HEGEL AND THE SCIENCES

BOSTON STUDIES IN THE PHILOSOPHY OF SCIENCE

EDITED BY ROBERT S. COHEN AND MARX W. WARTOFSKY

VOLUME 64

HEGEL AND THE SCIENCES

Edited by

ROBERT S. COHEN

and

MARX W. WARTOFSKY

D. REIDEL PUBLISHING COMPANY

A MEMBER OF THE KLUWER ACADEMIC PUBLISHERS GROUP

DORDRECHT / BOSTON / LANCASTER

Library of Congress Cataloging in Publication Data

Main entry under title:

Hegel and the sciences.

(Boston studies in the philosophy of science ; v. 64)
Includes bibliographical references and index.
1. Science—Philosophy. 2. Logic. 3. Hegel, Georg Wilhelm Friedrich, 1770–1831. I. Cohen, Robert Sonné. II. Wartofsky, Marx W. III. Series.
Q174.B67 vol. 64 001'.01s [509'.2'4] 83–13911
[Q175]
ISBN 90–277–0726–X

Published by D. Reidel Publishing Company,
P.O. Box 17, 3300 AA Dordrecht, Holland.

Sold and distributed in the U.S.A. and Canada
by Kluwer Academic Publishers
190 Old Derby Street, Hingham, MA 02043, U.S.A.

In all other countries, sold and distributed
by Kluwer Academic Publishers Group,
P.O. Box 322, 3300 AH Dordrecht, Holland.

Printed in The Netherlands

TABLE OF CONTENTS

III. DIALECTICS AND LOGIC

PREFACE

To the scientists and philosophers of our time, Hegel has been either a neglected or a provocative thinker, a source of irrelevant dark metaphysics or of complex but insightful analysis. His influence upon the work of natural scientists has seemed minimal, in the main; and his stimulus to the nascent sciences of society and to psychology has seemed to be as often an obstacle as an encouragement. Nevertheless his philosophical analysis of knowledge and the knowing process, of concepts and their evolutionary formation, of rationality in its forms and histories, of the stages of empirical awareness and human practice, all set within his endless inquiries into cultural formations from the entire sweep of human experience, must, we believe, be confronted by anyone who wants to understand the scientific consciousness. Indeed, we may wish to situate the changing theories of nature, and of humankind in nature, within a philosophical account of men and women as social practitioners and as sensing, thinking, feeling centers of privacy; and then we will see the work of Hegel as a major effort to mediate between the purest of epistemological investigations and the most practical of the political and the religious.

This book, long delayed to our deep regret, derives from a Symposium on Hegel and the Sciences which was sponsored jointly by the Hegel Society of America and the Boston University Center for Philosophy and History of Science a decade ago. To the Symposium papers we have added several others, and we are grateful to Dietrich von Engelhardt of the University of Heidelberg and Hermann Ley of the Humboldt University of Berlin for their agreement to contribute. We are also grateful to Carolyn Fawcett for her help with the difficult tasks of editing and for the beautifully organized and detailed bibliographical apparatus.

Where shall the philosopher of science see Hegel's contributions most clearly? To some, this question must lead to his dialectical logic; to others, it must suggest the methodology of the social sciences, especially via the philosophical analysis of history; yet another approach will look to his theory of knowledge; another his critical epistemology of religious claims; another his metaphysics of nature, particularly in its analysis of incompleteness and creativity; another his early *aperçus* concerning labor and cognition; yet others his inspiration for the historical dialectic of science, technology and society in the thought of Karl Marx. Some of these ways to Hegel will be found in this book. Perhaps there is no direct entry to Hegel, and perhaps there will never be a supplementary Section to Hegel's *Phenomenology* devoted to the stage of scientific consciousness, that extraordinary stage of Western, and now universal, civilization. We are pleased that the authors of these essays have so intelligently, lucidly, judiciously given us their signposts on the way.

October 1983

Center for Philosophy & History of Science
Boston University
ROBERT S. COHEN

Department of Philosophy
Baruch College,
The City University of New York
MARX W. WARTOFSKY

ACKNOWLEDGEMENTS

We gratefully acknowledge permission by the respective publishers to reprint substantial quotations from the following publications. Hegel's *Science of Logic* [1929]. Trans. W. H. Johnston and L. G. Struthers. 2 vols. London, Allen and Unwin, 1951. Hegel's *Science of Logic*. Trans. A. V. Miller. London, Allen and Unwin. New York, Humanities Press, 1969. Hegel's *Philosophy of Nature*. Trans. A. V. Miller. © Oxford University Press 1970.

INTRODUCTORY REMARKS TO THE SYMPOSIUM ON HEGEL AND THE SCIENCES

PETER BERTOCCI

THE SCHOLAR, THE LIBERAL IDEAL, AND THE PHILOSOPHY OF SCIENCE

I

Benedetto Croce was to live much of his life during the turbulent period in the history of Italy which had far-reaching consequences for the history of Europe and the world. At the end of World War II, this Senator of the Kingdom of Italy and Minister of Public Instruction, owing to his age, refused the request of Albert Einstein and others to offer himself for the leadership of a war-weary, war-torn Italy. Croce's whole life might be seen as an attempt, in the midst of the authoritarianism of the right and of the left, to discover what is living and what is dead in the philosophy of Hegel. Writing in 1940 on 'The Roots of Liberty', he said:

> The liberal ideal is a moral ideal, expressing an aspiration toward a better humanity and a higher civilization. The new ideal that is to triumph should therefore present itself with the promise of a newer, richer, deeper humanity and civilization.[1]

And he adds:

> As regards our scholars and thinkers of the present time, it is their task to keep the concept of freedom precise and clear, to broaden it and work out its foundations.

The act of thought, he urged, "is at the same time the act of willing."[2]

This theme is at the heart of Croce's famous statement: "History is the history of freedom." History is the effort of human beings to create political, religious, scientific, social and artistic modes of being that allow persons themselves to define the forms in which the different dimensions (*distinzioni*) of their lives will find fulfillment.

I happen to think that Croce saw, even more clearly than Hegel, what the dialectic of freedom involves. He saw, if I may put it this way, that even the dictum: " . . . Ye shall know the truth, and the truth shall make you free", can destroy the free search for truth, goodness, beauty, and social organization. There is not one of us in this gathering tonight who has not seen the search for truth cramped by some orthodoxy whose purpose was to protect *the* truth. In the name of humanity, in the name of the true God, in the name of a particular ethical code, of a specific aesthetic style, and of the true

R. S. Cohen and M. W. Wartofsky (eds.), Hegel and the Sciences, 3–10.

science, some conscientious truth-seeker has been warned, suppressed, destroyed in spirit and in body – in the name, once more, of some paradigm of truth and goodness. The true Church, the true State, the true science are to keep men free from ignorance, from superstition, from moral wickedness, from soul-destroying pornography, from allowing freedom to become license. Who of us assembled here is not experiencing the conflict between what he believes to be the quality of freedom – the way *really* to free the human spirit – and the belief that persons ought to be free to search for that quality of freedom?

In a moment I shall focus on some basic grounds of this conflict as experienced by Hegel himself. But perhaps I may first indicate one way in which this conflict between 'being free' and 'being really free' faces philosophers of science as it has, and does, philosophers of religion. In the realm of religion there are believers who think of faith as *the* way to truth and reality. They stipulate a particular method by which Reality manifests itself, and with such certainty that rationalistic or reasonable speculation may indeed explain, but may not call into question, the method itself or its fundamental revelation. The theological dogmas developed by this route are beyond reason, and beyond aesthetic and moral categories; man is to reason, *yes*, but not to reason about this basic method, epistemology, and the assumptions presumably consistent therewith. Freedom, *yes*, but not freedom to ask whether the method, faith or revelation, is correct; otherwise scepticism and moral chaos will overcome men. Nevertheless, in our own day the formulations of religious truth and theological truth are being questioned by many, within and without the religious-theological realm – by men who feel cramped by these formulations and by the institutions inspired by them.

It is not far-fetched for me to point to a faint parallel, at least, in the development of scientific method and dogmas. From many sides we hear complaints today that scientists are, wittingly or unwittingly, claiming access to the highway to reality; that, whatever the mysteries of being, there is no security apart from the scientific ascent to Being; that all other approaches, so-called, to reality are acceptable provided they do not question the basic possibilities of the method, epistemology, and assumptions presumably consistent with scientific effort and morals.

The response to these complaints often has a religious fervor about it. There can be no compromise with the autonomy of science; no appeal to moral, aesthetic, or religious considerations must be allowed to stain the purity of the fundamental method of science or of the world-view proclaimed in its name. Freedom within the basic 'presuppositions' of science, yes, but no freedom to question the fundamental validity of certain hard-won truths, or of appropriate method.

At the moment I am concerned with no more than the parallel between the claims and counter-claims made in the name of religious and of scientific world-views. What concerns me is the larger issue that lies within or beneath the claims and counter-claims. May it not be that these warisome claims and counter-claims need to be examined, especially in our day, within a larger phenomenology of spirit? Hegel was to place the different forms of human experience, such as the scientific, the aesthetic, the religious, within the *Idee*; in our day, the debate still goes on as to which of the sciences – physics, chemistry, biology, psychology, sociology, history – is to provide the sovereign paradigm in the kingdom of science. We need not despise such debates provided we hearken also to the cry in our day of the human spirit which asks, with Croce, that every segment of human experience be related to the human venture as a whole. When Hegel said: The true is the whole, he was reminding us that the true will not be found in less than the whole of human experience. Can the scholar be satisfied with less than a total phenomenology of the human spirit? This is our continuing problem, especially if we follow Croce's concern that the scholar "keep the concept of freedom precise and clean . . . broaden it and work out its foundations."

It may not be amiss, then, if in this context we remind ourselves once more of Hegel's own struggle to be clear about the nature of freedom and its foundation. Brevity forces me to put a complex issue over-simply, and largely in my own way, but, I hope, without losing the essential Hegelian thrust. My own response to Hegel calls for a dialectic of freedom that Hegel could hardly accept, but his struggle finds a response that should not be repressed.

II

The French Revolution broke out while Hegel was a student at Tübingen University; the singing of the 'Marseillaise', the anti-establishment speeches, were not events that were taking place in France alone; the new ferment expressed and inspired Rousseau's *Emile* and the *Social Contract.* "It is Rousseau," Hegel was to write in his *History of Philosophy*, "who has inaugurated the Absolute as Liberty."[3] But as he was to say in *The Phenomenology of Spirit*, "Universal Freedom can produce neither a positive achievement nor a deed . . . it is merely the rage and fury of disappearance and destruction." Freedom was no unmixed blessing, and for Hegel, the search for 'true freedom' was a continuing concern as he tried to understand the nature of personal and social experience.

In particular, Hegel was not satisfied with a doctrine that nevertheless influenced him greatly, namely Kant's concept of moral freedom. Kant, we recall, removes inclination and desire from the definition of a moral act and of moral worth. Kant insists that it is experientially false, and in any case reflectively spurious, to suppose that morality is based on anything less than a will that can choose among alternatives. As he says in Section I of *The Fundamental Principles of the Metaphysics of Morals*, that good will is "not . . . a mere wish, but the summoning of all means in our power". There is no doubt that a person's activity of willing is not to be reduced to any other mode of personal experiencing.

But this is only the beginning of Kant's view. Hegel was to brand it *Moralität* and move on to *Sittlichkeit.* Yet Kant himself realized that a person can not only will from the wrong motive but he can also be mistaken in his reasoning. Whatever else we may think about Kant's ethics, I think that there is no denying his insistence that the will is itself not an isolated part of a person's being. The willing person is a person capable of reasoning, and his actions take place in a social context affecting himself and others. There is no doubt in Kant's mind that a person who reasons about his choices ought to will by a maxim which his universalizing reason can approve.

I am not suggesting that Kant's categorical imperative is not by itself open to the charges of an empty formalism. But I am suggesting that there is open to us a not unreasonable interpretation of Kant's further meaning. Since the categorical imperative itself expresses what it means to be a reasoning person, persons ought to be treated never as means only but always as ends. In a word, the twisting of arms, the appeal to feeling (to 'gut-reactions'), to non-rational pressures, may bring conclusions; but persons who are not convinced through reasonable persuasion will not only remain unconvinced, but they will be degraded as persons.

If I am right, Kant's own basic effort, no less than Hegel's, is to reject ethical atomism, if only because a reasoning person is obligated to respect other persons and their capacities for freedom and reasoning. Even if a 'stepmotherly Nature' makes the realization of a will for reasonable interaction all but impossible, Kant had no doubt that the good, reasoning, will is a jewel shining by its own light – the one absolute, obligatory, moral good. I repeat that in all consistency Kant could hardly allow freedom and obligation to end in ethical atomism; for each free moral agent confronts another free and responsible moral agent. In this matrix of moral choice, whatever the specific good in question, each reasoning person is to respect other reason-capable moral agents. Duty for duty's sake, yes, but the duty cannot exclude others and their conscientiousness as persons. As I see it, it is not surprising, therefore, that Kant envisions a social moral order as a Realm of Ends in which persons respect each other's sense of duty and the freedom that inspires it.

I myself do not believe that Hegel would be in essential disagreement with such a view of Kant. But even such a conclusion is not penetrating enough for Hegel. For reasons I cannot pursue here, he questioned the atomic ontological view of the self that this view, as he saw it, presupposed. In any case, what he saw as the inadequacy of such Kantian moral autonomy (*Moralität*) led him to a view of the ethical ideal (*Sittlichkeit*) that, as I see it, handicaps and even threatens the Kantian autonomy of the moral agent. Yet in his inability to accept the Kantian realm of ends as concrete enough, Hegel puts his finger on the conflict that I think goes on in all thoughtful minds today. Let us follow his own reasoning for a moment.

Granting Kant's universalizing individual conscientiousness, we cannot find in it the concrete ethical ideal. Hegel is correct in noting that the very taking of the conscientiousness of other persons into account may well lead us to question whether Kant's initial thesis, that the individual conscience is an autonomous jewel, is a stable description of the actual situation in which persons find themselves. There are reasoning, conscientious believers; but there are also reasoning, conscientious unbelievers! The fundamental, and in some respects tragic, fact is that the conscientiousness of any free person may be a barrier to that of another free person.

Thus, Hegel reasons, if conscientious persons ought to take account of each other, the fact that they are persons and conscientious does not illuminate the choices they need to make *as* they disagree with each other. It is not enough, therefore, to seek a realm of ends, as if we could solve concrete issues on those terms. The fact with which we must begin is that as persons we need to see ourselves and our conscientiousness as intrinsically imbedded in a social whole without which we cannot exist as moral creatures. Mature ethical awareness, therefore, must resist the conception of monadic, self-determining persons linking themselves together into what would be a collection of morally righteous (self-righteous) persons. For, to press Hegel's objection, from such an atomic, moral, collection of beings one could never get to the concrete ethical community without which the moral consciousness remains an empty ideal.

Accepting this objection to Kant, does Hegel himself leave us with a tenable solution? As he says, the self-determining conscientious individual is self-defeating if he does not think out who he is and what he is to choose. A person's very attempt to think out what to choose does force upon him the realization that his freedom remains sterile until he finds concrete modes of realizing *what this freedom is for. Freedom to* involves some pattern of organization that will keep freedom from remaining a self-protecting *free from.* A self-righteous conscientiousness may break no windows, but neither will it light any candles. To paraphrase James, the road to hell is paved with a sense of moral purity that leaves freedom romantically idle. The way out is to deny any latent atomicity of the individual, to place him, his freedom and his conscience, squarely within a qualitative structure of reality and a qualitative social matrix that keeps freedom from being capricious and wicked, and also prevents conscientiousness from being sterile and self-righteous.

What has happened here? A conception of *qualitative freedom* becomes the only acceptable meaning of freedom. Kant's *formal freedom*, the freedom of conscience of the individual, succumbs to the dialectic of the moral person's search *for quality* which cannot avoid, which ought not to avoid, the demands of *Sittlichkeit*, or a well-organized society. What comes to mind here is Hegel's treatment of the dialectic of master and slave in the *Phenomenology.* The master, Hegel argued, comes to be so dependent upon the slave that he is no longer master. So here, Kantian formal freedom comes to be so dependent on qualitative ethical freedom, that both are unacceptable and must find their fulfillment in the large unity, the State, what Hegel called "the reality of the ethical idea",[4] or the "self-conscious ethical substance". That is, for Hegel, reason, inspired by the vision of *the* truth that shall make men free, confines the moral will to that quality of freedom at the cost of denying the validity of the autonomous free conscience as Kant saw it. The good of the State becomes at once the ideal of freedom, and the embodiment of reason, which, presumably cannot in fact be alien to the individual.

One cannot but sympathize with Hegel's conviction, as expressed in the third volume of his *Lectures on the History of Philosophy,*[5] that men find it hard to believe that a common reason exists, and can grow; and that they consequently clothe their vanity in conscientiousness. Hegel was aware, as Croce is, that freedom is man at war with himself. And Hegel found it necessary to keep freedom from destroying itself – in the name of a reason that could keep it qualitatively free! And each of us follows him in knowing that persons must find some way of rendering their consciences concrete, some laws and institutions, some economic order, and some educational ideal that will assure men that their freedom will be self-fulfilling.

III

But, I suggest, even this formulation of the solution is inadequate. For once men think they have found the way to keep men *really* free, be it in religion, in philosophy, or in science, they invite the alternative vice of forgetting that freedom and reason live in persons who are never complete either in their vision of truth or their own self-fulfillment. Thus I find Croce more accurate when he says:

> But the liberal mind . . . does what the heart of mankind cannot do: [it] regards the withdrawing of liberty and the times of reaction as illnesses and critical stages of growth, as incidents and means in the eternal life of liberty.[6]

And, in Kantian vein, I find Croce closer to the truth than Hegel when, in another context, he says:

> Liberty is not dependent on any particular economic systems [and I add political, theological, or scholarly structures] . . . . It calls all systems to the bar of judgment.[7]

May I, then, finally suggest that as philosophers of science – and not the least as scholars and as persons – we keep the Kant-Hegel (the *Moralität-Sittlichkeit*) tension before us as a struggle that does not have *a* or *the* historic solution? Can we avoid the great seduction, the belief that there is some specific system that will solve the dialectic opposition between formal and qualitative freedom? Is it not our task to keep the tension between formal and qualitative freedom alive? As Heraclitus said, "Strife is justice", a justice that never allows the opposition between formal and qualitative freedom to fade out.

As persons and scholars we conscientiously continue to confront each other in the search for truth and goodness. We know that every Eden will have its serpent, that every specific ideal of fulfillment may endanger the freedom to say *yes* or *no*. On the other hand, we must insist that, if persons are to be free to find full freedom, they must be free to be wrong in so doing. At the same time we must insist that every claim to *actual* freedom, *actual* truth, face the bar of reasoned judgment. Indeed, must we as scholars not be aware that the guide of reason is indeed at work in the realm of freedom? To be free is to seek qualitative freedom in some method or mode of life. Since every ideal, personal or communal, will be inadequate, persons will be imprisoned by their ideal unless they are responsibly-responsive to those who challenge their ideal. F. H. Bradley was right when he said that "there is no sin . . . which philosophy can justify so little as spiritual pride."[8] To say this is to underscore the reminder of Pico della Mirandola: we are neither earthly nor divine, neither mortal nor immortal, but we do have the freedom to shape ourselves in accordance with such truth as is open to reasonable and free men. Free men are not outside of reason; they are reasoners who are courageous enough to remain free – and responsible.

Boston University

NOTES

[1] Anshen, R. (ed.), *Freedom, Its Meaning* [by Croce and others] (New York, Harcourt Brace, 1940), p. 24.
[2] *Ibid.*, p. 27.
[3] I owe this reference to T. N. Munson, *Monist* **48** (1965), 100.
[4] *Ibid.*, p. 105.
[5] See edition translated by E. S. Haldane and F. H. Simson [1892] (New York, Humanities Press, 1968).
[6] Croce, B., *Politics and Morals*, tr. S. J. Castiglione (New York, Philosophical Library, 1945), p. 121.
[7] Anshen, *op. cit.*, p. 38.
[8] Bradley, F. H., *Appearance and Reality* (Oxford, Oxford University Press, 1897), p.7.

PART ONE

THE SCIENCES

GERD BUCHDAHL

CONCEPTUAL ANALYSIS AND SCIENTIFIC THEORY IN HEGEL'S PHILOSOPHY OF NATURE (WITH SPECIAL REFERENCE TO HEGEL'S OPTICS)

I

In a popular textbook on physics, written with an historical and philosophical bias, when turning to the chapters devoted to the subject of 'colour', we all of a sudden light upon a section entitled 'The Estrangement between Science and Philosophy'.[1] Prior to this, the reader has been taken through the properties of the refraction and reflection and propagation of light, lenses, mirrors, rays of light, angles of incidence, image formation, and so on. Here the text interrupts, and the author turns to Hegel, quoting from the *Zusatz* to § 275 of the *Philosophy of Nature* the 'definition of light':

> Light is the enclosed totality of matter, only as pure force, the intensive life sustaining itself within itself, the celestial sphere which has withdrawn within itself, whose vortex is just this immediate opposition of directions of the self-relating motion, in whose flux and reflux all distinction is extinguished. As existent identity, it is a pure line which refers only to itself. Light is this pure existent force of space-filling, its being is absolute velocity, present pure materiality, real existence which is in itself, or actuality as a transparent possibility (*PN*, § 275, *Z*, p. 87).[2]

The author of our textbook thereupon makes considerable play with the hostile reaction that this approach received from the men of science in Hegel's time, which, he says, was part of the cause of "the emotional drive behind the scorn which the scientific world came to hold for philosophy".[3] Now it is true that no greater contrast could be imagined between the poetic lines quoted above and the down-to-earth procedures of the sober scientist, with his concern for observations held together by a network of theoretical conceptions. I think, however, that where we come face-to-face with such a total intellectual, not to say emotional, clash of temperaments, it is likely that there is some grave misunderstanding concerning intentions. More specifically, Hegel's purposes in all probability have here been both misconceived and torn from their natural context.

For what Hegel tells us at the start of the passage from which I quoted is that he is attempting to provide an '*a priori* notional determination [*Begriffsbestimmung*] of light'. By this Hegel means a conceptual formulation which defines light as a state of matter, instead of as an independent element

R. S. Cohen and M. W. Wartofsky (eds.), Hegel and the Sciences, 13–36.

(as was still done in his day); furthermore, as a formulation which seeks to furnish an explication in terms of the categories and concepts of Hegelian dialectical logic.

Assuming here some familiarity with the outlines of Hegel's logic,[4] let me, in order to exemplify the process of notional determination, give a brief outline of Hegel's progressive elucidation of the concept of light, as it emerges from the structure of that logic. Let us take for granted an understanding of the place held in the logic by the overarching index of 'the Notion' [*Begriff*]. And singling out some of the details, we find, to start with, that the Notion is determined 'in itself' as pure Being, and in particular, as 'Being-for-itself'. Then, when specified further, the latter is conceived both as 'a one' and as 'a many'. This one-many relation, Hegel claims, further can be shown to involve a 'process' which may be labelled as 'repulsion-attraction', concepts which characterise Hegel's idea of 'matter'.[5] Now let us also note that all these logical articulations have their parallel in Hegel's philosophy of nature, where the logic of Being turns up more concretely interpreted as the subject of 'Mechanics', and in particular (for the above example of matter), of 'Finite Mechanics' (*PN*, § 262, p. 44; more of this in section III).

We cannot (and need not) – at least for the moment – enter into the details of this account. We only note that we meet here, in a nutshell, the kind of consideration of a concept – in our case, 'Matter', explicated in terms of Hegel's logical categories (all specifications or stages in the development of the so-called 'logical Idea' as 'Nature')[6] – which is what Hegel understands by an '*a priori* notional determination'. In a similar way, as we proceed up the ladder of the logic with our example, where the Logic deals with 'Essence', and in particular, with the categories of reflection, so in parallel, in the *Philosophy of Nature*, we likewise "enter logically into the sphere of Essence" (*PN*, § 274, *Z*, p. 86) which exhibits matter as a kind of 'reflection-into-self', interpreted by Hegel here as a kind of 'manifestation'. Where matter at first was purely inert, at best determined gravitationally and only 'externally', it has now gained a sort of 'independent existence', whereby it manifests itself as what Hegel calls 'light', but which – considering Hegel's characterisation – is perhaps on the whole more adequately labelled a field of energy, appearing only in certain instances as a 'singularity' (Hegel says 'individuality'); for example, as the stars; and especially, as the sun (*PN*, § 275, p. 87).

Now I want to claim that the general significance of this procedure amounts to an attempt to see certain very general scientific concepts articulated within a logical framework, to which they become thereby tied, in order to discover how much can be said about a given concept within such a local context. And this is not absurd. We have in our own day become familiar with the idea that the broad formative concepts of a science govern its development, and that a considerable part of their intellectual articulation relates to what we may vaguely call a 'metaphysical dimension'. And we have learnt that revolutionary changes in science frequently parallel, not to say are triggered off by, categorial changes. It is in this way we can understand Hegel when he writes in his Introduction to the *Philosophy of Nature*:

All revolutions, in the sciences no less than in world history, originate solely from the fact that Spirit, in order to understand and comprehend itself with a view to possessing itself, has changed its categories, comprehending itself more truly, more deeply, more intimately, and more in unity with itself.[7]

Similarly, we find that in our own day we have less difficulty with a typically Hegelian formulation, according to which 'nature is posited by spirit', though, to be sure, as a natural process the latter emerges from the former – remembering that a characterisation of this 'process' itself involves scientific and philosophical theorising. And equally we can say, with Hegel, that

spirit is not the mere result of nature, but in truth its own result; it produces itself out of the presuppositions which it creates for itself, – out of the logical idea and external nature (*PG*, § 381. *Z*, p. 23).

When we speak of Hegel's Logic, we need to keep in mind of course the dynamic character of that logic. Its very nature, as Hegel tells us, is best comprehended as a kind of 'serial progression' or 'evolution' (*PN*, § 249, *Z*, p. 21), from the abstract to the concrete, from the basic categories of pure Being to the most complex organisation of animal nature, of life and of cognitive thought.[8] By making the considerations of the philosophy of nature run in parallel with this serial process, it gets caught up in it, and thus gains its place as well as its significance within the whole.

II

So far, however, I have talked as though 'notional determination' was at best only an extra, altogether standing outside the scientific process, with its special concepts, empirical observations, unifying laws and theories. But clearly, unless there were some connection between these two elements, the whole procedure would lose its interest. In particular, we want to ask, what is the relevance of Hegel's notional determination for particular scientific theoretical articulations? And as a special case we must pose the question: what place in all this has the specific, contingent individual sensory fact?

Now, that Hegel does not ignore the observational and experimental detail of science, emerges from every page of his philosophy of nature; we need only mention his unequivocal declaration that

> not only must philosophy be in agreement with our empirical knowledge of Nature, but the origin and formation of the Philosophy of Nature presupposes and is conditioned by empirical physics (*PN,* § 246, *Remark*, p. 6).

I think this point need no longer be laboured, although it is necessary always to keep in mind that the science here involved is that of Hegel's time, of the first two decades of the nineteenth century. Moreover, in the Hegelian scheme these detailed facts, as pure contingencies, to some extent remain standing; the 'external material' here confronts 'spirit' (i.e. both scientific theory and notional determination); it remains 'indifferent' to it, and only accepts the idealisations of science 'on sufferance' (*PG,* § 381, *Z*, p. 22).

Hegel is so deeply aware of this confrontation of the notional by the empirical detail of nature, that he characterises it as a veritable 'contradiction of the Idea'; the 'necessity' of the forms generated by the Notion is faced by the 'indifferent contingency and indeterminable irregularity' of the empirical detail (Cf. *PN,* § 250, p. 22). It is a received tradition that Hegel claims to deduce empirical facts from the categories of his logic; which (if true) would of course strike an empiricist logician as a logically absurd undertaking, colliding head-on with their strong conviction of such a mistaken jump from the realm of logic to that of substantial fact, particularly when such facts happen to be different from those now recognized by modern science! Quite to the contrary, however, Hegel expressly denies deducibility of contingent facts:

> This confusion of contingency . . . , this impotence of Nature, [he writes], sets limits to philosophy and it is quite improper to expect the Notion to comprehend – or as it is said, construe or deduce – these contingent products of Nature (*PN,* § 250, *Remark*, p. 23).

In a similar way, the notorious Hegelian legend of the *a priori* deduction of the number and distances of the planets, on closer inspection turns out to be no more than a harmless insistence that we should try always to discover a plausible 'empirical formula' (as this would now be called) covering the independently and empirically ascertained observational planetary data (PN, § 270, *Remark, Z*, p. 82). There may be "traces of the determination by the Notion", but even these enter into science in a very complex way, to be considered presently. This is not to say that notional exploration may not on occasion yield some astonishing insights and anticipations, for instance those concerning the relations of energy and space. Thus, in accordance with the 'notional determination' mentioned before, light, Hegel writes, cannot properly be said to 'fill space', but only to be "present in space, and then not as something individual". Indeed, since under the Hegelian scheme, space turns out to be – as he puts it – "only abstract subsistence or virtual being", whilst 'light' is the "existent being-within-self" [*daseyendes Insichseyn*], there is an aspect of light according to which, in its 'abstract manifestation' it is as such primarily "spatial, an absolute expansion in space". And, Hegel concludes, "light is infinite spatial dispersion, or rather it is an infinite generation of space" (*PN,* § 275, *Z*, pp. 87–88).

We may interpret this as a conceptual exploration, according to which we cannot speak of empty space as such, but only of a spatial field of energy, not unlike ideas of more recent science. In such terms, field and space are more closely interwoven than in the traditional Newtonian context, though Hegel's ideas here go back perhaps also to older medieval or Renaissance or seventeenth-century ideas.

At any rate, the question of the relationship between such general ideas and the presence of empirical phenomena is easily misunderstood. For instance, one of the 'singularities' in our field of energy is a concentration of 'matter' which we call 'the sun'. Now on the one hand, that there *is* such a singularity, constitutes a fact which, so Hegel expressly notes, "is empirically ascertained" (*PN,* § 275, *Z*, p. 89). Here, seemingly a complication arises. For about this fact he expresses himself in an apparently contradictory way. At first he tells us that the body of the sun is "the primordial, uncreated light" (*PN,* § 275, *Z*, p. 89). However, immediately afterwards, when enquiring into "the finite causes of the existence of that which shines in this way", the account which he gives, in terms of 'internal friction' due to rotation of the sun, seems to clash with this claim to the light being 'uncreated'.

Now his answer to this problem tells us something further about the relation between scientific theorising and the Notion. The theory, Hegel implies, simply states the physical conditions incident upon the production of light in the finite contingent body of the sun. But this says nothing about the notional status according to which the sun is a singularity in a spatial field of energy, regarded as a tense-neutral manifestation. It is only this notional determination with which we are there concerned. And from this, 'matters of empirical fact', such as the account in terms of friction, cannot be deduced (*PN,* § 275. *Z*, p. 90). Evidently, the Notion supplies the conceptual framework; causal investigations at best are only *guided* by the conceptions thus explored.

III

In general, we can see that the relations between notional investigations, scientific theory, and empirical fact are rather complex, and to appreciate the variety of relations involved we must now explore these in more detail; partly, because, without such a survey, the chief aim of this essay, which is to characterise the method of Hegel's optics, must remain incomprehensible. Concerning the problem of deducibility of empirical data from the Notion, compared with the situation as it obtains in the theoretical approach of science, Hegel is in fact quite explicit. In an important passage, which appears at the end of his extended comparison between the approach of Kepler and Newton towards gravitational theory, he says:

> Philosophy has to start from the Notion, and even if it does not assert much, we must be content with this. The Philosophy of Nature is in error when it wants to account for every phenomenon; this is what happens in the finite sciences, which try to trace everything back to general conceptions, the hypotheses. In these sciences, the sole verification of the hypothesis lies in the empirical element and consequently everything has then to be explained. But what is known through the Notion is clear by itself and stands firm; and philosophy need not feel any embarrassment about this, even if all phenomena are not yet explained. I have therefore set down here only the rudiments of a rational procedure in the comprehension of the mathematical and mechanical laws of Nature as this free realm of measures (*PN,* § 270, *Remark, Z*, p. 82).

This passage makes it very clear that Hegel's philosophy of science has a two-tier structure. We must distinguish, he says, between the science, including its theoretical formulations, on the one hand, and "the course of its origin and of its preliminary investigations" (*PN,* § 246, *Remark*, p. 6). It is these preliminaries that are governed by notional determination, or – as we might say – conceptual explication. On the other hand, and this raises the problem, Hegel also maintains, in the same passage, that not only must experience "name the empirical appearance corresponding" to the result of conceptual explication; we must also show "that the appearance does, in fact, correspond to its Notion" (*PN,* § 246, *Remark*, pp. 6–7).

Is Hegel saying that there should, and that there should not, be deducibility between Notion and phenomena? The answer must be that this, for Hegel, is a matter of degree. The more 'brute' the empirical data, the less grip may be expected from the side of the Notion. But then, scientific theorising itself already banishes much of the purely qualitative and specific in its investigations: the more theoretical thought "enters into our representation of things, the less do they retain their naturalness, their singularity and immediacy. . . . The rustle of Nature's life is silenced in the stillness of thought", in one of Hegel's truly memorable formulations (*PN*, § 246, *Remark*, *Z*, p. 7).

Now the more theoretical our approach – and it is a commonplace for Hegel, that the purely immediate sensory datum cannot maintain itself intact as such – the logically 'softer' the constructions. Here, on the one side, hypothetical and constructive thinking always allow of alternatives. Not only that, but – and this is important – on the other side, science itself will always attempt to tie its fundamental constructs back to still more basic experience, and for Hegel this means: to the categorial determinations of the logical Idea. So it is here that we may expect Hegel, when confronted by scientific constructions and positions, usually and quite properly surrounded by much that is still problematic, to offer alternatives, or to reject hypotheses inconsistent with the development of his Logic. The variety of relations, either agreements or disagreements, between the developing Notion and the scientific laws and constructs, is however not so simple, and in this section I shall attempt to distinguish several different cases. Studying these in turn is perhaps the best way of coming to grips with Hegel's intentions.

I have elsewhere argued that there are some clearly definable groups of criteria for the acceptance of scientific theories, including conceptual or metaphysical, inductive, systemic, and regulative or architectonic components of appraisal.[9] Such a stratification seems to me to have had its first clear expression in the writings of Kant, from whom – as will be found presently – Hegel, too, appears to have received his prime inspiration. Hegel's development of this theme in rather an extreme fashion presents an opportunity to study at close quarters the range and limits of such a methodology, quite apart from the incidental gain of offering the modern student of the philosophy of science an entry to Hegel's philosophical intentions which has been somewhat overlooked.

1. Consider, as a first instance, Hegel's attitude towards atomistic theories in chemistry. In his *Science of Logic* we find a sustained discussion of Berthollet's law of mass action, Berzelius's development of this topic and the whole theory of chemical affinity. In particular, Hegel discusses Berzelius's account to be found in his *Textbook of Chemistry*, published contemporaneously with the composition of the *Logic* (Hegel was nothing if not

up-to-date), of the atomistic interpretation of what happened in chemical combination.

Now as is well known, Hegel has fundamental objections to any atomic theory which exhibits the essential features of the Democritean-Newtonian type of approach. For this allows the atomic unit to stand independently of and side-by-side with the void; and quite contingently, and only subsequently, adds forces of repulsion and attraction, in order to explain phenomenal effects. Hegel's rejection of Berzelius's form of this atomism occurs in two ways:

(i) He emphasises its hypothetical status, by complaining that "the theory goes beyond the limits of experience" (*SL* [M], p. 359). Furthermore, it is merely *ad hoc*, and "not corroborated in any other way", i.e. independently (*SL* [M], p. 360).

(ii) The conception of the atom as here used requires "corroboration . . . from metaphysics, which is logic" (*SL* [M], p. 360). And of course, Hegel's Logic is opposed to the atomist-mechanist metaphysics of the Newtonians.

In this example, then, we note that logic is not opposed to the empirical phenomena, but to their hypothetical explanation, and this for two reasons: (a) because they *are* hypothetical, as going beyond experience; (b) because in so far as they do transcend the phenomena, they are underpinned by a metaphysics inconsistent with that expounded by Hegel himself.

These are, it will be seen, two quite different objections. The first expresses a preference for a special type of scientific theory, labelled by subsequent philosophers of science in various ways: generalising, mathematical, external, abstractive, reticular, phenomenological or nomological; as contrasted with what are called explanatory, internal, hypothetical, analogical, dynamic, or aetiological theory-types. And it seems that Hegel usually sides with the first of these, as explained in the *Encyclopedia* section of his *Logic* which discusses the relations of whole and part, force and its expression, and the contrast between inner and outer.[10] Hegel thinks this distinction quite misplaced, and both here and in the Introduction to the *Philosophy of Nature* he quotes with approval from Goethe's poem '*Zur Morphologie*': "Nature has neither core nor shell, but is all at once".[11] In other words, he shares Goethe's hostility to the scientists' method of theorising in terms of entities regarded as concealed, underlying, hypothetical, espousing instead a formal approach which steers close to the pure phenomenon.

However, and this is the second objection, no doubt this hostility to aetiological theory-types is fed by an aversion to the *kind* of hypothetical entities espoused by the scientists, above all atoms and forces; and to the realisation (of which Hegel is well aware) that this involves a metaphysics of mechanism which forever seeks to absorb the realm of the organic into itself.[12]

2. The same opposition to existential or independent reification of forces (centripetal force of attraction, centrifugal force of repulsion) we meet with in the section on gravitational theory, already alluded to. Gravitation is there explicated essentially in phenomenological or nomological terms, as the Notion of a universal corporeality which partitions itself into particular bodies whilst at the same time these bodies close up again to what has an aspect of singularity, manifesting itself in motion, through which the several bodies make a *systematic* whole (*PN*, § 269, p. 62).[13] This is a Notion, says Hegel, which manifestly includes two moments, the being-for-self of the particular bodies, and the suspension [*aufheben*] of this as the continuity of the system of bodies. But these two moments, the aspect of continuity and the aspect of particularity, have been interpreted quite wrongly as two separate forces (*PN*, § 269, *Remark*, p. 63).

If we follow the bidding of the Notion, we must therefore confine ourselves to the nomological theory of Kepler, who restricted himself to an elucidation of the laws of the system, even though this does not very easily cope with perturbation, as Hegel has to admit (*PN*, § 270, *Remark*, p. 68). *Per contra*, we must oppose the reification of gravitation as an independently existing force, involving as it does "an unspeakable metaphysics which, contrary to experience and the Notion, has the said mathematical determinations alone as its source" (*PN*, § 270, *Remark*, p. 67). Basically he seems to have thought that – as he puts it in the *Encyclopedia Logic* – "the explanation of a phenomenon by means of a force is . . . an empty tautology".[14]

3. In the instance just mentioned, we have a case where notional determination (i.e. conceptual explication), though still not involving any deductive claims, does affect the form of presentation of a scientific theory. As a footnote to § 267 makes clear, Hegel preferred the analytical approach of Lagrange.[15] Sometimes, however, he goes further; we then get the notional approach to scientific theory in what we may call a 'degenerate' form. Optics (the laws of geometrical optics) provides an example; to this we shall return in the last section. Galileo's law of free fall is a second example. The empirical law with which we are here concerned is expressed in quantitative terms as $s = 1/2gt^2$, where s is the displacement, t the time and g the constant of

acceleration. Now, Hegel freely agrees, the "*a priori* proof independent of empirical methods" of this expression has "been furnished by mathematical mechanics" (*PN*, § 267, *Remark*, p. 57). It simply follows from the Galilean law that $v = \mathrm{d}s/\mathrm{d}t = gt$ by integration.

However, contrasted with this there is also another, logically otiose Newtonian proof, which involves the "conversion of the moments of the *mathematical* formula into *physical* forces", a force of inertia, responsible for uniform motion of falling, and a force of acceleration which adds continuous increments, and thus involves the *sum* of two independent elements, when the result is a relation of *powers*. (This is, of course, a distortion of the Newtonian argument, though not too uncommon in the period immediately following Newton.) Hegel's basic objections here are, first, that we introduce the law of inertia as such, a law which he holds to be an unreal abstraction, and a case that never occurs in nature; a taking in separation of the principle of identity, and thus simply reducing to the tautologies that motion is motion, and rest is rest.[16] Secondly, in the Newtonian proof we make an unreal transition from uniform velocity to uniformly accelerated velocity.

And it is this, what Hegel considers a confused Newtonian gloss of a proof, involving a confused 'metaphysics', that requires replacement by something more in line with his own notional approach. We must therefore, Hegel says, "against the abstract, uniform velocity of lifeless, external determined mechanism", seek a genuine proof for the "law of descent of a falling body", a law which is a "free law of Nature". As such, its basis must involve something notional, and it follows that "the law must be deducible from this Notion" (*PN*, § 267, *Remark*, p. 58).

The proof that follows is rather bizarre; but in any case, Hegel is concerned only with what he calls "the qualitative relation" of "powers" [*Potenzen*]. The argument (which I will put in my own words) seems to involve the contention that in free motion, space and time form an intrinsic whole. If we now plot the axes of space and time, we obtain a space-time diagram in which we can picture time as a second dimension, the diagram resulting in a plane figure whose simplest representation is a square. ('Raising to a power', it will be remembered, is the third determination of numbers, which are a development of 'the quantum'; and here the simplest case, according to Hegel, is 'the square'. This third determination is regarded as the "equality of the multiple and the unit [*Anzahl und Einheit*]". 'The square' has thus a definitive place in the development of the logical Idea.)[17] And this is the notional explication or 'proof' of why the *spaces* traversed when a body is falling *freely* with uniformly accelerated *motion* are developable as the *square* of the times.[18]

If punning were taken further, we might say that this is perhaps a species of proof now-a-days called 'a dimensional argument', except that Hegel would protest that his development, despite the involvement of the quantum, is concerned with only the purely 'qualitative' aspect of the interconnection of the concepts of space and time. So despite talk about "proof of the law of descent of the falling body as derived from the Notion of [this kind of] event" (*PN,* § 267, *Remark*, p. 59), we notice considerable looseness of fit between the argument and the mathematical conclusion. In any case, we have to keep in mind Hegel's complete awareness of the competing existence of the mathematical proof. We may say that Hegel's intentions are to let notional considerations provide a form of putatively-clinching plausibility; demonstrating a 'possibility' that Newton's exposition does not offer us. Above all, there lies behind this the assumption that a genuine support for the Galilean law, as expressed in the formula, needs more than merely mathematical demonstration.

4. Probably the prime inspiration for Hegel's approach, which centres on the need for a notional determination of the elements of natural science, was Kant's attempt to construct the metaphysical foundations of the basic conceptions of dynamics, especially of matter and force. It is not too much to say that, as for Schelling, so for Hegel, this was in all probability the most important source for his whole approach, with Hegel sticking more responsibly than Schelling to the basic intentions of Kant's procedure. Kant's metaphysical exposition, Hegel says in the *Science of Logic*

> is noteworthy, ... because as an experiment with the Notion it at least gave the impulse to the more recent philosophy of nature, to a philosophy which does not make nature as given in sense-perception the basis of science, but which goes to the absolute Notion for its determination (*SL* [M], p. 179).

Those who have difficulty in unravelling Hegel's views about the relation between notional determination and empirical science may therefore hope to gain some light from Kant's procedure in the *Metaphysical Foundations*, provided *that* has been properly understood.

Let us briefly consider Hegel's treatment of the subject of matter and force. Here again, Hegel follows the twofold path of empirical and notional determination. Whilst empirical science operates with the concept of matter regarded as subject to the twin forces of attraction and repulsion, these forces, says Hegel, in so far as they are regarded as forces of empirical matter, are also based on the pure determinations here considered, of the one and the many and their interrelationships, which, because these names are most obvious [the *Encyclopedia* describes them as 'metaphorical expressions' – *bildliche Ausdrücke*],[19] I have called repulsion and attraction (*SL* [M], p. 179).

Now what is here called 'pure determination' of matter had been described by Kant as an *a priori* elucidation of its concept; an elucidation which was supposed to establish the 'possibility of matter' by 'constructing' the concept through its reduction to the just-mentioned fundamental forces. The methodological approach here followed by Kant involves a sharp distinction between establishing the 'possibility' of a scientific construct, and the inductive 'probability' of its application.[20] In the present context, it is of course the establishment of 'the general possibility of the concept of matter'[21] with which we are concerned. Actually, Kant's procedure, 'constructive' from one point of view, from another was basically 'analytical' (as Hegel rightly points out (*SL* [M], p. 179)); it departs from an analysis of the meaning of matter as impenetrable, by interpreting this property as repulsive force. He then proceeds to show that matter cannot be conceived as existing with repulsion alone without attraction; which completes the exposition.

This, then, is an extremely clear case of conceptual analysis or explication; it seeks to show how much of the concept of matter is already contained in our basic modes of categorial thinking; and on the other side, how much of the fabric employed in empirical science can be shown to answer to this conceptual explication. There is no attempt to 'deduce' attractive or repulsive forces, let alone their laws of action.

This looseness of fit is also quite apparent in Hegel's treatment of repulsion and attraction, emerging as they do from within his own notional analysis as metaphorical expressions – evidently thus borrowed from the domain in which they originally have their empirical home.

The details of Hegel's account would take too long to unravel at any length. Of Kant's analysis he complains that although it has the merit of making the two forces internal to matter, deriving – unlike in the Newtonian procedure – matter from the forces, it still leaves them standing side by side, independent of each other (*SL* [M], p. 181), instead of treating them as "moments which pass over into each other" (*SL* [M], p. 182), an instance of the process of the Hegelian dialectical evolution of the logical Idea. The argument, as already mentioned at the start, is that Being, *quâ* being-for-itself, is a one; yet as such, at the same time, it relates itself to itself as an 'other'. Hence the one is incompatible with itself; it is "that which thrusts itself off itself; and what it thus posits itself as, is the many. Hence we may designate this side of the process of being-for-itself through the metaphorical expression, repulsion".[22]

The intention of the argument, if not its logical formulation, at least may be clear; when reflecting on matter as one, this turns into many; but the conclusion can maintain itself only *via* a process which bears similarity to what in mechanics is called 'repulsion'. Or rather, that concept, in its notional determination, boils down to this fundamental aspect of the being-for-itself of anything in respect of its material aspect. Similarly, each of the many is itself a one, and in this way turns round into its opposite, viz. attraction. Attraction and repulsion, on this view, are a dual aspect of the one-and-the-many.

Modern (i.e. Newtonian) physics, comments Hegel, still operates, through its conception of the molecule, with the picture-aspect of the old atom, even though admittedly – unlike original atomic theory – it has added the force of attraction to that of repulsion. "But their mutual relation, what constitutes their concrete and true nature, must be torn from its opaque confusion, in which even Kant's *Metaphysical Foundations of Science* has still left it."[23] In other words, attraction and repulsion must not be introjected *ab externo*, but need to be explicated as *logical* aspects of the Notion involved in the very consideration of matter *quâ* matter, under its aspect of the one and the many.

Hegel's analysis has thus a dual aspect. On the one hand, it seeks to incorporate these forces into the intimate process of the logical Idea, and thus project what Kant had called 'the possibility' of the concept. On the other hand, the method is so contrived as to deprive forces of their quasi-independent reality, or as merely borrowed from the theory of science. At the same time it is clear that Hegel's derivation is meant also to support his preference for the use of phenomenological theory-types, as already noted. For his 'deduction' is incidentally meant to have the effect of de-substantialising the forces of Newtonian physics, and thus to point to the adoption of the Continental 'mathematical' approach.

IV

With these explanations, we may now hope to gain some understanding of the significance of Hegel's philosophy of optics for his general philosophy of science. We shall be prepared for a phenomenological approach to that theory. We shall expect a rejection of any realist interpretation of the explanatory terms of the optical theory of traditional science. Further, we shall expect Hegel seeking to provide a notional determination (conceptual foundation) of the prime terms of optics. Finally, we shall have to see how, with the apparatus that results, Hegel approaches the laws of geometrical optics. And once again we will want to seek a satisfactory interpretation of those 'quasi-deductive' procedures of Hegel, which to the outsider so easily look like a competing scientific explanations of the laws of optics.

We began this essay with some comments on Hegel's notional determination of light, so that we may take these here for granted. It is of course an analysis of a purely qualitative kind. Like the conception of Essence (whose "standpoint is altogether that of reflection"[24]) which, as I said, it parallels, light is firstly – if we allow once again for a certain amount of punning, indicative of analogical insight – matter *quâ* 'unity of reflection-into-itself';[25] it becomes 'manifest', so to speak, just like opaque or mere Being, when reflected upon by ourselves;[26] when Being is lit up through consciousness. In fact, Hegel once likens light to what 'is simply Thought itself, present in the natural mode' (*PN,* § 276, *Remark, Z.* p. 93). Similarly, he calls light also 'the abstract self of matter', from which he seems to draw the questionable conclusion that it is 'absolutely light' (in the sense of being opposite to the absolutely heavy); nor is it therefore separable into parts; it is simple self-externality – remember the relation between matter and space, alluded to at the start.

The two aspects of interest are, (i) the parallel with the logic of Essence, and one of its models, reflecting consciousness; (ii) the attempt to make a characteristic play on the distinction between the 'passiveness' or 'lifelessness' of matter and the free activity of what we might call energy of radiation. The main point is the contention that the second cannot be notionally grasped without the logic of the first. This means that the 'manifest' continuity, and undifferentiated simplicity, of light, by contrast with the opposite characterisation of 'inert' matter (roughly falling within the domain of Being), becomes a notional necessity. But this at once leads to a difference, in respect of the relation between notional determination and scientific theorising, from the cases we have so far considered. For now the weight of the notion is such as seemingly to destroy the theoretical fabric of the optical science of Hegel's time altogether. In the case of gravitation, the impact from the notion resulted only in Hegel's preference for purely 'mathematical' (i.e. phenomenological) representations of dynamics, as contrasted with that of the Newtonian form. It is the kind of preference which Hegel thought he had found realised in French eighteenth-century dynamics, but which was later made explicit in Hertz's *Mechanics*, and which is reflected in the standpoint of General Relativity theory.[27] Again, in the case of the law of free-fall, although we saw Hegel offering what I called a 'dimensional argument' from the Notion, the procedure of mathematical mechanics was not rejected. It is apparently quite otherwise when we turn to optics. Newton's corpuscular theory, the wave theory, Euler's aether; indeed, the whole conception of "simple rays of light, and of particles and bundles of them", are all rejected, being "of a piece with those other barbarous categories for whose prevalence in physics Newton is chiefly responsible" (*PN*, § 276, *Remark*, p. 92; *Z*, p. 94).

Of course, once again these rejections are not thought of as being in opposition with the empirical phenomena; quite to the contrary. Since Hegel regards them as purely theoretical, "since [as he says] nothing empirical obtains here" (*PN*, § 276, *Remark, Z*, p. 94), he feels at liberty to condemn, but the opposition is primarily due to the fact that he assumes – quite correctly – that the scientists of his time regard their theoretical constructs as "materialistic representations"; and it is as such that they are "quite useless for the comprehension of light" (*PN*, § 276, *Remark, Z*, p. 94). Naturally, Hegel does not object to the geometrical representation of light-rays, as purely mathematical devices, witness the diagrams that accompany his account of refraction (*PN*, § 318, *Remark, Z*, p. 188). It is when these constructions are interpreted materialistically that they involve a break in the "abstract ideality of light", and hence, thinks Hegel, "its absolute continuity". Falling back again on the analogy of mind, he sarcastically remarks that the expression, 'bundles of rays' is meaningless; light can "no more be divided into bundles of rays than can the Ego or pure self-consciousness" (*PN*, § 276, *Remark, Z*, pp. 93–94).

It is of course possible that had there been a mathematical theory of optics, such as was developed during the nineteenth century, e.g. by Fresnel, Hegel would have opted for this alternative, and allowed some use to expressions such as wave-length, aether, light-ray. As it is, from the standpoint of his time, we have the result that the notional analysis tends to destroy the science of optics altogether. This is unusual. It is more normal for philosophers to underwrite the science of their time, and to let their metaphysics issue in an underpinning of the foundations of that science. True, even Hegel cannot quite stand up to the pressure. In the case of Newton's theory of gravitational force, he had admitted that that theory had a superior power because it accounted as precisely and elegantly for planetary perturbations; but he still rejects it, giving instead a purely qualitative, and indeed purely descriptive alternative to that of Newton (*PN*, § 270, *Remark*, p. 68). In the case of light, we meet with a similar awkwardness, and momentary anxiety. Here, the notional account for him implies that 'light has immediate expansion' (i.e. is propagated instantaneously). Yet he admits that astronomical observation of the solar system shows that it has a finite velocity. To meet the contradiction, he falls back on an auxiliary explanation, which connects the finiteness of velocity with the existence of a medium. On the other hand, he contends, we are not automatically entitled to extrapolate to empty space; and he finds "something eerie" in Herschel's reckoning that some of the light from the stars must have left them "about 500 years ago". But never mind "these far-fetched conclusions" (*PN*, § 276, *Remark, Z*, p. 94): words which betray a clear expression of unease; whether they mark honesty or dishonesty on the part of our philosopher, I will not decide.

Reflection, in all this, is – as we have seen – of central importance. Indeed, without reflection light is not visible. So light requires, as its opposite, the darkness of matter: "Light and darkness have an external relationship to each other; it is only at their common limit that light attains to existence, for it is in this being-for-another that something is lit up" (*PN*, § 277, *Z*, p. 95). And there is not here, he adds, "as yet any question of colour" – foreshadowing the theory of colour, which in opposition to Newton's will insist that coloured light is as it were a *modification* of light proper, through the interference from darkness or opacity. This aspect of the theory of colours we shall have to ignore here for lack of space. But we can perhaps understand better now how a coalescence of the deductions from the Notion, the rejection of the Newtonian theoretical constructions, and Hegel's preference for phenomenological approaches could lend force to his audacious attempt of total contradiction to the Newtonian theory of the dispersion of light. No doubt, this opposition to the Newtonian theory of colour was fed by psychological, not to say nationalistic considerations. And when we read on, and find Hegel exclaiming that in the Newtonian account "trivial phenomena obtained by intricate experiments are used as arguments against [Goethe's theory]; [that] Newton's experiments themselves are complicated, bad, pettily done, mean, and dirty" (*PN*, § 320, *Remark, Z*, p. 211), we must admit that Hegel often did lack a grasp of the circumstantial procedures of theoretical science. At least, his feel for these was sometimes so weakly developed that it yielded to the stronger pressure of the notional account, fed by the psychological motivations just mentioned.

V

Let us, however, now briefly consider the laws of rectilinear propagation, reflection and refraction. Since any theory of reflection in terms of the motion of luminiferous particles or the propagation of waves is rejected, only notional considerations are left, mixed up in a peculiar manner with a purely geometrical approach, perhaps somewhat akin to the procedure of Ptolemy's theory of reflection.[28]

The law of rectilinear propagation, according to Hegel, is implied by the fact that the manifestation of objects to each other is a purely spatial relationship; furthermore, since no other conditions obtain bar this, it must be "*direct* or rectilinear". Presumably this conclusion assumes some principle of simplicity, though Hegel does not make this explicit. Besides, it is not clear whether 'rectilinearity' has any meaning without the conception of a *ray*; but as already remarked, Hegel did not object to the use of this conception, provided it be treated as a purely geometrical model.

As regards reflection, Hegel's account is extremely brief, implying that we must not regard it as a form of deductive derivation. All consideration of material categories of transmission, he tells us, must be excluded. We are in these considerations simply to confine ourselves to the definition of light, "as abstract ideality, as inseparable self-externality", as "intrinsically spatial" (*PN,* § 278, p. 96). Once again, use is made of the basic notion of reflection: one object can be manifest to another, and that to a third: this is a basic aspect of light. Hegel now at once concludes – hardly a deductively mediated inference – that

> in these spatial determinations the law of manifestation can only be that of *equality* – the equality of the angles of incidence and reflection, and the *unity* of the plane of these angles. Nothing whatever is present which could in any way alter the identity of the relation (*PN,* § 278, p. 96).

Evidently, the 'can only be' of the text, and the appeal to simplicity consideration, shows that Hegel is not providing a deductive account but a notional contribution to what he knows is supported on experimental grounds. But such looseness pays of course its price. For as soon as we turn to more complex phenomena, e.g. of the polarization of light, the notional account can get us no further. Hegel, instead of admitting this, however, simply falls back altogether on to the pure *description* of a limiting case: the mere determination of the angle at which a second reflection occurs does not lead to any diminution of the light intensity.

Still, for us the moral is interesting: between notional determination, theoretical processing, and experimental description, any number of emphases are possible. At no point does Hegel flagrantly conflict with the observational phenomena; at no point is there an inkling of that famous '*a priori* deductive' approach of which he is usually accused, especially in the context of his optics. So his various escape routes can only be conducive to teaching us something about the stresses and strains that persist between the different components of scientific reasoning. If anything, Hegel – like Goethe – had too little appreciation of the importance of exceptions to any simple 'empirical' approach. Time and again we are told that a certain empirical phenomenon is in conflict with predictions from Newtonian theory, when in fact the 'conflict' could easily be resolved if Hegel allowed Newtonian constructions their rightful place. Instead, he sees his 'empirical phenomena' as backing up – though usually only illustratively – his notional interpretation.

Hegel's treatment of refraction is, however, perhaps one of the severest tests for his whole approach; where we can, so to speak, put the method under a critical microscope. For since he has abandoned the concepts of the physicists, i.e. their theoretical constructions such as those of Descartes or Newton, of Huygens or Young, which all employ 'deficient' 'materialistic representations' of light, and yet has no other theoretical ideas to put in its place, we may expect him to make a wholesale notional determination take the place of theoretical explanation; though once again, as he expressly says, merely "stating and sticking to general points of view" (*PN*, § 318, *Remark*, *Z*, p. 192).

Theories of refraction have often been a touchstone and symbol for metaphysical standpoints. Descartes's methodological as well as metaphysical (mechanistic) views are perfectly represented in his theory of refraction, just as Leibniz's teleological idea of 'least impediment' was the focal technical model for his all-embracing metaphysics of finality, of physical contingency being an expression of an optimum choice.[29]

What were some of the basic phenomena? When you fill a kettle with water, and view it from above, the bottom appears raised, and – as an immediate inference – the apparent volume is diminished. Geometrical optics explains this phenomenon by appeal to the refraction of the light rays, in accordance with Snell's law of sines; qualitatively, as the light passes from the water into air, it is bent away from the normal to the interface, the 'angle of incidence' increasing with increased angles of refraction. Note, moreover, that if the angle of incidence is zero, the angle of refraction is zero likewise, and the resulting phenomenon ought not to appear: a coin viewed vertically from above ought not seem to be raised. Yet it is! (*PN,* § 318, *Remark, Z,* p. 191).

This is a prominent specimen of one of Hegel's 'counter-examples', to 'prove' that even the physicists' own theory leads to a conclusion inconsistent with observation! I mention this to show that Hegel does not oppose physical theory merely on ideological grounds. Unfortunately, the counter-example involves a confusion: since both the eye of the observer and the coin are objects extended in space, Hegel's assumption of zero-angle of incidence does not apply. Still, as usual, we should note the intention, and the methodological implications, which make good sense.

What of the 'positive', notional 'explanation' of refraction? Let me state it in brief outline. We need to stick entirely to the phenomenological situation, employing solely concepts that have emerged in the course of the increasing specification of the logical Idea. The concepts involved are light, the transparency of the two media, and one primary property – there are others, but they have less relevance in the context – viz. 'specific gravity', or the relative density of the media. Then, Hegel holds, the only property which has any relevance to the passage of light from one medium to the other must be the specific gravity.

Hegel now proceeds to a move, very difficult to grasp for any quantitative, let alone mechanistically orientated approach, although it bears a curious likeness to one of the many hypothetical accounts of refraction – though unlike Hegel's, framed in purely quantitative terms – which Kepler conjectured in his treatise on optics of 1604, the *Ad Vitellionem Paralipomena.*[30] Whether Hegel, an avid student of Kepler's writings, was influenced by this work is not known. At any rate, Hegel talks as though the specific gravity of one medium could be 'active'; as though, for instance, the specific gravity of the air could become *affected* by the specific gravity of the water (*PN,* § 318, *Remark, Z,* p. 190); specific gravity being regarded here as a hybrid quantitative-qualitative concept, defined as 'space-determining form'; as though it were a kind of activity whose effects can spread.[31] Suppose now the eye, starting from the visual medium of the air, to reach a certain volume of the water, say V_w; we may then regard the situation as though a corresponding volume V_w of air had been impregnated with the specific gravity of water, contained in that volume, the latter resulting in the determination of a new (apparent) volume. Since the imaginary mass of air in the volume V_w is constant, to the greater specific gravity there must therefore correspond a smaller volume; the volume of 'air' (i.e. water) will seem diminished, which was what we tried to explain. (Remember: density $= \frac{\text{mass}}{\text{volume}}$.)

"This is the sort of way in which this phenomenon must to be grasped", Hegel remarks, "it may appear artificial but there is no alternative" (*PN,* § 318, *Remark, Z*, p. 190). We may wonder. Since about 775 cc of air at N.P.T. weigh 1 gram (its sp.gr. taken as 0.00129 g/cc), a very large diminution in apparent volume ought to result! This simple fact is overlooked in Hegel's reasonable (though incorrect) point that "the refractive power depends on the specific gravity of media" (*PN,* § 318, *Remark, Z*, p. 192). Besides, Hegel has to admit that even this is not always true; whereupon he falls back on auxiliary saving devices, introducing *ad hoc* the relevance of additional properties. It is clear that, being intent on operating with notional considerations, from their nature 'qualitative', the rejection of the scientist's quantitative approach has turned into a degenerate device; "the driving energy of Reason", of which Hegel once speaks (*PN,* § 316, *Remark*, p. 179) has spilled over into the theoretician's domain where it does not belong; the stratification of the metaphysical and inductive levels is insufficiently maintained.

Of course, this idea of a qualitative property impregnating another substance belongs to an earlier period, and it is indeed frequently the case that we find in Hegel such reversions to Renaissance and even earlier conceptions. The most interesting aspect here is his account of specific gravity as an 'intensive' quantity. The readiness with which he employs this device derives its 'justification' (in Hegel's eyes) from its definite position in the evolution of the logical Idea. In the relevant paragraph of the *Philosophy of Nature*, where he notes that the purely quantitative ('extensive') approach to specific gravity does "not express any reality" (*PN,* § 293, *Remark*, p. 128), he refers us back to the section of the *Logic* which deals with the doctrine of Being, under the heading of 'Quantity', and more specifically 'Degree' (*PN,* § 293, *Remark*, p. 127); and this itself is viewed as a transition to the next section, 'Measure' regarded as the 'qualitative quantum',[32] a determination of the Notion which, as he explains in the *Zusatz* to Sect. 111, immediately leads to Essence.[33] This approach to specific gravity as an intensive quantity is (as Hegel notes) due to Kant, where, in the section on 'The Anticipations of Perception' of the first *Critique*, it supposedly receives its transcendental justification (*CPR*, esp. pp. 206–207 [A173–5/B215–16]). In Hegel's hands this treatment of specific gravity is simply an attempt to replace the otiose atomistic approach of traditional mechanistic theory by a new hybrid quantitative-qualitative conception. It usefully links with his idea of light as something inward, since intensive magnitude for him "points to the category of Measure and begins to hint at an inwardness which, as a determination of the Notion, is an immanent determinateness of form, which only appears as a quantum by way of comparison" (*PN,* § 293, *Remark*, p. 127). In this way – I confess that Hegel's reasoning is almost oppressively opaque at this point – a connection is made between transparency of media and the relationship of two transparent media, whereby only the different specific gravities "bring about a particularisation of visibility". And so we get the transition to the final view that

> the second medium is, so to speak, infected by the immaterial density of the first one posited in it, so that the space in which the image is seen in the second medium shows the same contraction as is suffered by this medium itself (*PN,* § 318, *Remark, Z,* pp. 187–88).

Perhaps the fascination of the whole story for us *is* just that it is so vastly at variance with the methodological constraints of the scientific tradition which has come down to us these last 300 years. It provides us with a *reductio ad absurdum* of that view which I already mentioned when quoting from the Hegelian introductory passage to the *Philosophy of Nature*, where we are told that scientific revolutions originate solely from the change in categories through which Spirit seeks to comprehend itself more truly (*PN,* § 246, *Remark, Z,* p. 11).

> Light must not be represented as a material propagation: the water, as visible, is ideally present in the air. This presence is a specific gravity with this specific determinateness alone, the water preserves and exerts itself in that into which it has been transformed, and so transforms this its transformation into itself (*PN,* § 318, *Remark, Z,* p. 190).

Clearly, this is a different world. Hegel appreciates this; he offers us certain operative analogies, for example, that of the petty soul which "measures the grandeur of a great man's heroic deed by its own dwarfish standards and reduces it to its own level" (*PN,* § 318, *Remark, Z,* p. 190).

> As the hero I imagine is actively present in me, if only in ideal fashion, so too does the air receive within itself the visual space of the water and dwarf it to itself (*PN,* § 318, *Remark, Z,* p. 191).

Evidently, this is 'a different way of seeing' the physical phenomenon. The fact that the natural manifold needs interpretation through spiritual articulation has become transformed into a method of envisaging the phenomena themselves as spiritual. It has become fashionable recently to emphasise that Hegel's Logic is not the logic of Russell or Frege, but that its roots lie in the mystical visions of a Jacob Boehme or a Meister Eckhart, and that his influence must be sought in the direction of the poets, both from whom he received so much (Goethe, Hölderlin) and to whom he bequeathed at least his language, if nothing more (e.g. the influence on Mallarmé in recent times). For the philosopher of science, the value of a study of Hegel's logic lies in its clear characterisation of the structure of scientific theory, of its conceptual component, its theoretical dimension, and its empirical data. And the value of such a study is increased, not diminished, when we find it taken to extremes, transcending the boundaries of conventional methodology, for this teaches us something about the limits of its usefulness. In many cases, as remarked before, Hegel has to abandon the articulative power of his Logic; the empirical sensory material then serves only as illustration of the stages of the Notion. In this essay I have tried to light up those corners in the Hegelian edifice where we may glean a more definitive structure of scientific reasoning. Whether the specific abandonment of theoretical mechanistic approaches, in favour of the more plastic notional imposition, of which the theory of refraction gives such a bizarre example, might ever find application in other more naturally adapted regions of scientific enquiry, we must leave to subsequent speculation.

University of Cambridge
England

NOTES

1 Lloyd William Taylor, *Physics. The Pioneer Science* (Boston, Houghton Mifflin, 1941), p. 490. (New York, Dover, 1959, 2 vols.)

2 I have occasionally (as here) modified Miller's translation. Taylor himself actually quotes from the summary given by J. B. Stallo, *General Principles of the Philosophy of Nature* (London, 1847).

3 Taylor, *loc. cit.*

4 I cannot do better than refer the reader to J. N. Findlay's excellent modern introduction, *Hegel. A Re-examination* (London, George Allen and Unwin; New York, Humanities Press, 1958).

5 Cf. *Logik*, §§ 96–7, pp. 189–92. I have made my own translations of relevant passages from the *Logik*, Wallace's rather free translation being often too unreliable (*The Logic of Hegel*, second edition (Oxford, Clarendon Press, 1892)). I have, however, given the corresponding page references from Wallace, thus here: *L* [W, 1892], pp. 179–81.

6 *PN*, § 86, *Z*, 2, p. 166; cf. *Logik*, § 131, *Z*, p. 261; *L* [W, 1892], pp. 159, 240.

7 *PN*, § 246, Remark, p. 11. Cf. also *PG*, § 381. *Z*, pp. 21–3, for a summary of Hegel's position.

8 Cf. *Logik*, § 86, *Z*, 2, pp. 168–8; *L* [W, 1982], pp. 159–61. Cf. also *Logik*, § 160, *Z*, p. 315; *L* [W, 1892], p. 287.

[9] Cf. Gerd Buchdahl, 'History of Science and Criteria of Choice', in *Historical and Philosophical Perspectives of Science. Minnesota Studies in the Philosophy of Science*, vol. 5 (ed. R. Stuewer, Minneapolis, Minnesota University Press, 1970), pp. 204–30. I have used a similar criterial scheme for a more general characterisation of Hegel's Philosophy of Nature in 'Hegel's Philosophy of Nature and the Structure of Science', *Ratio* **15** (1973), 1–27.

[10] *Logik,* § 136, pp. 269–75; *L* [W, 1892], pp. 246–51.

[11] *Logik,* § 140, p. 276; *L* [W, 1892], p. 253; *PN,* § 246, *Remark, Z*, p. 12.

[12] Cf. Hegel's remarks in *Logik,* § 195, *Z*, p. 369 (section on 'Mechanism'); *L* [W, 1892], p. 337.

[13] I have paraphrased this, since neither the German text, still less the translation, give much inkling of what Hegel intended.

[14] *Logik,* § 136, p. 270; *L* [W, 1892], p. 248.

[15] *PN,* § 267, *Remark,* p. 58n. Cf. also p. 67, the reference to Francoeur's *Traité élémentaire de mécanique* (Paris, 1798). Hegel goes on to say that Newton's presentation is such that 'what have been adduced are not so much propositions as bare facts; and the requisite reflection is only this, that the distinctions and determinations brought forward by mathematical analysis, and the course it has to follow in accordance with its method, are wholly distinct from what is supposed to have physical reality'.

[16] *PN,* § 266, *Remark*, p. 52; cf. *SL* [M], pp. 379 ff.

[17] Cf. *Logik,* § 102, pp. 203–4; *L* [W, 1892], pp. 190–92.

[18] *PN,* § 267, *Remark*, p. 59. For a somewhat fuller development, see my *Ratio* article, note 9 above, sect. vii, pp. 17–21.

[19] *Logik,* § 97, *Z*, p. 192; *L* [W, 1892], p. 181.

[20] Cf. Gerd Buchdahl, *Metaphysics and the Philosophy of Science. The Classical Sources: Descartes to Kant* (Oxford, Blackwell; Cambridge, Mass., MIT Press, 1969), ch.8, sect. 4c(iv), pp. 512–16.

[21] Cf. Kant's *Prolegomena and Metaphysical Foundations of Science* (tr. E. B. Bax, London, Bell, 1883), ch. II: 'General Observation on Dynamics', p. 200; *Schriften* (Ak. ed., Berlin, 1911), vol. 4, p. 524, line 24. Also Buchdahl, *Metaphysic . . . .*, pp. 567–8.

[22] *Logik,* § 97, *Z*, p. 192; *L* [W, 1892], p. 181.

[23] *Logik,* § 98, p. 193; *L* [W, 1892], p. 182.

[24] *Logik,* § 112, *Z*, p. 224; *L* [W, 1892], p. 208.

[25] *PN,* § 275, p. 87; § 277, p. 95.

[26] *Logik,* § 112, *Z*, p. 224; *L* [W, 1892], p. 208.

[27] Cf. E. Nagel, *The Structure of Science* (New York, Harcourt, 1961), ch.6, especially sect. 4; P. Duhem, *The Aim and Structure of Physical Theory* (tr. P. P. Wiener, Princeton, Princeton University Press, 1954), Part I; N. R. Campbell, *Foundations of Science* (New York, Dover, 1957), ch.6. For an earlier source, E. Mach, *Die Principien der Wärmelehre* (Leipzig, Barth, 1896): 'The contrast between mechanical and phenomenological physics', pp. 362–4.

[28] For Ptolemy, see M. R. Cohen and I. E. Drabkin, *A Source Book in Greek Science* (New York, McGraw Hill, 1948), pp. 268–71.

[29] Cf. Buchdahl, *Metaphysics* . . . ch.3, sect. 2c–d, for Descartes's Optics, and ch.7, sect. 3b, pp. 425–34, for Leibniz's use of his theory of refraction.

[30] Cf. Johannes Kepler. *Gesammelte Werke*, vol. 2 (ed. W. v. Dyck and Max Caspar, Munich, Beck, 1938 –), ch.2, pp. 85–86. See my 'Methodological Aspects of Kepler's

Theory of Refraction', *Stud. Hist. Phil. Sci.* **3** (1972), 265–298, especially pp. 283–86.
[31] *PN*, § 318, *Remark*, *Z*, p. 190; cf. § 293, *Remark*, p. 127.
[32] *Logik*, § 107, p. 215; *L* [W, 1892], p. 201.
[33] *Logik*, § 111, *Z*, p. 221; *L* [W, 1892], pp. 205–206.

JOHN J. COMPTON

A COMMENT ON BUCHDAHL'S 'CONCEPTUAL ANALYSIS AND SCIENTIFIC THEORY IN HEGEL'S PHILOSOPHY OF NATURE (WITH SPECIAL REFERENCE TO HEGEL'S OPTICS)'

In his 'Foreword' to the English translation of Hegel's *Philosophy of Nature*, J. N. Findlay confesses that the section devoted to physics is the most difficult part of the work and "requires a detailed, paragraph-by-paragraph commentary by one as much versed in physical science and its history as in Hegelian concepts" (p. xviii). Fortunately for us, Gerd Buchdahl is just such a one. His lucid and subtle comments illuminate what are indeed very dark corners of the Hegelian edifice. My intent in these brief comments on the comments is simply to lure him on to further, more intensive and extensive luminosity.

At several junctures in his paper, Buchdahl suggests that his purpose is to explicate what he regards as a general Hegelian philosophy of science – particularly its "clear characterization of the structure of scientific theory, of its conceptual component, its theoretical dimension, and its empirical data." (p. 34) He does this by considering cases. He carefully examines Hegel's treatment of gravitation, free-fall, matter and force, and most notably of optical effects, in order to explore the relationships among (i) the 'notional' meaning of the related concepts, (ii) the prevailing theoretical constructs used by the physicists to deal with them, and (iii) the observed phenomena. Buchdahl effectively disposes of the hallowed misunderstandings which have Hegel either totally *ignoring* empirical facts and regularities or else claiming somehow to *derive* them deductively from the notion of Nature as manifestation of the Idea perfecting itself in self-conscious spirit. And he points out the complex relations of agreement and disagreement between the developing Notion and the scientific laws and constructs. However, from this the implied (if not promised!) general methodological account does not emerge. *Does* Hegel have a statable, general theory of science, and particularly a theory of the logical relationships between his philosophy of nature and empirically based, scientific theorizing? And if so, what is it? Putting Hegel's specific, contentual claims aside – including the claim that there *is* only one coherent philosophy of nature, namely his own – what I should like to know is whether one could take *his philosophical style* as a paradigm and go out into the world, the world of contemporary physics, say, and *do* philosophy of nature after him? What would one do and how would one understand the task?

R. S. Cohen and M. W. Wartofsky (eds.), Hegel and the Sciences, 37–40.

Now I do not even *hint* that I am able to answer these questions. I believe that Buchdahl can do so and I profoundly hope that he will. However, I can suggest a possible context for an answer – one which is already implicit, I believe, in Buchdahl's paper, and one which is by now familiar in any case.

Consider, for example, that there are three distinct levels of scientific thought – and their correlative concepts of nature: (1) The level of *observed and experimentally tested regularities* in the behavior of things; (2) The level of *physical constructs or theoretical explanations*, hypothetically offered to account for those regularities – leaving the crucial expression "to account for" undefined for the moment; and (3) the level of *metaphysical or conceptual models* which represent what is thought to be intrinsically intelligible or self-explanatory in nature, and which function in a regulative way to suggest hypothetical constructs for possible scientific use.

Now consider further that there are competing master, metaphysical models – among them one given by the tradition of atomistic materialism, suitably amended by adding various sorts of forces, and another based on the notion of a self-alienating and self-fulfilling Absolute Spirit. And suppose, still further, that while most of the tested and well-accepted physical constructs have been suggested by the atomistic-mechanical model, one has convincing reasons to believe that this model is inferior in systematic coherence to that of the Absolute Spirit. Then it seems clear that what one should rationally set out to do is precisely what Hegel did set out to do in the *Philosophy of Nature*: One should try so to develop and apply the model of nature based on the notion of Absolute Spirit that it would give sense to the given phenomena of level (1) but without using the physical constructs of level (2), which are based on the competing, atomistic-mechanical picture.

But "give sense" or "account for" in what sense? What precisely should be the logical relations among the three levels of scientific thought? At this point, as Buchdahl writes the story, Hegel seems to have failed clearly to distinguish two quite distinct alternatives or, if he did, he seems to have taken them both.

I. One option is a *conventionalism* of scientific theory. It is to treat the (largely atomistic-mechanical) physical constructs of level (2) as *merely* hypothetical, conventional, or instrumental, but not true, while taking the metaphysical model (of the self-exteriorizing Spirit) as truly referential. Then the philosophical task is simply to *eliminate false reifications* of forces and to *re-interpret* the given phenomena of level (1) directly in notional terms drawn from the metaphysical model (3). This is the view that seems to be implicit in Hegel's analysis of gravitation and of free-fall, where he expresses strong preference for formal, mathematical laws only and seeks to supplement such laws with a qualitative demonstration of their inner meaning and basis. Similarly, he appears willing to allow talk of light rays and of their rectilinear propagation so long as this is understood to refer to purely geometric, heuristic representations. On this option scientific agnosticism (*à la* Kant) allows metaphysical fulfillment (*à la* Hegel).

II. On the other hand, a second option is a *realism* of scientific theory. But this implies a new and more adequate scientific theory, the constructs of which will be based on the genuinely referential model of the Absolute Spirit. Here, the philosophical task is not only to criticize materialistic representations and to re-interpret observed phenomena in notional terms, but more: it is to use the notional determinations to *generate new theoretical hypotheses* which might afford discovery and explanation of *new phenomena*. While Hegel never explicitly espouses this view, perhaps it is nascent in his reformulation of the Kantian dynamic theory of matter – which could be used to suggest a field theory of matter-energy. And it seems obvious in his treatment of the relations between light and space which, as Buchdahl points out, yields some astonishing insights such as that of the dependence of space upon the dispersion of light-energy. Even, and perhaps especially, Hegel's 'excessive' spiritualization of light in his account of refraction, suggests this more constructive role for metaphysical models – for we should not expect a metaphysical model invariably to yield positive results!

Now let me put my basic puzzlement this way: Why is it that Hegel held back from this second, more constructive view of the role of philosophy of nature? Why did he seem to favor the first, more cautious option – conventionalism and formalism of scientific theory coupled with metaphysical redescription of the already given empirical data? To be sure, it is far less risky; but it is also far less dialectical, less open to permitting new phenomena to bear upon and perhaps force modifications in what we had *thought* to be the categorial features of nature, and in being far less dialectical it is less true to Hegel's own vision. The second option would have more adequately fulfilled Hegel's aim to reveal the drama of the Idea not only as *consistent* with known phenomena but as *requiring* new ones – not deductively, but in the way in which a rich and fundamental model affords new analogies for examination, quantified formulation and test.

Perversely, I suppose, I find myself wishing that Hegel had indulged in some more 'bizarre' excesses like that of "the ideal presence of water in air"; he might then have found some more astonishing successes as well.

Vanderbilt University
Nashville, Tennessee

DIETRICH VON ENGELHARDT

THE CHEMICAL SYSTEM OF SUBSTANCES, FORCES AND PROCESSES IN HEGEL'S PHILOSOPHY OF NATURE AND THE SCIENCE OF HIS TIME

In the history of Hegel research his philosophy of nature has been unjustly neglected. Only some fifty studies of this subject have appeared since 1830; and the greater part of these are devoted to the philosophy of nature as a whole. On chemistry as a special domain of physical nature no more than perhaps five treatises are to be found. Academic concern with Hegel's philosophy of nature has not been less insignificant. In the years from 1945 to 1970, 15,500 seminars and lectures were held at the universities of the Federal Republic of Germany. Of these only twelve treated Hegel's philosophy of nature, only two exclusively so. This low esteem seems strange if we consider that the second part of Hegel's system is formed entirely by the philosophy of nature and recall furthermore how much attention the philosopher had devoted to the natural sciences throughout his life. Indeed, the treatment of philosophy of nature was so important for Hegel that he did not hold back from polemical commentaries dealing with the entire formal-analogical speculation about nature of his time, and he did not shrink from allowing his friendship with Schelling to be broken off because of these.

The rejection which his philosophy of nature has met, is based above all on Hegel's alleged contempt for and neglect of both empirical work and modern research. The general indifference displayed toward the philosophy of nature after 1830, however, cannot adequately be explained by these objections, because they were raised only among natural scientists and by few of them even though with a vehement and offensive tone; and they found but meager resonance. The reason for the indifference lies rather in the polarization of the sciences of nature and the sciences of the mind [*Geisteswissenschaften*] during the nineteenth century, which Hegel had foreseen; indeed he had seen one essentially important significance of philosophizing to be the overcoming of that polarization. The positivist sciences' understanding of themselves, separating nature and mind [*Geist*], and apparently legitimated by impressive technical achievements, has become the epitome of the dominant attitude toward nature after the idealist period; it has confirmed a self-centered and self-satisfied readiness to leave the world of nature to the empirical approach and knowledge, and in this way has freed the natural scientist from the philosophic requirement to establish the grounds for the forms and materials of nature.

R. S. Cohen and M. W. Wartofsky (eds.), Hegel and the Sciences, 41–54.

The study of Hegel's philosophy of nature offers the possibility of allowing reflection about the falling apart of nature and mind to begin again. Hegel's philosophy of nature can be investigated from three perspectives: (1) according to its place in the system as a whole, its relation to the logic and the philosophy of mind; (2) with respect to the contents and structures themselves; and (3) in its relation to empirical studies. However, while a critical interpretation of the idealist philosophy of nature will receive little consideration and support by those philosophers who are scarcely interested in nature, it could much more readily awaken the unrestricted interest of historians of natural science and medicine if it were to take on the form of a confrontation between empirical science and philosophy.

We will seek to give an example of such a method of interpretation in what follows, and this with respect to chemistry alone. To a limited attitude toward nature, such a restriction may seem questionable, which is incontrovertible in the domain of the mental [*Geistige*]; for who would expect of an investigation in philosophical psychology at the same time an evaluation of the philosophy of right? The scientific system of chemistry will be presented first, and after that the philosophical; and finally we will set forth the essential 'moments' of their agreement and difference together with several general provisions [*Bestimmungen*] of the philosophy of nature.

I

To unite system and empirical work was the general goal of natural science in the second half of the eighteenth century. To meet the demand of empirical research and at the same time to proceed systematically, this was to initiate a new phase of research. In addition, mathematics was declared to be the model of scientific method, the possibility of mathematization held to be proof of the scientific (d'Alembert, 1751: *Discours prèliminaire*; Kästner, 1768: p. 2, 5; Erxleben, 1772: Preface; Condillac, 1780). Both parts of natural science – theoretical knowledge of nature and natural history – sought to achieve the principle of the empirical combined with the systematic. In natural history, the description and classification of phenomena of the three natural kingdoms was required, which meant the development of a natural order instead of an artificial system or any arbitrary preoccupation with some specific subject-matter; in the natural sciences explanation of the causes and effects of phenomena was demanded, the uniting of separate phenomena and their manifold conditions into a natural order. Theoretical natural science was subdivided into physics and chemistry, natural history into mineralogy, botany and zoology. However, these fundamental criteria of research were by no means met consistently, neither in the theoretical sciences nor in natural history. This shortcoming also applied to making mathematics absolute, which in any case found persuasive critics.

Chemistry, the science of elements, compounds, forces and processes, was also guided by the general goal of the natural sciences (Erxleben, 1775: Preface and p. 5ff; Lavoisier, 1789: *Discours prèliminaire*; Berzelius, 1820: p. 19ff), but it too was not capable of complete success. For here too mathematics served as certification of scientific character. The conception of the elements and their combination was transformed in the closing years of the eighteenth century by the controversy between phlogistic and antiphlogistic chemistry. Around 1800 the oxygen theory had established itself. It, and the doctrine of forces and processes developed mainly by the proponents of the phlogiston theory, was combined during the nineteenth century with the atomic and electrical theories to form a system of chemistry.

The concept of the element fulfilled the requirements of empirical investigation, insofar as it was bound to the observable. In this way the absolute concept of the element was replaced by a relative one. Element is no longer, as formerly, simply and entirely indivisible, but an undivided substance whose future dissociation [*Disoziirung*] could not be excluded. Instead of the designation 'element', the term 'undivided substance' [*unzerlegte Substanz*] became current (Trommsdorff, 1800, p. 22ff). This change of meaning took place in phlogistic as well as antiphlogistic chemistry (Erxleben, 1775: Preface; Elliott, 1784; p. 7; Lavoisier, 1789: *Discours prèliminaire* and p. 193ff). The absolute conception of element of the past did survive, however, in the confrontation of simple, undivided and compound substances, in which, however, the older conception of three or four elements was not taken up again, but instead all those things were included among the simple substances, whose further division was considered impossible. Although, with the principle of quantitative measurement, antiphlogiston chemistry had replaced phlogiston by oxygen, it propounded a view of elements which was indifferent as to weight, for it accepted light, heat, magnetism and electricity among the elements. Only after the turn of the century did the conception gradually prevail that chemistry was the science of substances which could be weighed; with that the so-called imponderable substances were dropped from the class of elements. The vacillation of the concept of element, and the discovery of new elements and compounds, which till then had been considered as basic substances, led to the virtually unchanged number of about fifty elements.

To the concept of element corresponded the concept of compound. By this was understood the union of two or more basic substances. A compound was considered to be what could be dissolved into simple substances, and then be put together again from these. Although synthesizing organic substances was not accomplished until the second decade of the nineteenth century (Wöhler, 1828), their status as compounds was established earlier. Physical mixtures and chemical compounds were not always clearly distinguished, especially during the eighteenth century. A contributing cause to the imprecision of the concept of compound was the indecisiveness about the elementary nature of the substances that could not be weighed – light, heat, magnetism and electricity. Around 1790 the number of possible compounds was determined by arithmetic means to be about 1200; however, this was a hypothetical number that could not be verified empirically, for its reality was far from being scientifically attested (Lavoisier, 1789: p. 182).

But chemistry was not regarded as an exclusively geometrical and static science of the elements and their compounds; rather, it was above all a theory of forces and processes. Basically, the chemical processes were divided into synthetic and analytic (also diathetic), the former producing compounds from the basic substances, the latter extracting basic substances from the compounds. The theory of the forces which underlie chemical changes was also called the theory of relationships [*Verwandtschaftslehre*], the theory of attractions and affinities. Generally a combinatory affinity and a simple and dual elective affinity were distinguished; the former designated the striving of heterogeneous or differentiated substances to amalgamate to form homogeneous or undifferentiated substances, the latter the tendency of homogeneous substances to separate or to exchange their components. In the last third of the eighteenth century, this theory was decisively completed by the phlogistonists, while at first the antiphlogiston chemists neglected this domain (Lavoisier, 1789: *Discours préliminaire*). According to the theory of affinities, it should not only be possible to specify in which proportion elements and compounds would combine, but also which distinguishable degrees of affinity exist among the substances. With the exploration of preferences, the theory of chemical forces had already made its first beginnings shortly after 1700 (Geoffroy, 1718), but only in the last third of that century was it extended to the whole of chemistry and an attempt made to bring chemistry under mathematical categories (Bergman, 1775; de Morveau *et al.*, 1777; Wenzel, 1777; Richter, 1789, 1972–4). The achievement of the mathematization principle first raised chemistry from the status of an art to the rank of a science, as was emphasized everywhere (de Morveau *et al.*, 1780: vol. 2, Preface; Richter, 1792: vol. 1, pp. 1 ff). This transformation appeared all the more justified, the more the demonstration of the constancy of the proportions of substances within compounds succeeded, and where these proportions were variable, to show them as multiple – which means to show that they stood in a simple relation to each other. The theory of forces was questioned at first, before its final adoption due to the observation and investigation of such factors as heat, physical conditions, and mass, which influenced the constant and multiple proportions; but finally the theory was refined and extended by these insights (Berthollet, 1803). But the combination of the theory of attraction with electrochemical theory (Volta, 1792, 1800; Davy, 1807, 1812; Berzelius, 1803, 1819) and atomic theory (Higgins, 1789, 1814; Dalton, 1808–1810) proved much more problematical. According to the electrochemical view, also called Galvanism, substances were positive and negative electrically charged matter; the chemical processes were the uniting and the separating of the two kinds of electricity. This interpretation reduced the theory of attraction to the two electrical poles, and beyond that made the validity of atomic theory appear questionable; for with the tendency then prevailing to regard electricity – as well as heat, light and magnetism – not as substance, but as force or manifestation of forces, required that a dynamic view of nature become more probable again. To be sure, the controversy, which was not to be resolved empirically, was then decided in favor of atomism. The dogmatic and, for the time, still ambiguous character of the system of chemistry that was linked to this was certainly noticed by scientists, as in general it had to be admitted that science and its progress was unthinkable without hypotheses and speculation (Gmelin, 1817: vol. 1, pp. 54ff; Berzelius, 1820, p. 19ff).

Elements, compounds, forces and processes were the constitutive 'moments' of chemistry between 1780 and 1830. A closed system was not, however, produced from these moments by the scientists of those decades. Fundamentally a system of chemistry could have been constructed from two principles and their combination: first, according to a principle of substances, and this either in the form of a mathematized classification of the elements and their compounds, those of a simple and a higher order, or in the form of a natural history taxonomy; the other according to the principle of forces and processes. Neither of the two possibilities nor their integration into a comprehensive unity was realized. The system of chemistry remained disjointed among the varied presentations because of the irregular connection of the two fundamental possibilities, as well as due to the special preeminence accorded to specific substances or forces, as for instance oxygen or electricity. To be sure, proceeding logically from its scientific premisses (Condillac, 1780), antiphlogiston chemistry had developed an analytic method for the language of chemistry (de Morveau *et al.*, 1787) and for the system of elements and compounds on various levels, but then, due to the emphasis given to oxygen and an orientation reflecting the three kingdoms of nature, it had arrived at an arbitrary classification (Lavoisier, 1789). Furthermore, the theory of forces and the theory of processes on the whole remained excluded from the antiphlogiston theory, although they were mentioned here and there. This situation did not change in the subsequent decades. About 1800 a system began with a conceptual explanation of the elements, compounds, forces and processes, proceeded to the examination of fire, water and air, then went on to the acids, alkalis, oxides and chlorides, and metals and ended with an account of the mineral, vegetable and animal substances (Trommsdorff, 1800–1804). The organization of chemistry reached a certain consistency around 1815, since after the weightless substances the ponderable elements were presented, these divided into non-metals and metals; and to the description of the individual metals, that of the compounds of each was added. However, the theory of forces and processes was expounded independently of the elements and compounds (Gmelin, 1817). Around 1830, the non-metallic fundamental substances and their combinations stood at the beginning of the system; air and water and the theory of attraction followed; to these finally the acids, bases and metals were added (Mitscherlich, 1829–1830). All the chemical systems were characterized by the fact that the theory of forces and processes had no influence on the order of the substances. The theory was treated quite arbitrarily, now at the beginning, now at the end or at some point within the account of the elements and their compounds, and at times not at all or only in part. Substance, force and process remained apart. Beyond that, the theory of forces, of atoms and of electrochemistry were not and could not be brought into correspondence. By proceeding neither from one of the basic possibilities by themselves nor from their combination was a consistent order of chemical substances and processes developed. The universal demands placed upon science during the eighteenth century were not fulfilled.

II

Hegel's philosophy of chemistry is a unified system of empirical chemistry; substance, force and process are combined to form a real and conceptual unity. Without resorting to all those logical categories that are decisive for the study of chemistry, and without taking too much into consideration those physical phenomena and their philosophical interpretation that precede it and form its foundation, the general concept of chemistry can be represented and clarified in the following manner: chemistry is the 'indifferentiation' of differing substances and the 'differentiation' of indifferent substances. The first process refers to the uniting of opposed substances into neutral products, the second, their separation into substances that are again opposed. A third process is the combination of these two [*Vereinigung*]; it is the neutral process, the production of neutral substances through dissolution of other neutral substances. It neither begins with opposing substances, nor does it lead to opposing substances; it can be called the 'indifferentiation' of indifferent substances, and is called by Hegel the 'total process'. However, the chemical substances are not subjected to these changes by external, mechanical means – this would only lead to mixtures, aggregations, amalgamations; for Hegel this is "the *formal* process which is a combination of merely different, unopposed bodies" (*PN*, § 327, p. 236).

Instead, these changes are the expression of forces inherent in the substances, their striving to unite and to separate. Acids and lyes are in reality themselves the force of neutralization. The attractive forces belong to the individual substances; these are to be characterized according to the principles of the theory of attraction: every individual substance is a specific combination of bodily-individual phenomena and equivalent forces. As every substance is phenomenon as well as force, and the force does not coincide precisely with the phenomenon, but is instead directed toward its changing in chemistry, we can speak of a falling apart of concept – the goal of the substance's force – and reality – the phenomenal state of the substance. The process overcomes [*aufhebt*] this falling apart. Forces are the tendencies of heterogeneous substances to become homogenized, of neutral substances to separate again into heterogeneous substances or merely to exchange their components. In the then contemporary terminology which Hegel too employs, combinatory affinity and simple and double elective affinity are the conditions of the three processes. Such a philosophy cannot share the usual understanding of element and compound, which however, it well knows. In philosophical chemistry the element is simple, being the resultant substance of the synthetic process. On the other hand, by its different nature, namely as substance and force in one, the element is also a compound, while the compound, through its quality of indifference, its neutral character, which means the negation of the opposition of substance and force, is also elementary. This complex definition of element and compound cannot be reconciled with a purely atomic interpretation. But according to Hegel, the question of atomism or dynamism will not be decided within chemistry, but already in the derivation of matter.

The class of the simple product, the resultant substances of the synthesis, which are heterogeneous because of an affinity directed toward neutrality, is thus to be determined in a twofold respect: as heterogeneous substances with their striving toward homogenization they are in reality different, but as their striving is directed toward this homogenization, as it anticipates it, they are at the same time conceptually indifferent. Overcoming and transcending [*Aufhebung*] this contradiction between concept and reality is the synthetic process. Conversely the class of substances of the analytic process are indifferent in reality and different conceptually, as they are indeed neutral, but still consist of opposing substances to which separation will again restore them. The class of neutral substances, the formation of neutral compounds from neutral compounds, the process of exchange solely of the foundations of their composition, all remain on the level of neutrality or indifference. The substances different in reality, resultant substances, products of the synthetic process, are: metals, metal hydrates and metal oxides, non-metallic and combustible substances like sulphur, phosphorus, and finally the acids and lyes; the substances indifferent in reality, resultant substances, products of the analytic process, are: the salts. The subdivision of the synthetic and analytic processes into a number of subprocesses forms the basis of a detailed grouping of the simple and compounded substances. To list and describe all of the substances in detail, according to Hegel, is not the task of philosophy. Its task is to structure the manifold conceptually, but not to prepare a handbook for it.

The division of the chemical substances is philosophical; it must, however, be documented by experience: substances manifest their nature as phenomena, forces and processes in reality, and this nature establishes their place in the system. We cannot mention nor evaluate here the numerous empirical results, which Hegel selects from the literature to show what substances are qualified to represent the different or indifferent classes. The philosopher who, with the science of his time, has gone beyond the alternative of phlogiston or antiphlogiston chemistry, compiles natural perceptions and experimental observations from worldwide research, compares them, notes contradictions and incoherencies, and above all discovers the hypothetical character of many empirical statements and the metaphysical within natural science. Pre-eminently in the theory of forces and processes, of constant and multiple proportions and also with respect to the table of equivalents, Hegel proves remarkably knowledgeable. The history and the most up-to-date state of research are equally familiar to him.

The philosophical system of chemistry is not consistent solely because substance, force and process are brought into an immanent order, which furthermore is verified in empirical work, but also because a foundation is established for conceptually concrete connection with physics and organic science, and a derivation is provided for the specific independence of the various domains of nature. Reduction and the production of neutral compounds do not pass over from the one to the other without external mediation. The neutral process of double elective affinity is, to be sure, separation and compounding at the same time, but does not take place in the real difference between reality and concept, which is the distinctive mark of synthesis and analysis: in chemistry, there is no immediately coherent cycle of synthetic and analytic changes. The unity of the two processes is only a conception in thought without any corresponding reality. Just as concept and reality fall apart for the compounding and separating substances, so the concept of the unity in thought of the synthetic and analytic processes also does not correspond to the chemical reality. The first two incongruities between simple and compounded substances are overcome [*aufheben*] by the chemical processes, and the resolution of the second incongruity is achieved by a new process, the process of the organic. The existent combining of separating (analytic) and uniting (synthetic) acts is the distinctive mark of living individuality. In Hegel's words:

this *concrete unity* with self, self-produced into unity from the particularizing of the different corporealities, a unity which is the activity of negating this its one-sided form of reference-to-self, of *sundering* and particularizing itself into the moments of the Notion and equally of bringing them back into that unity, is the *organism* (*PN* § 336, p. 270).

More decisive than the deduced emergence of the living from chemistry is the real-ideal genesis of chemistry from the phenomena of physics that precede it. The concept of chemical reality follows immanently from the processes of magnetism and electricity; the concept of chemistry is the union, conceived in thought, of two aspects of magnetism and electricity, namely the conceptual indifference of a polar body – that is, magnetism – and the real indifference of two polar bodies – that is, electricity:

In magnetism, the difference is manifest in one body. In electricity, each difference belongs to a separate body; each difference is self-subsistent, and the whole shape does not enter into this process. Chemical process is the totality of the life of inorganic individuality; for here we have whole, physically determined shapes. . . . The two sides into which the form sunders itself are thus whole bodies, such as metals, acids, and alkalis; their truth is that they enter into relation (*PN,* § 326, *Zusatz*, p. 233).

Just as the organic realizes a concept that follows from chemistry, so chemistry realizes a concept that arises from magnetism and electricity. And beyond this, a class of physical elements is constitutive for the chemical substances and their transmutations; these elements are: air, fire, water and earth, or, as one could also say, airiness, fieriness, wateriness and earthiness. The physical elements are fundamentally distinguished from the chemical ones in that they are formless general materials and display no chemical affinity toward each other, indeed are in reality indifferent to themselves. Actually, earth, water, fire and air find each other without a chemical process, with neutral products resulting from this, even though they stand in changed relationships to each other; however, these are solely physical, as, for example, meteorological processes. Now the processes of chemical substances take place within the medium of the physical elements. The salts exchange their constituents in the medium of water. In chemical actions, physical elements too are decomposed into a further class of substances, called abstract chemical substances; abstract, as they are the decomposition products of physical elements. They are: oxygen, hydrogen, nitrogen and carbon. They unite with the chemical substances and multiply the forms of possible combination. From the immanent union of substance, force and process and the distinctiveness of their physical and their chemical mode comes the complete philosophical system of chemical substances, which differs considerably from the scientific one. Hegel states his view with respect to this in the following manner:

In empirical chemistry, interest mainly centres in the *particularity* of the *substances* and *products* grouped together according to superficial, abstract determinations in such fashion that no order is brought into their particularity. In this grouping, metals, oxygen, hydrogen, etc., metalloids (formerly called earths), sulphur, and phosphorus, are placed side by side as *simple* chemical bodies and on the same level. The great physical diversity of these bodies is such that it straightway arouses opposition to such grouping; no less varied, too, is their chemical origin, the process from which these bodies result. But in equally chaotic fashion, the more abstract processes are put on the same level as those that have more reality. If scientific form is to be introduced into this sphere, each product must be defined according to the stage of the concrete, fully developed process from which it essentially results and from which it has its peculiar significance; for this purpose it is equally essential to distinguish between the stages of abstraction or reality of the process.... A further matter is the empirical, quite special particularity of reaction shown by bodies to all other particular bodies; a knowledge of this involves going through the same litany of reactions to every re-agent (*PN* § 334, Remark, pp. 264–265).

Although the physical elements participate in the chemical processes and share in determining them, they are not properly chemical substances; the methods of physics are also transferrable to other domains of nature, but they do not lose their own proper meaning. Rejecting the intermingling of the domains of nature and their categories – an identification, which, according to Hegel, much more deserves a reproach than that which identity philosophy has received – also forbids inclusion of imponderable substances such as light, heat, magnetism and electricity within chemistry, since science has reached agreement concerning their physical character only hesitantly. Though electricity – like magnetism too – has a part in chemistry, and above all in the syntheses of metals, for which Hegel also adopts the term Galvanism, the category of polarity can by no means adequately explain the qualitative characteristics of substances, their forces and processes:

> Let philosophy no more be reproached with 'its ignoring of particularity and its empty generalities', when all the physical properties just mentioned can be disregarded in favour of positive and negative electricity. A former style of Philosophy of Nature 'potentialized' – or rather dissipated and attenuated – the system and process of animal reproduction to magnetism and the vascular system to electricity: such schematism was not more superficial than this reduction of concrete, corporeal opposition to electricity. In the former case, such a summary method of dealing with concrete phenomena to the neglect, by abstraction, of their characteristic features, was rightly rejected. Why not also in the present instance? (*PN* § 330, Remark, pp. 248–249).

According to Hegel it is also not legitimate to equate mathematizability with scientific status. The conditions for scientific character lie in the inner connectedness of conceptual categories with the phenomena, and this cannot be achieved only quantitatively, for the relation of quantity and quality in nature, apart from individual domains, has an indeterminate measure. An arithmetic classification of the chemical substances constructed with the usual scientific conception of element and compound is inadequate. However, the application of mathematics is not rejected in any fundamental fashion. To comprehend the measures of the constituents quantitatively is a justified and extremely significant goal, from the perspective of constant, as well as multiple and equivalent proportions. Hegel did not turn solely against the absolutized transfer of mathematical-physical methods to chemistry – according to him, it is just as impossible to explain the organic entirely chemically, and yet it is still permissible to subject the realm of the living to chemical analysis.

III

To unite system and empirical work was not the universal goal of science alone, for philosophy too took itself to be subject to the demand for a natural order. An adequate conceptual grasp and explanation of nature is, however, for Hegel, only possible for a science that does not deny resting on metaphysical conditions, that does claim to be without presuppositions, but is conscious of its metaphysics, tests it in its empirical work and orients its experience by it. The inadequacy with respect to nature, and the inconsistency of the chemical system are the result of a lack of metaphysical reflection, of the admixture of unrecognized speculative presuppositions with hypotheses and empirical theories; they are the result of the isolated treatment of chemical phenomena that destroys their natural unity, of neglect of the specific methodologies of individual domains of nature, and especially of the internal disconnectedness of the chemical substances, forces and processes. However, several other attempts in the natural sciences to pursue correspondence and lawfulness in the scattered and contingent sensuous world meet with Hegel's approval: the ordering of plants according to the principle of mono- and dicotyledons, the conception of the metamorphosis of plants, the division of animals into those with and without vertebrae, and, more specially, according to their teeth and claws, these are in accord with nature and not artificial; for teeth and claws are the instruments with which the animal itself confronts its environment, and plants too arise in the beginning as the unfolding of one or of two germinating leaves; the leaf is the basic form of the plant's form, and the vertebra the archetype of the animal bone structure. However, these attempts, too, are still inadequate for Hegel; they fail to see the one-sidedness of their category, do not comprehend its conceptual character, tear apart the unity of nature.

If for the formation of natural order, science must depend on philosophy, so philosophy must also depend on science. The conceptual categories are extracted from description and observation, the speculative operations that make a formal-material derivation of nature have to be verified in experience. The concept of chemistry is the result of the reality of magnetism and electricity – and chemical reality makes a concept emerge whose reality is the organic. The concept of chemistry and the genesis of a new concept must be capable of confirmation in empirical research. But, philosophy is not research, it must presuppose it, it must allow itself to be instructed by it: "Not only must philosophy be in agreement with our empirical knowledge of Nature, but the *origin* and *formation* of the Philosophy of Nature presupposes and is conditioned by empirical physics" (*PN*, § 246, Remark, p. 6).

Hegel refers to about 100 natural scientists; one third of these are chemists. The observations and insights incorporated in indirect and direct quotations, which – as a consequence of inadequate attribution – to some extent appear to be Hegel's text, testify to his complete familiarity with the fundamental traits, divergent theories and essential facts of the chemistry of his time, and even of its history.

The reproach that Hegel showed contempt for empirical study and neglected it, was the predominant justification for science to disparage Hegel's philosophy of nature; the reproach is indefensible. Three further accusations: low esteem for mathematics, scorn for experimentation, and rejection of technology, these too cannot withstand even a casual examination of his philosophy of nature. The utility of mathematics is emphasized, the necessity of its application to the disciplines of physics, chemistry, mineralogy, botany and zoology is demonstrated – but at the same time the independence of these domains is underlined. Experiment is recognized as a procedure – but beyond this the demand is made not to overlook the implications of artificial observation. The technological relationship to nature receives Hegel's approval, and is even justified as an integral 'moment' of the philosophical relationship to nature – but it is rejected as the sole guideline for research.

To place philosophy in contradiction to empirical science is, according to Hegel, one of the 'evil prejudices' and 'violent bisections'. The characterization of Hegel's philosophy of nature as closed off to empirical research must be abandoned; a critical exegesis will continue to investigate the conditions of agreement and difference between science and the philosophy of nature. To confront philosophy and science with each other facilitates access to speculation about nature, "a subject-matter that in any case is abstruse", as Hegel confesses; and furthermore, it aids in the clarification of other obscure passages of Hegel, as for instance the chapters 'Force and Understanding' and 'Observation of the Organic' in the *Phenomenology of Spirit* and 'measure' and 'objectivity' in the *Logic*. To relate philosophy and science as opposed to each other, allows us to comprehend the claim and the meaning of a philosophy of nature, and to recognize its transience and its present relevance.

Ruprecht-Karl-University
Heidelberg

BIBLIOGRAPHY

Alembert, J. L. R. d'. 1751. 'Discours préliminaire' of the *Encyclopédie ou Dictionnaire raisonné des sciences, des arts et des métiers.* 17 vols. Paris, 1751–1765.

Bergman, T. 1775. 'Disquisitio de attractionibus electivis' in *Nova acta Regiae societatis scientiarum upsaliensis* **2**, 159–248.

Berthollet, C. L. 1803. *Essai de statique chimique*. 2 vols. Paris.

Berzelius, J. J. and Hisinger, W. 1803. 'Versuche betreffend die Wirkung der electrischen Säule auf Salze und auf einige von ihren Basen', *Neues Allgemeines Journal der Chemie* **1**, 115–149.

Berzelius, J. J. 1819. *Essai sur la théorie des proportions chimiques et sur l'influence chimique de l'électricité*. Paris.

Condillac, E. B. de. 1780. *La logique*. Paris.

Dalton, J. 1808–1827. *A New System of Chemical Philosophy*. 3 vols. London.

Davy, H. 1807. 'The Bakerian Lecture on Some Chemical Agencies of Electricity' in *Philosophical Transactions of the Royal Society* **1**, 1–56.

Davy, H. 1812. *Elements of Chemical Philosophy*. London.

Elliott, Sir John. 1782. *Elements of the Branches of Natural Philosophy Connected with Medicine* . . . London.

Erxleben, J. L. P. 1775. *Anfangsgründe der Chemie*. Göttingen.

Erxleben, J. L. P. 1772. *Anfangsgründe der Naturlehre*. Göttingen.

Geoffroy, I. S. C. 1741. 'Table des differens rapports observés en chimie entre differentes substances' in *Mémoires de l'Académie des sciences* (Paris), année 1718, pp. 202–212.

Gmelin, L. 1817. *Handbuch der theoretischen Chemie*. 2 vols. Frankfurt a. Main.

Guyton de Morveau, L. B., Lavoisier, A. L., Berthollet, C. L. and Fourcroy, A. F. de. 1787. *Méthode de nomenclature chimique*. Paris.

Guyton de Morveau, L. B., Maret, H. and Durande, J. F. 1777. *Élémens de chymie, théorique et pratique* . . . 3 vols. Dijon, 1777–1778.

Hegel, G. W. F. 1958. *System der Philosophie*. Part II: *Die Naturphilosophie*. (*Jubiläumsausgabe* **9**).

Higgins, W. A. 1789. *A Comparative View of the Phlogistic and Antiphlogistic Theories*. London.

Higgins, W. A. 1814. *Experiments and Observations on the Atomic Theory, and Electrical Phenomena*. London.

Kästner, A. G. 1768. *Anzeige seiner nächsten Vorlesungen über Mathematik und Physik*. Göttingen.

Lavoisier, A. L. 1789. *Traité élémentaire de chimie*. Paris.

Morveau *et al*. See Guyton de Morveau *et al*.

Mitscherlich, E. 1829–1830. *Lehrbuch der Chemie*. Berlin.

Richter, J. B. 1789. *De usu matheseos in chymia*. Königsberg.

Richter, J. B. 1792–1794. *Anfangsgründe der Stöchyometrie*. 2 vols. Breslau.

Trommsdorff, J. B. 1800–1804. *Systematisches Handbuch der gesammten Chemie*. 7 vols. Erfurt.

Volta, A. 1792. 'Observationum circa electricitatem animalem specimen' in *Commentarii de rebus in scientia naturali et medicina gestis*. Leipzig. Vol. 34, pp. 685–688.

Volta, A. 1800. 'On the Electricity Excited by the Mere Conduct of Conducting Substances of Different Kinds' in *Philosophical Transactions of the Royal Society* **90**, 403–431.

Wenzel, C. F. 1777. *Lehre von der Verwandtschaft der Körper*. Dresden.

Wöhler, F. 1828. 'Über künstliche Bildung des Harnstoffs, *Annalen der Physik und Chemie* **12**, 253–256.

HENRY PAOLUCCI

HEGEL AND THE CELESTIAL MECHANICS OF NEWTON AND EINSTEIN

My object here is, first, to review Hegel's criticism of Newton's 'system of the world' and then to examine critically the many aspects of it that seem to anticipate the approach to mathematical physics, which is today associated with the name of Einstein.

One must emphasize that Hegel's criticism was well informed. Certainly he knew the *Philosophiae naturalis principia mathematica* and *Opticks* first hand and had the requisite training in mathematics to comprehend what he read. Through hundreds of well-documented pages of his *Science of Logic* (large and small) and *Philosophy of Nature*, Hegel explores the meaning of Newton's fluxional calculus, his concepts of space, time, mass, inertia, centripetal and centrifugal forces, his laws of motion, his gravitational world-system, and, finally, his theory of light and colors. Particularly under the headings 'Quantity' and 'Measure' in the *Logic* and 'Mechanics' in the *Philosophy of Nature*, Newton's doctrine provides much of the empirical datum upon which the Hegelian philosophical dialectic operates. And that represents no small tribute to the achievement of Newton, especially when we bear in mind that, according to Hegel, "without the working out of the empirical sciences on their own account, philosophy could not have reached further than with the ancients" (*HP* **3**, p. 176).[1]

I

Unfortunately for Hegel's reputation, some of his pages on Newton have a vein of rather heavy-handed humor running through them that can easily be mistaken for *lèse-majesté* or, worse, an ethnic slur against the British people. Such passages are few, but ethnic pride and sensitivity have sought them out; and, at least among sanguine admirers of the greatness of Newton – *qui genus humanum ingenus superavit*[2] – their effect has been to draw upon Hegel an imputation of gross scientific ignorance and presumption.

Even at its worst, however, Hegel's criticism nowhere matches the sustained sarcasm of the anti-Peripatetics of the century before Newton – the Galileos and Bacons, for instance – who pilloried Aristotle for his 'bad' science, regardless of the fact that he had been revered for centuries as, in

R. S. Cohen and M. W. Wartofsky (eds.), Hegel and the Sciences, 55–85.

Dante's phrase (*Inf.* IV, 131), the "teacher of those who know" – *maestro di color che sanno.* Still, there is no denying that the Newtonian *sacramenta* are subject to much abuse in the pages of the German philosopher; and, despite the reigning relativism of our age, it is a bold scholar who will presume, as J. N. Findlay has, to qualify it as "much *fine* abuse."[3]

Admittedly hard to take, on a first reading, is Hegel's comparison of Newton with Molière's *bourgeois gentilhomme* – all to the advantage of the latter! Molière's M. Jourdain was "surprised to learn that he had talked prose all his life, not having had any idea that he was so accomplished"; but he, at least, learned – Hegel observes – whereas Newton seems never to have realized "that he thought in and had to deal with notions of the understanding, while he imagined he was dealing with physical facts" (*HP* **3**, pp. 323–324). On the same level is Hegel's discussion of what the English generally understand by the term Philosophy and why it is that among them "Newton continues to be celebrated as the greatest of philosophers." In England, he explains, not only the empirical sciences but also the mechanical arts are confounded with philosophy; "and the name goes down as far as the price-lists of instrument-makers." With grudging admiration, as well as irony, Hegel in the same passage commends the heirs of the tradition of Bacon and Newton for at least taking the name of philosophy seriously enough to apply it to the study of matters of the greatest importance to them – like political economy, free trade, and the imperial administration, as well as chemistry, mineralogy, natural history, agriculture, and the arts – whereas in most other lands "the name of philosophy is now generally used only as a nickname and insult, or as something odious" (*L* [W, 1975], § 7, pp. 11–12).[4]

More abusive, from a Newtonian standpoint, is Hegel's rejection of the *Principia*'s 'proofs' of Kepler's laws – proofs which he characterizes as "demonstrational jugglery and counterfeiting" (*SL* [J/S] **2**, p. 290); and most offensive of all, perhaps, is his assessment of the *Opticks*, where he speaks not only of *ineptitude, incorrectness*, and *thoughtless inconsistency*, but also of *stupidity, blind prejudice*, and *dishonesty* (*PN,* § 320, *Remark*, p. 199).[5]

It is no wonder, therefore, that, for longer than a century in Newton's English-speaking world, most scholars have responded by refusing to take Hegel's *Naturphilosophie* seriously, or even to read it at all – as, it is said, the seventeenth century Aristotelians of Padua refused to take Galileo seriously, or even to look through his optical tube. The fact is that, whereas the French and Italians, less sensitive to criticism of Newton, have had translations of the complete *Encyclopedia* of Hegel, including the *Naturphilosophie*, since the 1860s, the English-speaking academic community, on the contrary, has delayed until this bicentennial anniversary of Hegel's birth the labors of filling the gap between the excellent versions of the *Logic* and the *Philosophy of Mind*, which Professor Wallace provided three generations ago.

But this year is destined evidently to mark a turning point in the fortunes of Hegel's *Philosophy of Nature*. Thanks largely to the sustained interest of Professor Findlay, we now have an excellent translation – that of A. V. Miller – presented to us by Findlay with a challenge that the Anglo-American academic community will sooner or later have to accept. Back in 1958, Professor Findlay had written of the *Naturphilosophie*:

This part of the system is one that many Hegelians have thought fit to ignore entirely, mainly on account of the outmoded character of the science on which it reposes. Nothing can, however, be more unfit than this ignoring, and, in view of Hegel's undoubted greatness, more impertinent. The *Philosophy of Nature* is an integral part of Hegel's system, and one can no more understand that system without taking account of it, than one can understand Aristotelianism while ignoring the *Physics* or the *History of Animals*, or Cartesianism while ignoring the physical portions of the *Principles of Philosophy*. In Hegel's theory of Nature, as in the parallel theories of Aristotle and Descartes, one sees the philosopher's principles at work, casting their slant upon our talk and thought about the world around us. The complete misunderstanding of Hegel's idealism by British philosophers, and its reduction to a refined form of subjectivism, are probably due to their ignoring of the *Naturphilosophie*.

. . . Hegel's grasp of contemporary science was, moreover, informed and accurate: the reading of the *Naturphilosophie* is made easy by its wealth of experimental illustration, and by its long citations from contemporary treatises. Hegel *gives* one the sciences of his own day, together with the interpretations he puts on them. [His views remain] as worthy of study, and of detailed scholarly comment, as are the views of Aristotle, or (in recent times) of Whitehead.[6]

In his foreword to the Miller translation, Professor Findlay says that his object in publishing it has been primarily utilitarian, to make its thought accessible "to students and teachers, particularly in regions where prejudiced simplifications might otherwise be their only route of access to it." The work itself is proof, he asserts, "that Hegel, like Aristotle and Descartes and Whitehead, is one of the great philosophical interpreters of nature, as steeped in its detail as he is audacious in his treatment of it." Praising its "thoroughgoing realism," Findlay concludes that what is to be admired most in it is Hegel's manifest "willingness, unusual in philosophers, to read, digest, and take full account of so much detailed scientific material, a willingness which puts him on a pinnacle of scientific information and understanding shared only by Aristotle" (*PN*, pp. viii–ix, xxv).

Thus, under unprecedentedly favorable auspices, we now have the

Philosophy of Nature before us in a thoroughly English context. There can no longer be, therefore, any excuse among us for the 'impertinence' of ignoring its doctrine – including the details of its criticism of Newton. Still, it remains to be seen how that criticism will fare under the closer scrutiny it is now certain to receive. Studied popularly in its fulness, it may appear to be all the more presumptuous and ill-informed and lead to a general discrediting of the rest of the Hegelian System – the *Philosophy of Mind* as well as the *Logic* – precisely on the grounds that the System is an integrated whole. "Nature, and the world or history of spirit, are the two realities" – says Hegel in the closing pages of his *History of Philosophy*; and the "ultimate aim and business of philosophy is to reconcile the thought of the Notion with reality" (*HP* **3**, p. 545). If the Newtonian perspective, which is also the Darwinian, is true, and the Hegelian perspective false, what is the use, apart from antiquarianism, of a renaissance of interest in the *Naturphilosophie*? If Hegel's notion of the reality of nature is inadequate, then there can be no truth, no *adaequatio intellectus et res*, in his general synthesis. And, except for truth's sake, why should the Anglo-American academic community take Hegel seriously when it can get as much of German dialectical philosophizing as it apparently needs in the doctrine of Karl Marx, which leaves the greatness of Newton and of Darwin intact?

II

Before focusing on Hegel's systematic elaboration of his criticism of Newton's science which is developed in the larger *Science of Logic* and culminates in the *Philosophy of Nature*, we would do well to consider briefly the condensed expressions of it in scattered pages of his *Phenomenology of Mind, History of Philosophy*, and smaller *Logic*. Newton is not named in the *Phenomenology*, but his chief contribution to the development of scientific thought – the concept of force – is taken up as part of the phenomenology of consciousness under the heading "Force and Understanding, Appearance and the Supersensory World (*Kraft und Verstand, Erscheinung und übersinnliche Welt*)." In attempting to understand the perpetual flux of the world of appearances (*phenomena*), consciousness labors to look through the phenomena to see what, in reality, sustains them. In the process of 'thinking through' appearances, the understanding, Hegel says, discovers or fashions for itself a supersensory model which stands for it in place of the complexities of the restless sensory world. As Hegel expresses it, continued reflection develops that supersensory model into a "kingdom of laws," which is "no doubt beyond the world of perception – for this exhibits the law only through incessant change – but likewise present in it, and its direct immovable copy or image" (*Phen* [B], p. 195).[7]

In the first stages of this reduction of phenomenal complexity to the simplicity of reflection, laws of the understanding are discovered or formulated singly, to account for a relatively narrow range of phenomena. But the understanding longs for unified comprehension; and its tendency is, as Hegel says,

> to let the many laws coalesce into a single law, just as, e.g., the law by which a stone falls, and that by which the heavenly bodies move have been conceived as one law. When the laws thus coincide, however, they lose their specific character. The law becomes more and more abstract and superficial, and in consequence we find as a fact, not the unity of these various determinate laws, but a law which leaves out their specific character; just as the one law, which combines in itself the laws of falling terrestrial bodies, and of the movements of celestial bodies, does not, in point of fact, express both kinds of laws. The unification of all laws in universal attraction expresses no further content than just the bare concept of law itself, a concept which is therein set down as existing. Universal attraction says merely that all things retain their difference with respect to one another. The understanding presumes that in this it has found a universal law that expresses universal reality as such; but in fact it has merely found the notion of law in itself, though in such a way as to permit it to assert that all reality, in itself, is subject to law (*Phen* [B], pp. 196–197).

The important point to stress here is that Hegel exhibits this activity of the understanding as an essential moment in the phenomenological process. What Newton did *had* to be done by human consciousness. The idea of universal gravitation "is therefore of the greatest importance," Hegel concludes, "because it is directed against the unthinking way of representing reality that makes everything appear to have happened by accident, and for which qualitative distinctiveness has the form of merely sensory differentiation" (*Phen* [B], p. 197).

Newton thus comes off rather well in the *Phenomenology* – in marked contrast with the caustic treatment we get of him in the lectures on the *History of Philosophy*. There Hegel calls the author of the *Principia* an intellectual "barbarian" for having treated the basic concepts of his natural science as if they were physical facts – sensuous things, to be dealt with "as men deal with wood and stone." Especially barbaric, according to Hegel, was the Newtonian attitude summed up in the maxim, "Physics, beware of metaphysics," which amounts to saying, "Science, beware of thought." The worst of it is, Hegel continues, that physical scientists since Newton's time have for the most part faithfully observed this precept, inasmuch as they have not entered upon an investigation of their conceptions, or thought about thoughts . . . . And this is even now the case. In the beginnings of physical science we read of the power of inertia, for instance, of the force of acceleration, of molecules, or centripetal and centrifugal forces, as of facts which definitely exist; what are really the final results of reflection are represented as their first grounds (*HP* **3**, p. 323).

Yet here again, Hegel acknowledges the value of Newton's scientific work in "introducing to physics the determinations respecting forces, which pertain to reflection." By setting "the laws of forces in the place of the laws of phenomena," Newton "raised science to the standpoint of reflection," and for that he deserves high praise. He is to be blamed, according to Hegel, only for imagining that he is still functioning on the level of sensory perception, free of metaphysics, when, by undertaking to 'compose' the empirically-derived laws of phenomena out of the interrelations of 'component' forces, he has obviously slipped back into metaphysics without knowing it – which is to say, ignorantly. The harshest part of Hegel's judgment here is his prophecy that physical science will make no significant advance in the theoretical sense until it gives up the naive Newtonian dogmatism that confounds abstract concepts of the understanding with the reality of nature (*HP* **3**, pp. 322–324).

That surely was a 'hard saying' in the first decades of the nineteenth century. Today, however, it has become the prevailing view of the most expert practitioners in the field. In his *Physics and Microphysics*, for instance, the Nobel Prize-winning physicist Louis de Broglie writes:

For scientists, and in particular for the theorists, there is a certain danger in trying to ignore the efforts of philosophers and especially their work as critics . . . . Thus many scientists of the present day, victims of an ingenuous realism, almost without perceiving it, have adopted a certain metaphysics of a very materialistic and mechanistic character and have regarded it as the very expression of scientific truth. One of the great services that the recent evolution of physics has rendered contemporary thought, is that it has destroyed this simplified metaphysics, and with the same stroke has caused certain traditional philosophic problems to be considered in an enterely new light . . . . For the development of science to continue, we must embark on, or at any rate touch upon, questions of philosophic import and sometimes consider their new and very original solutions.[8]

Discussing the reluctance of the Newtonians to abandon their long-cherished concepts of space, time, motion, and force, the same author observes that too many physicists even today would no doubt "prefer merely to perfect and amend the existing theories rather than be obliged constantly to reconstruct them." But a willingness to abandon long-cherished concepts is, he concludes, "the condition and ransom of scientific progress."[9]

In the smaller *Logic*, Hegel does not explicitly discuss the Newtonian celestial mechanics. He does, however, discuss the logical presuppositions upon which any analysis of laws of phenomena (such as Kepler framed) into laws of forces (such as Newton framed) ought to be based, if it is to make philosophical sense. And this he does at considerable length under the sub-headings 'Repulsion and Attraction', 'Quantity', 'Magnitude', 'Quantum', 'Number', 'Degree', 'Quantitative Ratio', 'Measure', 'Thing', 'Properties', 'Matter', 'Form', 'Phenomena', 'Forces', and 'Expression of Forces' – all of them terms we now familiarly encounter in contemporary treatises on post-Planckian physical theory.

Under the heading "Repulsion and Attraction," Hegel draws a parallel between Newtonian physics and the ancient atomism of Leucippus and Democritus. In both systems, particles of matter and an enveloping void are posited as the ultimate constituents of nature, with the void serving as a separating or repulsive principle. The Newtonian advance over the old doctrine consists in having posited "an attractive by the side of a repulsive force," which – Hegel acknowledges – "certainly gives completeness to the contrast" between the Democritean Full and Empty (*L* [W, 1975], § 98, p. 143). Still, what Aristotle criticized in ancient atomism remains to be criticized, according to Hegel, in the Newtonian doctrine.

The old atomists traced fixed patterns (equivalents of the Newtonian laws of forces) through all the manifold flux and multiplicity of the phenomenal world. But when required to say what, in the first place, caused the atoms to start bumping together, acting and reacting as they do in their great web of necessity, the reply of the Democriteans was simply: a primordial, fortuitous swerving, a whirl of indeterminate origin – which is another way of saying that, in the final analysis, like our modern Heisenbergians, Democritus ascribed all things to chance. ("*Democrito*," Dante wrote in the fourth canto of the *Inferno*, "*che 'l mondo a caso pone*" [who ascribes the world to chance][10] – and it can easily be demonstrated that, despite the popular view which holds that Necessity is the atomists' God, Dante's phrase is drawn straight out of the very precise Aristotelian analysis of the Democritean doctrine in Book IV of the *Physics* – an analysis upon which Hegel dwells in his Lectures on the *History of Philosophy*.) The Newtonians, when all is done, do precisely the same – Hegel argues; for, despite all their talk of matter and void, inertia and gravity, it is a *chance thrust* dating from some indeterminate moment of the remotest past that is assumed to have originally set the planets in course to be pushed and tugged as they are by centrifugal as well as centripetal forces.

Hegel's concern in all of this is to indicate that the Newtonian mechanics, despite pietistic and deistic protests to the contrary, is essentially a materialist doctrine. He is, of course, aware that, as a man of God, Newton strenuously resisted the assimilation of his doctrine to that of the avowed materialists. "I feign no hypotheses" about final causes, Newton had pleaded in the celebrated General Scholium of the *Principia*, where, indeed, he explicitly rejects as untenable the Cartesian theory of vortices on the grounds that no merely mechanical causes could give birth to so much orderly motion as we can trace in the heavens. By asserting that such orderliness presupposes an intelligent creation, Newton had hoped, as Hegel remarks, to leave unimpaired the "honor of God as the Creator and Governor of the World" while at the same time excluding him from consideration in the actual search for underlying forces (*L* [W, 1975], § 136, p. 195).

Under the sub-headings 'Force' and 'Expression of Force' in the smaller *Logic*, Hegel discusses at length why it is that Newtonian science would sooner or later have to give up its veneer of theistic piety. "Contrasted with its deinfinitized world of independent forces and matters," he writes, "the only terms in which it is possible to describe God will present him in the abstract infinity of an unknowable supreme Being in some other world far away." And that, Hegel continues, is "precisely the position of materialism," from as far back as Democritus and Epicurus down to the "free-thinking" deism and agnostic atheism of the Enlightenment. The medieval Church, confronted with the renaissance of ancient doctrines, was in that respect right, therefore, Hegel says, in resisting the "search for underlying causes" as impious, as tending to deny to God the things that are God's, which are assigned instead to indeterminate causes. From the vantage point of religious faith, no less than from that of philosophy, it must be acknowledged, Hegel says, that "the finite forms of the understanding certainly fail to fulfill the conditions for knowledge either of Nature or of the formations of the world of Mind as they truly are" (*L* [W, 1975], § 136, p. 195).

Yet while religion has a right to be dissatisfied with the results, it has no right to frustrate the labors of the understanding, whose 'finite forms' are – for Hegel – indispensable moments in the development of consciousness to the level of truly scientific comprehension. In words that anticipate the scientific posture of Pierre Duhem in the concluding sections of his masterful *Aim and Structure of Physical Theory,* Hegel writes against the false claims of religious piety:

On the other hand, it is impossible to overlook the formal right which, in the first place, entitles the empirical sciences to vindicate the right of thought to know the existent world in all the speciality of its content, and to seek something further than the mere abstract faith that God creates and governs the world. When our religious consciousness, resting upon the authority of the Church, teaches us that God created the world by his almighty will, that he guides the stars in their courses, and vouchsafes to all his creatures their existence and their well-being, the question Why? is still left to answer. Now it is the answer to this question which forms the common task of empirical science and of philosophy (*L* [W, 1975], § 136, p. 195).

Defending the practice against the piety of the Newtonians, Hegel asserts that when religion appeals to the 'unsearchableness' of the decrees of God, it is, in effect, aping the agnosticism of the enlightenment rationalists. "Such an appeal," he concludes, "is no better than an arbitrary dogmatism which contravenes the express command of Christianity, to know God in spirit and truth, and is prompted by a humility which is not Christian, but born of ostentatious bigotry" (*L* [W, 1975], § 136, p. 196).

III

The larger *Science of Logic*, which provides what amounts to a running commentary on the terse paragraphs of the *Logic* of the *Encyclopedia*, takes us to the philosophic core of Hegel's criticism of the Newtonian celestial mechanics. No one pretending to assess the adequacy of that criticism can afford to ignore, or skim over, what Hegel has to say about Quantity, Quantum, and Quantitative Ratio in his book-length discussion of 'Magnitude', or about Specific Quantity, Real Measure, and the Measureless (as transition to Essence) under the heading 'Measure', or about Mechanism, Chemism, and Teleology under the heading 'Objectivity'. As in the smaller Logic, Hegel is here examining the preconceptions and methods of study of all sciences that take not abstractions but the concrete realities of nature and mind as their qualitatively determined objects.

Under 'Magnitude' and 'Measure', Hegel examines, among other things, the basic notion of the Newtonian-Leibnizian calculus and traces the history of the mathematical infinite and infinitesimals upon which that calculus is built; he considers at length the difficulties of assimilating analytical calculus to analytical geometry and of applying both to the analysis of accelerated rectilinear and non-uniform curvilinear motion; and, anticipating things to come in our own time, he speculates on the possibility of developing a mathematics of qualitative quanta which would be a science of measures, competent to deal with the qualities as well as the quantities of existent things (as, for instance, Einsteinian and Planckian physics now deals with qualitatively determined quanta, or measures, of space, time, light, and a host of electro-magnetic phenomena).

The first explicit reference to Newton comes under the sub-heading 'Quantum' and it consists of singularly high praise. Considering the "chief determinations which have been offered in mathematics about the infinite," Hegel asserts unequivocally that "no correcter determination of the thought can be made than that offered by Newton." The problem of the mathematical infinite is ancient. It was first clearly defined by the Eleatics (notably Zeno) in their efforts to demonstrate the irrationality of the Heraclitean 'flux'; and Zeno's paradoxes have remained vital paradoxes for mathematical physicists to this day – Hegel insists – despite all the walking to and fro of old Diogenes and cynics of more recent times. That Hegel was competent to deal with the problem in its abstractest geometrical and algebraic as well as logical aspects, he has amply demonstrated not only in the *Science of Logic* itself, but also in the *History of Philosophy*, where he takes it up as often as it surfaces in his sources.

Newton tells us that he "invented the methods of series and fluxions in the year 1665," and his first published treatise on the subject, *Methods of Fluxions and Infinite Sequences*, appeared in 1674. But Hegel is concerned, as he says, with Newton's maturest thought on the subject, which is that of the *Principia*. Hinting at what he will later criticize, Hegel writes:

> I here set apart the determinations belonging to the idea of motion and velocity (from which latter chiefly he took the name of *fluxions*), for there the thought appears not in its due abstraction, but concrete and mixed with unessential forms.

In praise of the duly abstract thought, Hegel says:

> Newton explains these fluxions (*Princ. Mathem. Phil. Nat. L. i. Lemma XI, Schol.*) by saying that he takes them not as *indivisibilia* (a form used by earlier mathematicians, Cavalieri and others; it contains the concept of a Quantum determinate in itself), but as vanishing *divisibilia*; and, further, not as the sums and ratios of determinate parts, but as the limits (*limites*) of the sums and ratios. It will be objected, he says, that vanishing magnitudes have no final ratio, because the ratio before they vanish is not the last, and, after they have vanished, no longer exists. But by the ratio between vanishing magnitudes must not be understood the ratio that exists either before or after, but that *with which* they vanish (*quacum evanescunt*). And, similarly, the first ratio of becoming magnitudes is that *with which* they arise (*SL* [J/S] **1**, pp. 271–272).

After noting that the "magnitudes at the point of vanishing" are understood by Newton to be "quanta no longer," Hegel makes clear that, in his judgment,

Newton has here adequately expressed the notion of transition from quantity to quality which is so essential an element in the Hegelian dialectic. As much Newtonian as Hegelian is the statement that: "The limit of the magnitudinal ratio is that point where it is and it is not – or, more precisely, where the Quantum has vanished, and the ratio, therefore, is preserved only as qualitative quantity-ratio" (*SL* [J/S] **1**, p. 272).

But while Newton grasped the concept of the qualitative quantity-ratio and expressed it clearly, in practice – says Hegel – the author of the *Principia* let it slip through his fingers. And the same must be said of the other great mathematicians of the period, including Newton's master Barrow, as well as Fermat, Leibniz, Lagrange, Landen, Euler, and their successors. It seems, Hegel writes, that "when mathematicians turn to practice, the finite determinateness of quantity returns, and the operation can no longer do without the idea of a Quantum which is merely relatively small" (*SL* [J/S] **1**, p. 277). Indeed, the idea of the relatively small, of increment, of addition, "of growth of x by dx or i, and so on," Hegel concludes, "must be considered the fundamental evil inherent in these methods – as an enduring obstacle which makes it impossible to disengage the determination of the qualitative moment of quantity from the idea of ordinary Quantum" (*SL* [J/S] **1**, p. 274). In other words, the evil is that those "ghosts of departed quantities" (as Bishop Berkeley called them) are too easily mistaken for never-perishing *indivisibilia*, and the transition from quantity to quality is lost to thought.

If mathematicians could take another approach so as to develop a mathematics of qualitative quantity-ratios, Hegel observes, the result would be a magnificent instrument for the advancement of a truly physical as distinct from an abstractly mechanical science. Such a mathematics, he writes, would have to be in essence –

> a science of measures – a science for which much has been done empirically, but little in a truly scientific, that is philosophic, manner. Mathematical principles of Natural Philosophy – as Newton called his work – if they were to fulfill this determination in a philosophic and scientific meaning deeper than that which was reached by Newton and the whole Baconian generation, must contain quite other things in order to bring light into these regions, dark as yet, but most worthy of contemplation (*SL* [J/S] **1**, p. 361).

Later, in the *Philosophy of Nature*, Hegel will add on this theme:

> The truly philosophical science of mathematics as *theory of magnitude*, would be the science of *measures*; but this already presupposes the real particularity of things, which is found only in concrete *Nature*. On account of the *external* nature of magnitude, this would certainly also be the most difficult of all sciences.

Newtonian science, Hegel stresses, is not a science of measures. In fact, all that most needs to be measured in nature – the actualities of space, time, and the light by means of which spatial measurements are possible – is taken for granted by the Newtonians, exactly as defined in the abstract presuppositions of their calculus. That calculus was developed, it is true, to deal with motions of qualitatively determined phenomena; but, in the process, the traditional mathematical spirit of abstraction took over. Forgetting the qualitative forms of the phenomena of falling bodies and orbiting planets, the mathematicians focused on the terms of the ratios in the formulas – with the result that products of analysis (of the decomposition of the magnitude of a phenomenon, such as motion) there received, in Hegel's words –

> an objective meaning, such as velocity, accelerated force, and so on; according to this newly acquired objective meaning they were to produce correct propositions and physical laws; their objective connections and relations, too, were to be determined by analytic means; for instance, it was said that in a uniformly accelerated motion, there existed a special velocity proportional to the periods of time, while besides, an accretion was added uniformly from the force of gravity. In the modern analytical form of mechanics such propositions are regularly cited as results of the calculus (SL [J/S] **1**, p. 289).

Hegel was well aware that Newton had not arrived at the conclusions demonstrated in Book Three of the *Principia* (where the universal law of gravitation is expounded as the System of the World) by means of the analytical calculus and geometry he uses there. Newton himself says, at the beginning of that book, that the method of exposition there was an afterthought, and that he introduced it rather to prevent than to facilitate general understanding. He had at first, he says, written the entire book in a more direct, popular way; but afterwards, "considering that such as had not sufficiently entered into the principles could not easily discern the strength of the consequences, nor lay aside the prejudices to which they had been many years accustomed," Newton decided to avoid any direct confrontation with ignorance and prejudice by choosing "to reduce the substance of this Book into the form of Propositions (in the mathematical way), which would be read by those only who had first made themselves masters of the principles established in the preceding books."[11]

What might have become of the great law of universal gravitation of that Third Book, had Newton not *reduced* his exposition to the form of Propositions, in a mathematical way, one can hardly guess. But perhaps the Ernst Machs and Einsteins would have got at its absolute space and time, the essentially non-empirical character of its laws of motion, and its oddly convenient equivalence of inertial and gravitational mass, much sooner. In the light of what has since become the fate of Newton's mathematical 'proofs', it was eminently fair and extraordinarily perceptive for Hegel to have said over 150 years ago:

It will be impossible to deny that in this sphere much has been accepted as proof – chiefly veiled under the kindly mist of the infinitesimally small – on no other ground than that the result was already known beforehand, and the proof, which was arranged in such a manner as to produce the result, at least effected the illusion of a framework of proof – which illusion was preferred to mere belief or empirical knowledge. But I do not hesitate to regard this method as no better than demonstrational jugglery and counterfeiting; and I include even some of Newton's demonstrations, and especially such as belong to those just mentioned, for which Newton has been extolled to the skies and above Kepler, because what Kepler had discovered empirically *he* demonstrated mathematically (*SL* [J/S] **1**, p. 290).

Hegel then proceeds to explain that, by its very nature, the mathematics of Newton and his peers is unable to "prove the magnitudinal determinations of physics in so far as they are laws based upon the qualitative nature of the moments." The qualitative element is lost in the processes of the calculus; and it is rather a moral than a scientific question why, to many mathematicians, it has seemed against the honor of their discipline – so Hegel phrases it – to "acknowledge simply experience as source and sole proof of empirical propositions" (*SL* [J/S] **1**, p. 290).

Needless to say, Hegel has long since won his point on this score. Modern mathematicians and the best mathematical physicists of recent years are all of a mind in acknowledging – and indeed often boasting – that they no longer hold, with Galileo, Descartes, and Newton, that mathematics of itself, or mathematically conducted experiments, can lead to true knowledge of the realities of Nature. As Professor Tobias Dantzig expressed it in his *Number: The Language of Science:*

The mathematician may be compared to a designer of garments, who is utterly oblivious to the creatures whom his garments may fit. To be sure, his art originated in the necessity for clothing such creatures, but this was long ago; to this day a shape will occasionally appear which will fit into the garment as if that garment had been made for it. Then there is no end of surprise and delight! There have been quite a few such delightful surprises. The conic sections, invented in an attempt to solve the problem of doubling the altar of an oracle, ended by becoming the orbits followed by the planets in their courses about the sun .... The absolute differential calculus, which originated as a fantasy of Riemann, became the mathematical vehicle for the theory of Relativity .... Mathematics and experiment reign more firmly than ever over the new physics, but an all-pervading skepticism has affected their validity. Man's confident belief in the absolute validity of the two methods has been found to be of an anthropomorphic origin, both have been found to rest on articles of faith.[12]

Bertrand Russell, who (with Alfred North Whitehead) gave us an updated *Principia Mathematica* in 1914, has said to the same effect: "It is a curious fact that, just when the man in the street has begun to believe thoroughly in science, the man in the laboratory has begun to lose his faith"; and, again: "Science, which began as the pursuit of truth, is becoming incompatible with veracity, since complete veracity tends more and more to complete scientific scepticism."[13]

But the truly great scientists of our time, the Einsteins and Plancks, have not been content that mathematical science, which once boasted of its certainty, should now boast of its uncertainty. Einstein admits that, in its pursuit of the open questions of the Galilean and Newtonian mechanics, mathematical science has been led inevitably – as Hegel predicted – to Democritean, or Heisenbergian, indeterminism. But, according to Einstein, that is hardly an outcome to gladden the scientific heart. "Some physicists," he wrote in 1941, "among them myself, can not believe that we must abandon, actually and forever, the idea of direct representation of physical reality in space and time."[14] Hegel, too, refused to abandon that idea. Like Aristotle long before him, he sought, by means of a dialectical criticism of the limits of mechanistic science and its mathematical methods, to prepare the way for a truly philosophic science of nature – a science that can not only represent but also comprehend in its truth the qualitative reality of nature, from the merest externality and duration of space and time, up through the great chain of inorganic and organic being, to the animal existence – birth, life, health, reproduction, disease, and death – of thinking man.

IV

With one notable exception, all references to Newton in the *Philosophy of Nature* are confined to the first part – Mechanics – where the object of study is matter in the abstract, as contrasted with the qualified matter of Physics and the living matter of Organics. Under Mechanics, the sub-divisions are: 'Space and Time' (including the abstract concepts of Place and Matter); 'Matter and Motion, Finite Mechanics' (including the concepts of Inert Matter, Thrust, and Falling); and 'Absolute Mechanics' (with Universal Gravitation, Kepler's Laws, and Transition to Physics as sub-headings). The subdivisions and subheadings obviously represent a graded order, ranging upward from the simplest and most abstract object of study in mechanics (space) to what in Hegel's judgment is the most complex and least abstract (the solar system as described in Kepler's laws). The fact that Universal Gravitation is placed beneath Kepler's Laws in the graded order reveals at a glance that here, as elsewhere, Hegel will represent Newton's 'System of the World' to us as a 'reductionist' analysis of a higher into the terms of a lower determination.

In the whole range of Mechanics – Hegel writes – "bodies exist only as points; what gravity determines is only spatial interrelations of points. The unity of matter is only the unity of the place it seeks, not the unity of a concrete One. That is the nature of this sphere." And yet, on its highest level of development, Mechanics is able to transcend its own abstract character; for, in its descriptive laws of the solar system – tracing a pattern of elliptical orbits about an energizing center – it comprehends the notion of what matter, as manifest motion, must really be in itself. In a startling anticipation of the vision of Niels Bohr, Hegel writes:

> What the solar system is as a whole, matter is now to be in detail . . . . The determinations of form which constitute the solar system are the determinations of matter itself and they constitute the being of matter. The determination and the being are thus essentially identical; but this is the nature of the qualitative, for here, if the determination is removed, the being, too, is destroyed. This is the transition from Mechanics to Physics (*PN*, § 271, *Z*, pp. 83–84).

The whole of Newtonian science is comprehended within the abstract range of Mechanics, excepting only the theory of light of the *Optics* – Newton's one major attempt to make the transition from Mechanics to Physics (in Hegel's sense of the terms). Newton had confidently imagined that, with the method of the *Principia*, he could easily make that transition, and then move on from Physics into Organics to complete his Natural Philosophy. But in Hegel's judgment, as in the judgment of Einstein and his peers, Newton's attempt to comprehend scientifically the phenomena of light was a failure, and foredoomed to be a failure because of the inadequacies of its analytical, decompositional method. Just as Newton decomposed the ellipses of Kepler's laws into rectilinear components, thereby depriving them of their qualitative identity, so he decomposed the fluid transparency of white light into colored corpuscles, which he declared to be primary and simple; and, with that, the qualitative, distinctive actuality of light slipped out of his scientific grasp.

Hegel was aware that Light – which for him as for Einstein was the first physical manifestation of matter – united qualitatively the characteristics of abstract corpuscular and wave motions; he knew also that light, in reflecting, refracting, and otherwise manifesting itself, shows all the colors of the entire universe, including those revealed in the dark when a prism is set up, just right, as a divider between light and dark. But, about the corpuscular and wave theories, Hegel held that they were both one-sided, and therefore inadequate approaches to the dialectical notion of light in its qualitative unity. Hegel writes:

> Light is an interesting theme to treat; for we think that in Nature we have only the individual, *this* particular reality. But light is the very opposite of this . . . . In thinking of light, we must renounce all conceptions of composition and the like . . . . The expression 'bundles of rays' is merely one of convenience, it means nothing; the bundles are light in its entirety, which is only outwardly limited; it is no more divided into bundles of rays than is the Ego or pure self-consciousness. It is the same when I say: in *my* time or in Caesar's time. This was also the time of everyone else; but here I am speaking of it in relation to Caesar, and restrict it to him without meaning that he really had a separate ray or parcel of time. The Newtonian theory according to which light is propagated in straight lines, or the wave theory which makes it travel in waves, are, like Euler's aether or the vibration of sound, materialistic representations quite useless for the comprehension of light (*PN*, § 276, *Z*, pp. 93–94).

But the best, in this extraordinarily thorough anticipation of things to come in physical theory, is what follows – where Hegel rejects (as modern Quantum Physics rejects) the notion that the wave theory is superior to the corpuscular theory. According to the wave theory, he writes –

> the dark element in light [the color-producing element] is supposed to run through the movement in a series of curves which can be mathematically calculated; this abstract determination has been introduced into the theory, and is nowadays thought to be a great triumph over Newton. But this is nothing physical; and neither of these two ideas is in place, since nothing empirical obtains here. There no more exist particles of light or waved aether than the nerves consist of a series of globules, each receiving an impulse and setting others in motion (*PN*, § 276, *Z*, p. 94).

Hegel died, it is to be noted, more than a generation before the birth of the people who are usually honored in our catechisms of popular science for having discussed such matters for the first time. Also on the *speed* of light, Hegel anticipates the pioneers, writing:

> The propagation of light occurs in time, since, being an activity and an alteration, it cannot dispense with this moment . . . . The distances which light is supposed to travel involve time; for illumination, whether through a medium or by reflection, is a modification of matter requiring time (*PN*, § 276, *Z*, p. 94).

But the speed of light is not, according to Hegel, like the speed of other physically determinate manifestations of matter. In the sphere of *qualified* as distinct from abstract matter, light is primary – the first, and therefore *limiting* manifestation of matter in motion. The sun, says Hegel, does not pour out its light *incidentally* while serving as the central material point of the solar system. That system's orbital structure is the inner form of matter itself, as we have noted, and it is that circling about a center that shows itself physically as light. But, as each particle of matter is a microcosmic solar system, all matter is – for the eye that can discern its true, centralized orbital speeds – luminous. Again in a thoroughly Einsteinian vein, Hegel characterizes Light as the "self-contained totality of matter," explaining that, as an "existent, space-filling force," it is "absolute velocity, pure materiality which is everywhere present, real existence which remains within itself, or actuality as transparent possibility." "Light," he concludes, "brings us into the universal interrelation; everything exists for us in theoretical, unresistant fashion because it is in light (*PN*, § 275, *Z*, p. 88).

Einstein has told us that it was as a consequence of the labors of Faraday, Clerk Maxwell, and Hertz, but particularly of Maxwell, in the study of light and electro-magnetic phenomena, that modern science generally abandoned the established view that Galilean-Newtonian mechanics could stand as the "basis of all physics." Attempts to solve the difficulties of application of the Newtonian principles to the newly-developed spheres led first to formulation of the field theory of electricity, then to the effort to base all physics upon the concept of field, and finally to the theory of relativity, which is the "evolution of the notion of space and time into that of the continuum with metric structure,"[15] which is in our view then qualitatively determined. "The general theory of relativity," Einstein has said with his usual respectful regard for the overthrown Titan of classical mechanics, "formed the last step in the development of the programme of the field-theory. Quantitatively it modified Newton's theory only slightly, but for that all the more profoundly qualitatively."[16]

Already in Hegel's day, the phenomena of magnetism, electricity, and chemism, together with those of light, were calling into being a unified science very different in principle from Newtonian mechanics. Scientific 'law' in the mechanical sense means, he explains, "the combination of two simple determinations such that merely their simple interconnections constitute the whole relationship, and yet each must have the show of independence with regard to the other" (*PN*, § 270, *Z*, p. 72). But such a concept of law has, he insists, a very limited range – that of mathematical abstraction. It will not do in the study of bodies as they *actually* fall, or as they actually orbit around an energized center; but the inadequacy there may be obscured by assuming that the bodies in those qualitatively determined relationships are really nothing more than points in the geometric sense. It is otherwise, however, with the phenomena of light, magnetism, electricity, and chemism.

In magnetism, for instance, Hegel writes, "the inseparability of the two determinations is already posited"; consequently there is not a co-relationship of 'law' in the mechanical sense (*PN*, § 270, *Z*, p. 72). Magnetic phenomena, he explains, are not gravitational phenomena which mathematical analysis can, for its arbitrary convenience, separate into inertial as well as gravitational components. "Magnetism," he writes, "differs from gravity in this, that it forces bodies into a quite different direction from the vertical." Motion in the sphere, or 'field' of magnetism is not that of rectilinear attraction; but it is also "not rotary, not a curve upon itself, like the motion of the heavenly bodies, which is accordingly neither attractive nor repellent" (*PN*, § 313, *Z*, p. 171).

Summing up his case against scientific 'reductionism' in this sphere, Hegel continues:

> In all higher forms [as contrasted with the forms of abstract mechanics] the individualized whole constitutes the third in which the determinations are conjoined, and we no longer have the direct determinations of two things which are in relationship with each other.

Laws in the mechanical sense are possible for the planetary motions, he says, because those motions involve two distinct phenomenal elements – not to be confounded with the non-empirical centripetal and centrifugal forces. Those two phenomenal elements are "the form of the path and the velocity of the motion." But if such laws are to be comprehended in a higher principle so as to embrace the phenomena of light and magnetism as well, the "thing to be done," he concludes, "is to develop this from the Notion. This would give rise to a far-reaching science and the difficulty of the task is such that this has not yet been fully accomplished." (*PN*, § 270, *Z*, p. 72).

The men who have undertaken to develop that "far-reaching science" in our time have all started from the premise that it would have to be built up from the very bottom – from a rectification of the first 'reductionist' error of classical mechanics, which consisted in its positing space and time as only externally interrelated in the phenomena of motion. Identifying the essential element in Einstein's labors of rectification, Bertrand Russell wrote:

The scientific merit of Einstein's theory lies in the explanation, by a uniform principle, of many facts which are unintelligible in the Newtonian system. The philosophical interest lies chiefly in the substitution of the single manifold, space-time, for the two manifolds, space and time.[17]

How the original 'reductionist' error came to be made, Einstein himself has explained as follows:

The lack of definiteness which, from the point of view of empirical importance, adheres to the notion of time in classical mechanics was veiled by the axiomatic representation of space and time as things given independently of our senses. Such a use of notions – independent of the empirical basis, to which they owe their existence – does not necessarily damage science. One may however easily be led into the error of believing that these notions, whose origin is forgotten, are necessary and unalterable accompaniments to our thinking, and this error may constitute a serious danger to the progress of science.[18]

Literally billions of words have been written on this theme since Ernst Mach and then Einstein opened it up for the academic community of mathematical scientists. Yet in all that literary production, which still floods the markets today, there is, I am confident, no discussion of the reductionist error as profound or as philosophically consistent as Hegel's in the opening sections of his Mechanics, from which point he pursues the process of philosophical *rectification* up through the whole range of mechanics into those higher spheres of nature where to mathematize at all is to commit the gravest sort of reductionist error – unless one's purpose is avowedly reductionist and utilitarian.

In reviewing Hegel's approach to Newton in the *Philosophy of Nature*, we have started from the 'difficulties' of the phenomena of light and electromagnetism only because it has been from that standpoint that our contemporary science, always working 'reductionistically', has found its way back to the beginnings, ironically getting a reductionist satisfaction out of the effort to rectify the original reductionist error of Newton. Hegel pursued a 'reductionist' course in his great intellectual voyage of discovery, the *Phenomenology of Mind*; but in his systematic exposition of the doctrine of Mechanics, as in all the philosophical sciences of his *Encyclopedia*, he reverses that course, working his way up from the most abstract to the most concrete conceptions.

V

Although we have the assurances of the greatest practitioners that, in mathematical science, intellectual insight and discursive reasoning (*nous* and *episteme*) come first and mathematization follows, many of us are still reluctant to believe that we are dealing with truly scientific thought when its utterance is not in equations. For those of us who feel that way, the best approach to Hegel's Mechanics is, as we have suggested above, through close study of the several hundred pages on 'Quantity' and 'Measure' in the larger *Science of Logic*, which bristle with formulas and constructions of the analytical calculus and geometry. Hegel was, as we noted, well trained in the higher mathematics. But, as a great philosopher in the Aristotelian tradition, he knows that the place of mathematically-shaped abstractions lies between the purely logical and the sensory, and that such shaped abstractions can be used meaningfully in natural science only after first principles (in mechanics, the principles of space, time, matter, and motion) have been logically determined in thought.

Early in the Mechanics, Hegel sums up the range of concepts he is about to explore. "Self-externality," he writes (linking the *Naturphilosophie* with what precedes it in his System),

> splits at once into two forms, positively as Space, and negatively as Time. The first concrete thing, the unity of these abstract moments, is Matter; this is related to its moments, and these consequently are related to each other, in Motion. When this relation is not external, we have the absolute unity of Matter and Motion, self-moving Matter (*PN*, § 253, *Z*, p. 28).

When mechanistic physics speaks of material 'points' in Space (as Newton does), the concept of Time, Hegel explains, is incidentally introduced. For the mathematical point, having no dimension, is obviously a negation of Space, and that is precisely what Time is. And it is the point of Time in space that traces the 'lines' of spatial dimensionality as its *loci*. Anticipating the need for the Riemann-Einsteinian advance, he observes:

> There is no *science of time* corresponding to the *science of space*, to geometry. The differences of time have not this *indifference* of self-externality which constitutes the immediate determinateness of space, and they are consequently not capable of being expressed when the Understanding has paralyzed it and reduced its negativity to the unit (*PN*, § 259, *Remark*, pp. 37–38).

Here is where Hegel observes that a science of *measures*, competent to deal with time in its negativity, as the moving-point of space, "would be the most difficult of all sciences."

Hegel then points out that, in its traditional meaning (philosophically defined by Aristotle), the term Place has always expressed "a posited identity of space and time." That is a profound as well as obvious observation. The moderns who speak of the *space-time continuum* for want of a suitable single word, might do well to consider the long historical development of the doctrine of Place – a doctrine which, according to Pierre Duhem, holds the future of natural science. In our ordinary usage, Hegel notes, a thing in 'place' is understood to be a thing in space-time, here and now, or there and then. Zeno's paradox of motion, we are reminded, was defined as a space-time paradox of places. A thing at rest or in motion is in 'place' always, even though, in motion, we distinguish (as Hegel says) "three different places: the present place, the place about to be occupied, and the place which has just been vacated" – in which distinction the "vanishing of the dimensions of time is paralyzed." Yet it remains clear in all of this, Hegel continues, that "there is really only *one* place, a universal of these places, which remains unchanged through all the changes; it is *duration*, existing immediately according to its Notion, and as such is Motion" (*PN*, § 261, *Remark, Z*, p. 43).

Zeno's paradox shows that rectilinear motion, out of one place into another, no matter how far extended, contains no *measure* of itself, and is therefore not really motion. Having no *measure* means that it has no time, which is the measure of motion. Curvilinear motion that returns to its former places, on the other hand, contains its own measure. The lines of a space-time continuum (as contrasted with the timeless space of Euclidean geometry) must, therefore, curve back on themselves. With 'Einsteinian' insight, Hegel writes:

> This return of the line is the circle; it is the Now, Before, and After which have closed together in a unity in which these dimensions are indifferent, so that Before is equally After, and vice versa. It is in circular motion that the necessary paralysis of these dimensions is first posited in space. Circular motion is the spatial or subsistent unity of the dimensions of time .... It is motion in its essence, motion which has sublated the distinctions of Now, Before, and After, its dimensions or its Notion. In the circle, these are in unity; the circle is the restored Notion of *duration*, Motion extinguished within itself. There is posited *Mass*, the persistent, the self-consolidated, which exhibits motion as its possibility (*PN*, § 261, *Remark, Z*, pp. 43–44).

The so-called *mass*-point, in other words, is 'circling motion' in itself. I doubt that anything more profound has been written in modern times on the abstract foundations of mechanics. To sum up, and hurry forward, we may say: Anticipating the language of Ernst Mach, Einstein, and Planck, Hegel defines space and time as positive and negative determinations of Motion, which, in its self-circling, is Mass. Time is the restless, ubiquitous point of space that negates dimensionality in itself while generating the same as 'loci' of its Future and Past. Time gives 'place' to the lines, planes, and volumes of space. And Hegel dares to conclude that 'moving place' or, as we usually say, *loco*motion, is the sole constituent of Matter.

Every time-point is thus a mass-point circling in itself: an essentially *moving* place that tends restlessly to circle 'out of place' and back into it. Thus this universal gravitation of time-points is simultaneously a moving away and a moving toward, self-repulsion from one center and attraction toward another. As Hegel expresses it: "Matter is both moments [repulsion as well as attraction] and their unity, centralized in a point, is gravity." Matter itself is "tending toward a center, but – and here is its other determination – a center located outside itself." More precisely: Gravity is not mere attraction, which is the tendency to negate spatial separateness and produce continuity. On the contrary, "gravity preserves both separateness and continuity" as moments of its concept (*PN*, § 262, pp. 44–47). And these moments, apprehended abstractly as repulsion and attraction, are destined to be narrowly determined as independent forces, centrifugal and centripetal (*PN*, § 269, pp. 62–65).

Those who take space and time abstractly, who conceive of 'matter' as 'occupying' space and as being 'acted upon' by independent forces of 'inertia' and 'gravity' have, in other words, taken rectilinear uniform motion as simple, and accelerated and curvilinear motion as composed. From Hegel's standpoint these are to be reversed. And in what he has to say on this point, "there is undoubtedly [as Professor Findlay has said of his concept of light] a flavor of relativity-physics."[19] Hegel says:

That movement in general is movement that returns upon itself may be concluded from the determinations of particularity and individuality in a body, which give it an internal focus of separateness together with a tendency to join a center outside itself. These are the determinations underlying the concepts of centripetal and centrifugal force; these are then taken abstractly as independent vectors which, brought to bear accidently on the same inert body, give it the empirically observed motion. Thus you have the transformation of physical reality into lines that really serve only to facilitate mathematical expression (*PN*, § 270, *Remark*, pp. 68–69).

Before the development of a 'field' theory of gravitation in our own time, Hegel's application of the dialectic to analysis of physical phenomena had no counterpart in the mathematical thinking of physicists or astronomers. But now even children are familiar with a concept of geometric space suffused with a time that varies with each point of space, and with pictures of the circling centers of atomic particles. In terms less rigorous than Hegel's but not inconsistent with them, Einstein has written:

We have two realities: *matter and field* . . . . But what are the physical criterions distinguishing matter and field? Before we learned about the relativity theory we could have tried to answer the question in the following way: matter has mass, whereas field has not. Field represents energy, matter represents mass . . . . From the relativity theory we know that matter represents vast stores of energy and energy represents matter . . . . the division into matter and field is, after the recognition of the equivalence of mass and energy, something artificial and not clearly defined . . . . What impresses our senses as matter is really a great concentration of energy into a comparatively small space. We could regard matter as the regions in space where the field is extremely strong . . . . There would be no place, in our new physics for both field and matter, field [in Hegelian terms, 'moving place'] being the only reality. This new view is suggested by the great achievements of field physics, by our success in expressing laws of electricity, magnetism, gravitation in the form of structure laws, and finally by the equivalence of mass and energy.[20]

We noted in passing that it was on the basis of the field theory that the Einsteinians rejected the Newtonian principles of rectilinear inertial and gravitational motion, adopting a four-dimensional geometry in place of Newton's calculus to express mathematically their conception of an astronomical order in which inertial and gravitational mass are identified and the fundamental motion is curvilinear. And, while we cannot dwell on it, we should at least mention here that, at one point in his larger *Science of Logic*, Hegel speaks of the possibility of developing a four-dimensional space-time geometry to do justice to Kepler's laws. Such a geometry, Hegel says, "might prove powerful with regard to free movement, wherein one (spatial) side is governed by geometrical determination (in Kepler's law s^3:t^2), and the other (temporal) side by arithmetical determination." The mathematics of that geometry, Hegel suggests, might serve to comprehend the elliptical motions of planets in their uncomposed naturality, in a manner consistent with the Hegelian concept of matter and gravity (*SL* [J/S] **1**, p. 324).

VI

Hegel, as we said, places Kepler's laws of planetary motions at the apex of the grand pyramid of Mechanics. On that apex, looking beyond itself, mechanistic science transcends the abstract, and the transition is made into the sciences of qualified matter. Just below that apex Hegel sets the Galilean laws of falling bodies.

Of these laws, Hegel says: "they are immortal discoveries which redound to the greatest honour of the analysis of the Understanding." Still, in themselves they are not enough. "The next step," he continues, requires "their *proof* independently of empirical methods." And such proof, according to the Newtonians, has already been supplied – furnished, allegedly, "by mathematical mechanics itself, so that even a science based on empirically-ascertained facts is not satisfied with the merely empirically *pointing out* (demonstration)." What Einstein and Planck and their peers have undertaken to do mathematically, beyond the sphere of the empirically verifiable, Newton also attempted to do; but, according to Hegel, his method was inadequate. Newton's 'proof' of Galileo's laws, Hegel says, takes for granted what those laws describe: the empirically-established fact of the uniformly-accelerated motion of a freely falling body. Newton's analytical treatment of that given fact consists, Hegel says,

> in the conversion of the moments of the *mathematical* formula into *physical* forces, into an *accelerating* force imparting one and the same impulse in each unit of time, and into a force of *inertia* which perpetuates the (greater) velocity acquired in each moment of time – determinations utterly devoid of empirical sanction and equally inconsistent with the notion (*PN*, § 267, *Remark*, p. 57).

In other words, Hegel indicates that Newton treats a heavy, falling body, which *naturally* gravitates toward a center outside itself, as if it were an inert body responding inertly to an external thrust or force – as if it were *not* itself essentially gravitational. "The motion of falling," Hegel writes, "forms the transition and middle term between inert matter and matter whose Notion is absolutely realized, that is, absolutely free motion." Inert matter is the abstraction of a single material point, conceived as having no time in itself. It will remain at rest, or, if moved, will move inertially as if it were at rest, the two being equally timeless and therefore not really distinguishable as rest and motion. On the other hand, "heavy matter in the motion of falling," Hegel writes, is at least "partly adequate to its Notion, namely, through the sublation of the Many, as the effort of matter to reach one definite place, as center." The next step in the determination of matter is that 'falling *toward*' a center must also be seen as 'repulsion *from*' a center – which is what the so-called fixed stars manifest in their 'constellational' pattern. Though in themselves they are matter orbiting about an energized center, with respect to one another, they are simply held in a rigid equilibrium of attraction and repulsion. The stars, says Hegel, "belong to the sphere of dead repulsion . . . . Matter, in filling space, erupts into an infinite plurality of [luminous] masses, but this, which may delight the eye, is only the first manifestation of matter."[21]

Not in abstraction as inert, not in falling, not in the rigid equilibrium of stellar repulsion and attraction, but only in the absolute mechanics of universal gravitation – in the grand mechanics of the sun and planets, moons and comets – is the true Notion of matter realized.

As in his discussion of the Newtonian calculus and the concept of force, so here in his discussion of Newton's concept of universal gravitation, Hegel praises before he criticizes. "Universal gravitation," he writes, "must be recognized as a profound thought in itself, though it is especially by reason of the quantitative determinations bound up with it that it has attracted attention and credit." So far from criticizing the concept, Hegel takes it up as an essential moment of his system, which (it should be recalled) is, like Aristotle's, a system of gravitation, in which all components, those of the *Logic* and *Philosophy of Mind* as well as those of the *Philosophy of Nature*, are drawn together and sustained by a single principle that operates on or in them like a magnet on metal filings; or, better, like the command of a commander-in-chief to all his armies – a command that exists apart, perfectly, in the commander-in-chief, but which is also present in all the purposeful activities of the army on all levels. Hegel is more concerned than Galileo or Newton were with the *cause* of such universal gravitation; but, as we suggested earlier, he does not deny the right of empirical science to study the filings in a magnet field (or the moving multiplicity of nature as a whole for that matter) as if there were no underlying, magnetic mover.

The trouble with Newton's use of the notion of universal gravity in his treatment of Kepler's laws is the same we identified in considering his treatment of Galileo's laws. In Newton's "law," Hegel notes, "there are included, as we have shown, two moments" whose unhappy fate it has been "to be regarded as separate *forces*, corresponding to the forces of attraction and repulsion, more precisely determined as *centripetal* and *centrifugal* forces. These forces are conceived as *acting* on bodies, and as accidently and independently brought to bear upon a third something" – the material point of the pattern of motion under study. "In this way," Hegel goes on, "whatever deeper meaning might have been in the thought of universal gravitation is again lost"; and he predicts that science will not be able to "penetrate into the theory of absolute motion so long as the much-vaunted discoveries of *forces* prevails there" (*PN*, § 269, *Z*, p. 63).

To make clear the line of Hegel's argument, we should remark here on the great cultural paradox that presents Newton to us as the fulfillment of a scientific development that is usually traced to him from the labors of Copernicus through those of Galileo and Kepler. That development from its inception aimed at creating a single physics, made up of a single set of principles, to displace the old Aristotelian double physics with its double set of principles (celestial and curvilinear, terrestrial and rectilinear). Galileo, as any serious student of his works must know, opted for unity on the basis of curvilinear motion, so that all apparent rectilinear motion would be either an optical illusion (like that of falling bodies, where the rotational movement of the earth is shared by body and observer, and therefore cancelled out) or a resultant compounded of naturally curvilinear motions. Copernicus had abolished the entire realm of rectilinear motion by assigning celestial motion to the terrestrial orb. And Kepler, too, while laboring to eliminate the complexities of circles on circles in the Ptolemaic and Copernican systems, was none the less a champion of curvilinear physics. From Copernicus to Galileo and Kepler, the triumphant claim of science had been, therefore, that the physical universe is indeed One, and all celestial. But then came Newton to bring the whole business down with a Baconian thud: The universe is physically One, as required, but – according to Newton's doctrine – it is all earth, all rectilinear – so that all curves are either optical illusions or resultants of compounded, naturally rectilinear motions.

From Hegel's point of view both positions, the Galilean as well as the Newtonian, were one-sided abstractions of the Understanding – like the contradictory corpuscular and wave theories of light. Against both (and again in an Einsteinian vein), Hegel argues that the proper scientific course is to distinguish and preserve qualitative differences. Inert, falling, rigidly-fixed stellar, and the freely-moving matter of the solar system must be comprehended together in their qualitative differences. Altogether, Hegel concludes after a long discussion of principles,

> There exist three motions: (a) mechanical motion which is communicated from outside and is uniform; (b) the partly conditioned and partly free motion of falling, where the separation of a body is still posited contingently but where the motion already belongs to the gravity itself; and (c) the unconditionally free motion, the principal moments of which have been indicated, the great mechanics of the heavens. This motion is a curve. In it, the positing of a central body by the particular bodies, and conversely, the positing of these by the central body, occur simultaneously. The center has no meaning apart from the periphery nor the periphery apart from the center. This puts to rout those physical hypotheses which start now with the center and now with the particular bodies, sometimes making the former and sometimes the latter the original factor. Both points of view are necessary, but, taken separately, they are one-sided. The diremption into different bodies, and the positing of the moments of subjectivity, is a simple act, a free motion, nothing external like pressure and thrust (*PN*, § 269, *Remark Z*, p. 64).

That is a summation worthy of Einstein or Planck. On the lowest, abstract level are the inert, unorganized material entities: their motion comes to them from the outside, is uniform, according to the pressure or thrust, but essentially random. Focusing by abstraction on the inert particle, we cannot know what forces may act on it from unknown sources not yet posited; focusing on posited *forces*, we cannot know what particles, if any, will be acted upon, here or there. But on the level of heavy, organized matter, the motion is not only random or mechanical, but also determined, by its organization around a center of gravity, as a weighing down of one organized mass toward other organized masses. Higher still, however, is the gravitational relation manifested in the organized equilibrium of the solar system – and that, as Hegel says, has nothing random or forced about it. On that level, he writes,

> we must not speak of forces. If we want to speak of force, then there is but *one* force, and its moments do not, as two forces, pull in different directions. The motion of the celestial bodies is not any such pulling this way and that but is free motion; they go on their way, as the ancients said, like blessed Gods (*PN*, § 269, *Remark, Z*, p. 65).

That last phrase, it should be noted, was cited by Sir Arthur Eddington in *The Nature of the Physical World* to epitomize his own doctrine of the planetary movements, in the light of the Einsteinian general theory of relativity.[22] Einstein himself has observed that in the "translation of the law of inertia into the language of the general theory of relativity" we get a new law of motion which is not rectilinear at all, being the law of "a system of total differential equations, the system characteristic of the geodetic line" which curves back on itself. Einstein says further: "In place of Newton's law of interaction by gravitation, we shall find the system of the simplest generally covariant differential equations which can be set up for the $g_{\mu\nu}$-tensor. It is formed by equating to zero the once contracted Riemannian curvature tensor ($R_{\mu\nu} = 0$)" – which means that all apparently rectilinear motion is henceforth to be understood as the segment of a vast curve.[23]

But from Hegel's point of view, this Einsteinian development is also one-sided. All that Hegel says against Newton's 'proofs' would apply also, *mutatis mutandis*, to the Einsteinian-Riemannian formulas, were they offered as 'proofs'. The philosophic virtue of the latter, in comparison with the *Principia*, is their philosophic modesty. The Einsteinians acknowledge that their mathematical analysis, which permits them to speak of non-empirical clocks slowing down non-empirically and about non-empirical rulers shrinking, proves nothing. But, if more were claimed for that analysis, it could hardly expect to fare any better at the hands of another Ernst Mach than Newton's analysis has fared.

In his fullest commentary of the relation of Newton's mechanics to Kepler's laws, Hegel makes the following points:

1. It is a mere mathematical translation that turns the $\frac{A^3}{T^2}$ of Kepler's Third Law into a product ($\frac{A}{T^2} \times A^2$) of which one of the factors is the ratio of Newton's law of gravitation.

2. Newton's 'demonstration' that a body moving around the central body in subjection to the law of gravitation pursues a *conic section*, does not, by any logical means, show the necessity for its being an *ellipse*. And the worst is that

> the conditions which make the path of the body a *specific* conic section are, in the analytical formula, *constants*, and their determination is referred to an *empirical* circumstance, namely, to a particular position of the body at a certain point of time, and to the *fortuitous* strength of an *impulse* which it is supposed to have received in the beginning; so that the circumstance which determines the curve to be an ellipse falls outside the formula that is supposed to be proved, and no one has ever dreamt of proving this circumstance (*PN*, § 270, *Remark*, p. 66).

3. Newton's law of the force of gravity has only an empirical, not a philosophical basis. There is, for instance, no *reason* given as to why the centrifugal force manages to 'overcome' the centripetal force just when the latter ought, by definition, to be the greatest (perihelion), and then manages to let itself be 'overcome' in turn just when the centripetal force ought to be weakest (aphelion). The explanation is that, the empirical fact being known in advance (as summed up in Kepler's laws), the analytical calculus is conducted accordingly. One of the most thorough analyses of the relation between Kepler's achievement and Newton's, in recent times, is that of Pierre Duhem in his *Aim and Structure of Physical Theory*. Pursuing a course of reasoning that comprehends Hegel's arguments, Duhem concludes:

> The principle of universal gravity, very far from being derivable by generalization and induction from the observed laws of Kepler, formally contradicts these laws. If Newton's theory is correct, Kepler's laws are necessarily false.[24]

VII

Summing up, we may say that the great value of Hegel's criticism of Newton's celestial mechanics consists in its emphasis on the error of scientific 'reductionism' upon which that mechanics was founded – an error that still plagues theoretical physics, obstructing its advance, and making impossible a philosophically-integrated Natural Science. The popularity of Einstein has made this part of Hegel's criticism more acceptable now than ever before. Yet hardly less impressive than his criticism of Newton is Hegel's sustained polemic against the pre-Darwinian advocates of an evolutionary origin of species, which he conducts on the same grounds. Like the Newtonians in Mechanics, Hegel argues, so the evolutionists in biology are 'reductionist'. They 'decompose' the self-generative cycle of a species into two constitutive elements or vectors, one of which is the characteristic actuality of the species just below it in the unbroken chain of organic forms, while the other is a sort of fortuitous thrust (chance mutation) acting at right angles, so to speak, and thus deflecting the old species out of its old generative cycle into a new one. Summing up the lesson of his entire Philosophy of Nature from this standpoint, Hegel says:

> Now, here is what is really involved in this conception. What, speaking generally, we call inorganic nature is thought of as having an independent existence, while the organic is attached to it in an external fashion, so that it is a mere matter of chance whether or not the organic finds the conditions of existence in what confronts it . . . . The question is this: Is that the true concept of the inorganic, and do living things represent what is dependent? [On the contrary] this is the true relation: man is not an accident added on to what is first; the organic is itself what is first . . . . Regarded in this fashion, the universe is not an aggregate of many accidents existing in a relation of indifference, but is a system endowed, in its essential character, with life.[25]

Here we break off. The wealth of materials in Hegel's *Philosophy of Nature* that have a bearing on the theme of this paper is inexhaustive. As I have observed elsewhere, no scholar who has seriously occupied himself with the history of the sciences of celestial and terrestrial mechanics, and of inorganic and organic physics, can read Hegel on these subjects without a sense of awe; for that history, from its Greek origins down to the nineteenth century, lives in his pages. Those of us who have for decades poured over the sources, guided by Pierre Duhem and his peers, can hardly fail to recognize in Hegel an easy mastery of much of the relevant materials. For me, at any rate, the ten volumes of Duhem's *Le système du monde* serve as an historical commentary on Hegel's *Mechanik* and *Physik* which, in turn, serve to make philosophical sense out of the long development of scientific thought traced by Duhem through all its evolutions, revolutions, and involutions, through all its errors and aberrations, exhuming once-famous doctrines from long oblivion, showing how the science of each epoch is nourished by the systems of past centuries, indicating how often the scientific certainties of one proud age are laughed at as absurdities in the next, while as often today's absurdities are seen forcing themselves forward into acceptance as the high truths of tomorrow.

Duhem's history of science displays a *dialectical* development, and Hegel's *Philosophy of Nature* compels us to recognize the continuous presence of that dialectical past in contemporary empirical science. It is thus a work of permanent scientific as well as philosophic value.

St. John's University
Jamaica, New York

NOTES

[1] The translations cited here from Hegel's *Lectures on the History of Philosophy* have sometimes been adapted in accordance with the German text (Michelet, 1840).

[2] Inscription on statue in the ante-chapel of Trinity College, Cambridge.

[3] John N. Findlay, 'Hegel and the Philosophy of Physics,' in *The Legacy of Hegel: Proceedings of the Marquette Hegel Symposium, 1970*, ed. J. J. O'Malley, *et al*. (The Hague, Nijhoff, 1973), p. 83.

[4] The translations cited here from Hegel's *Logic* have sometimes been adapted in accordance with the German text (Henning, 1840, 1955).

[5] The translations cited here from Hegel's *Philosophy of Nature* have sometimes been adapted in accordance with the German text (Pöggeler, 1959; Michelet, 1847).

[6] J. N. Findlay, *Hegel: A Re-Examination* (New York, Humanities Press, 1958, 1976), pp. 167–8.

[7] The translations from Hegel's *Phenomenology of Mind* have been adapted in accordance with the German text (Hoffmeister, 1949).

[8] Louis de Broglie, *Physics and Microphysics* (New York, Grosset and Dunlap, 1966), pp. 238–9.

[9] *Ibid*., p. 75.

[10] See V. Cioffari, *Fortune and Fate from Democritus to St. Thomas Aquinas* (New York, 1935), *passim*, for a review of the commentary on Dante's Heisenbergian line (*Inf*., IV, 136) epitomizing the doctrine of Democritus.

[11] H. D. Anthony, *Sir Isaac Newton* (New York, Collier, 1961), pp. 140–141.

[12] Tobias Dantzig, *Number: The Language of Science* (New York, Macmillan, 1954), pp. 234, 335.

[13] Bertrand Russell, *The Scientific Outlook* (New York, Norton, 1932), p. 264.

[14] Albert Einstein, *Out of My Later Years* (New York, Philosophical Library, 1950), p. 110.

[15] Einstein, *Later Years*, p. 97.

[16] Albert Einstein, *The World As I See It* (New York, Covici Friede, 1934), p. 57.

[17] Bertrand Russell, 'Introduction,' in A. V. Vasiliev, *Space, Time, and Motion* (New York, Knopf, 1924), pp. xv–xvi.

[18] Einstein, *Later Years*, p. 69.

[19] Findlay, *Hegel*, p. 279.

[20] Albert Einstein and L. Infeld, *The Evolution of Physics* (New York, Simon and Schuster, 1961), pp. 256–58.

[21] *PN*, § 267, *Z*, p. 59; § 268, *Remark, Z*, pp. 61, 62.

[22] Arthur Eddington, *The Nature of the Physical World* (Cambridge, Cambridge University Press, 1948), p. 147.
[23] Einstein, *Later Years*, p. 80.
[24] Pierre Duhem, *Aim and Structure of Physical Theory* (Princeton, Princeton University Press, 1954), p. 193.
[25] *Hegel's Lectures on the Philosophy of Religion*, trans. E. B. Spiers and J. B. Sanderson (New York, Humanities Press, 1962), Vol. III, pp. 340–341.

JOHN N. FINDLAY

THE HEGELIAN TREATMENT OF BIOLOGY AND LIFE

I gave a paper on Hegel's treatment of the physical sciences at Milwaukee last June (1969), and another not entirely similar paper on the same theme at Stuttgart in July, where I also contended with the German language. On this occasion my theme is Hegel's treatment of Life and the Life-Sciences. I shall not, however, limit myself to the dialectical analyses which occur in the *Philosophy of Nature*, but shall go back to the far more fundamental treatments which are to be found in the *Science of Logic* and the *Encyclopaedia*. Life is for Hegel a category before it is a physical phenomenon in time and space, and we have to try to understand just what sort of a category it is, and where and how it has its place in the logical array of categories. I also wish to say something about the treatment of Life and Biology which occurs in the *Phenomenology of Spirit*, where Life is being set forth as one of the objects which the mind contemplates and investigates on its journey to Absolute Knowledge. After these two fundamental treatments I shall consider what Hegel has to say about Life and Biology in the *Philosophy of Nature*, enquiring, in the end, what he has to say *to us* in the present stage of development of the biological sciences. I am compelled to leave out the interesting discussions of Life in the various sets of Jena lectures, because I have never been able to give these lectures that complete sentence by sentence scrutiny which is an indispensable preliminary to understanding anything in Hegel.

It is all-important, in understanding any fundamental notion in Hegel, that we should approach it in a purely logical, as well as, and probably before, approaching it in an empirical manner. Hegel does not deny that certain fundamental categories may first present themselves to us in empirical instances: we may take note of various numerical assemblages and quantitative differences before we consider what it is to be numerous or quantified, we may observe interactions before we consider what it is for anything to be the cause of anything, and obviously we may observe and even formalize the behaviour of mechanically interacting bodies, or of self-active, living bodies, or the behaviour and inner activities of particular minds, before we ask what it is to be inertly mechanical, or what it is to be living, or what, lastly, it is to be conscious or self-conscious. But the logical treatment of these categories is not a mere elevation into generality of what has thus been empirically

R. S. Cohen and M. W. Wartofsky (eds.), Hegel and the Sciences, 87–100.

distinguished: it is an entirely new, much sparer treatment, a sort of reduction to a logical skeleton in which the empirical flesh vanishes altogether. Most Hegelian commentators have barely risen to the austerity of his logical strippings; they have generally retained a good deal of empirical flesh and have then been astonished how it got there. That Hegel's whole logical stock-in-trade consists of such severe abstractions as the positive and the negative, the one and the many, the universal, the species and the instance, the necessary and the contingent etc., they have been very far from realizing. I myself have only fully understood the purity of the basic Hegelian abstractions in very recent years, and I shall use this occasion in trying to communicate what I have come to see.

Let me first outline what I conceive to be the whole pattern of the Hegelian Logic: I may say that it is nothing but a series of stages in which the *Begriff*, the Principle of Active, Concrete Universality, becomes explicitly the *Begriff*, becomes, if one likes, the *Begriff* of the *Begriff*, or the *selbstbegreifende Begriff*. On Hegel's view what absolutely is, is simply Universality as Such, but a Universality which only is what it is because it includes Specificity and Individuation as a subordinate mode of itself, because it is a Universality endlessly specified and individuated. This absolute Universality operates in a pure and detached form in thought, but it also operates in an obscure, embedded form in external things, and it is the categories of such embedded Universality that are studied in the two earlier, 'objective' parts of the Logic, the Doctrine of Being and the Doctrine of Essence. It is the pure principle of Universality which appears in such inadequate guises as Indeterminate and Determinate Being, as Qualified and Quantified and Metric Being, as what is Identical-in-Difference and Different-in-Identity, as what is Underlyingly Essential and Superficially Manifest and the necessary union of both, as what is Necessary and Contingent and Possible. What releases itself from all this dance of forms is, however, the *Begriff* or Principle of Universality as Such, now seen, not as an external linkage of Species and Instances, but as the unifying principle without which specification or instantiation would not be possible at all. This *Begriff* first reveals itself in a system of deterministic mechanism where the behaviour and character of everything is rigorously determined on general lines by the behaviour and character of everything else, the Universal being the rigorous laws which pervade and link the whole system. But this rigorous *Wechselwirkung* has its 'truth' in a Teleology where it is a unifying End, rather than a set of compelling Laws, which pervades and works through the whole: in Teleology the mechanistic laws, which there still seem to be among mutually external things, must be brought together and connected. In thoroughgoing Teleology the whole is operative in all the parts, nothing is externally coerced by anything, and the End is distinguished from its logical ancestor, the Cause, in that it, and it alone, produces itself and only itself. Life is simply Teleology 'collapsed into immediacy', a system where wholeness and unity is everywhere at work, where the external becomes part of the internal, and where all sides of the internal are merely the same Universality differently but not separately operative. But the immediacy of Life necessarily leads on the still deeper and more magisterial unity which both posits and overcomes diremption, in which detached Universality in the form of subjective Thought and Will confronts embedded, materiate Universality in the form of objective natural and social being, and in which the former overcomes the latter in the activities of knowing and doing. In all this we have nothing but the Universal distinguishing itself from specifications and individuations, and also asserting its unqualified mastery over the latter, as that of which they are only the parasitic adornments, instruments and illustrations. This philosophy of dynamic Platonism will not recommend itself to all, and I cannot argue for it on the present occasion: I can only make lucid what I think it without question means and involves.

Let us now consider the dialectical phases which Hegel distinguishes in his treatment of Life, the immediate expression of the Absolute Idea. Life is, as we have said, a universality which absolutely dominates the specific and the individual, the former being a set of distinct organic functions quite inseparable from the total being to which they contribute, and which are constantly modified to enable that being to be continued, and the latter being the instantiations of that totality. In Life, as thus outlined, three interrelations of the moments of Universality, Specificity and Individuality call for consideration, which Hegel describes as 'processes', in some abstract, logical sense of the word, and also as 'syllogisms'. The first of these process-syllogisms is the immanent process of the living individual, conceived as in relation to an objective environment. This represents the lower mechanistic expression of the Idea, of which the living organism represents the transcendence, but not specifically seen in its interaction with the latter. In this immanent organic unity each organ and function, though specific, has something totalitarian about it, and uses every other organ and function for its own support and sustenance, *is*, in fact, the whole which it also specifies and instantiates. Hegel tells us expressly that this intra-organismic commerce is *not* to be understood in a purely mechanical or chemical fashion. The living organism, in so far as it has mutually external parts, is capable of such mechanico-chemical relations, but in so far as it functions mechanically or chemically it is also functioning inorganically, or in a dead manner. The inner teleology of Life does not therefore consist in the mere *use* of mechanism for a purpose: the organism is not a watch, though it may have watchlike aspects. In this internal teleology Hegel distinguishes three overall functions of Sensitivity or Receptivity, Irritability or Reactivity, and Self-Reproduction, and tries to give to each an *a priori* notional warrant.

But the living teleological system not only confronts a non-teleological, mechanico-chemical environment: it, as the higher expression of the *Begriff*, must necessarily endeavour to overcome and dominate the latter. It is aware of the gulf between itself and the non-teleological environment in the experience of need and pain, and it launches out from this to assimilate the environment to itself, to make it as it wants it to be. The precise modification inflicted on the environment depends on the instinctual make-up of the organism, but *that* it should have instincts directed to remoulding the environment is categorial, ineliminable, necessary. The living teleological system is, in the third place, concerned, as pure Universality, to transcend the mere contingency of its individual embodiment: instantiated it must be, but not necessarily or solely in *this* living body. It therefore extends itself as a genus of mutually external instances, and connects its instantiations by various genetic relations. What these are is essentially an empirical matter, but the generic and the genetic are categorial features of living universality. It is in fact at this point that living Universality passes over into the two-sided Universality of Subject and Object. There can never be an adequate individualization of living Universality in any number of living generations, however prolonged: it achieves, however, another sort of individualization in the thinking mind, in which guise it can stand over against the inadequate individualization it achieves in the outside world, and can take up the latter into the former. It is a very wonderful transformation when the endless stream of biological generations, whose serial form is given concreteness by, without being identical with, a series of generations in time, gives rise, in a logically grounding rather than a temporally causative sense, to the static, non-serial perfection of a species in thought, and shows us with what logical purity Hegel thought of the dialectical phases in question. They are called Life and Cognition, but they are only as close to Life and Cognition as we ordinarily understand them as an abstract equation is to the wealth of phenomena that it formalizes. Life is, as it were, Universality drowned in the Specificity and Individuality which it also needs, whereas Cognition is Universality which has come out of specific-instantial immersion, but which still drips with what it has come out of. My images will antagonize many, and seem anthropomorphic, but what they picture is the varied forms of that distinction-without-separation which analytic thought hates, and which is the central feature of Hegelianism.

I shall now go over to the treatment of Life and the Life-sciences we encounter in the *Phenomenology*. Here Hegel is dealing with Life as an object of Consciousness which emerges *after* Consciousness has been dealing with the world of phenomenal natural things, and seeing it as the theatre in which unmanifest powers and laws are forever displaying themselves. Contemplating Nature in this manner, it has gone beyond sundered moments of the *Begriff* to the *Begriff*: it has been the *Begriff* as a Universality which is forever specifying itself and individualizing itself, and in which none of the *Begriff*'s constituent moments, and their living interplay, can be separated from the others. This living *Begriff* is, however, as much Consciousness itself as Consciousness's object, and, in contemplating it, Consciousness, the faculty of liberated, active Universality, is in a sense only contemplating itself. The *Begriff, qua* object of Consciousness, is, however, first invested with traits reminiscent of the sensuous externality which has belonged to its former objects, and, seen in this manner, it is Life, the definiteness which is always melting into indefiniteness and the out-thereness which has also all the lack of out-thereness which marks inner-life interpenetration. As Hegel puts it:

The determination of Life as it arises out of the Notion or general result which we encountered on entering this sphere, is sufficient to characterize it without attempting to develop its nature further: it is a circle embracing the following moments. Its essence is infinity as the supersession of all differences, the pure revolution about an axis, the rest of its own absolutely restless infinity, the very self-sufficiency in which the differences of motion are dissolved, the simple essence of time which in this inner uniformity has the solid shape of space. Differences, however, are just as much present as differences in this simple universal medium, since this universal fluidity only has its negative nature in so far as it abolishes such differences: differences must be there if differences are to be abolished. But this fluidity, as self-sufficient uniformity, is itself the subsistence or the substance of these differences, in which they occur as distinct members and independent parts. Their being no longer has the meaning of being in the abstract, nor their pure essence the meaning of abstract universality: their being is just the simple fluid substance of pure movement in itself (*Phän* [H, 1937], pp. 135–136). [Trans. by J. N. F.].

The word that encapsulates the sense of this paragraph and of several others that follow it is the word *Flüssigkeit*, Fluidity: Life is Flux as a pure concept. It is a flux in which there are differences which assert themselves against the flux, which mean to be determinate, specific, separately individual. These self-asserting differences are, however, the mere instruments of a mightier self-assertion, one which has posited such differences only in order to abolish them, to make them pass into one another, to be the pure negativity that it is in the process of negating them all. The many differences involved in Life do not even pretend to the separateness and self-sufficiency of solid, inorganic things: they are there only to have their separateness and self-sufficiency and even clear character denied, to become mere points of passage in an unbroken motion. The teleology of the living organism simply consists in its denial of substantial separateness, even of determinate character to its parts: they are there only to be modified, to be bye-passed, to be swept through and swept away. Nowhere has anyone put forward so stably unstable and so purely logical an account of life in which structure rests on fundamental structurelessness: beside it the Platonic-Aristotelian accounts of Life as the Self-moving seem hopelessly gross and inadequate. But they make sense, for they constitute an image of that being-no-one-in-particular which is also the possibility of being-anyone-and-everyone which is somehow the deepest essence of our conscious personal being: life in the more definite medium of what exists out there is a monogram, an analogue, of that more ultimate indefiniteness that we experience as ourselves.

Hegel develops his account of Life at a later stage of the *Phenomenology* (*Phän* [H, 1937], pp. 193–221) where he is discussing the observation of organic phenomena. Here he maintains the interesting thesis that the teleology of life, where it is present, essentially transcends natural law: there is rigorous law connecting acids with bases, but there is no rigorous law connecting cold climates with hairy pelts, or fluid media with the shapes of fishes. Organisms do indeed adapt themselves to the environmental situation in which they arise, but this does not causally produce their peculiarities, and they evince their freedom from it by having many non-adaptive characters, some anticipating the adaptations of other types of organisms or harking back to the same. (Hegel does not believe in organismic evolution in time, but his logical arrangement of species none the less has an evolutionary cast.) The duck-billed platypus, which Hegel does not seem to have heard of, is a typical example of Hegelian freedom. And not only are organisms not products of the environment, but they produce no uniform products but themselves: it is they themselves which they alone produce and reproduce. An organism may be stimulated by the environment into doing this or that, but, unless this stimulation passes over into inorganic violence, what the organism does is solely what suits its own programme, what is part of its repertoire, what is in short itself. Organisms in short use the natural objects around them, they incorporate them and their dead properties to some extent into themselves, they feed on them, they walk on them, they play with them etc., but in all this they make external objects their own: the objects which surround organisms never use organisms, on the other hand, since they have no programmes, no directed procedures, into whose service organisms can be impressed. These are elementary points, but they seem, none the less, to have eluded most contemporary philosophers.

Hegel also thinks that, while we may properly distinguish the specific bodily structures of organisms, their brain-structures, muscular structures etc., from the universal organic functions which work in and through these structures, it is quite wrong to think of these as correlated aspects of organic being somehow connected by laws of nature. The structures of brain, intestines, muscles are nothing apart from the uses, the indwelling purposes to which they are put: we can no more study them to throw light on integrated organic functions than we can throw light on the semantics of sentences by a mere examination of their syntactical make-up. The uses to which such structures are put are, however, ineliminably universal, and this means that an organism can always shift its exploration or reaction or self-maintenance from one structure to another. It is indifferent, Hegel tells us, to the stream of life what manner of mills it drives (*Phän* [H, 1937], p. 211). The freedom of organic nature is in fact such as neither to obey the rigours which bind physical substances in physico-chemical necessity, nor the dialectical laws which govern the development of conscious spirit. Organic nature reveals only rudiments of laws, suggestions of order and sequence, amusing and specious relationships, in which the most serious statement that is ever made is that something has a 'great influence' on something else (*Phän* [H, 1937] p. 220f). Since Hegel's day there has been a great development in the life-sciences and social sciences, but what he says is as true today as it was in his time. There is not a single rigorous biological or psychological or psycho-physical or sociological law on record, only rudiments of all these things, given undue prominence by those who hope to assimilate the sciences of life to the sciences of lifeless things. When will scientists and philosophers give up this vain endeavour, and learn to study life as life and not as a peculiarly complex case of lifelessness? I believe myself that only a revolution in thought and life, one that radically demotes experimental science and manipulative technology will do the trick. We are in need of a radical de-Baconization of thought and life.

From the more fundamental treatments of Life which occur in the *Logic* and the *Phenomenology*, I pass on to consider Hegel's treatment of Life as a natural phenomenon. The 'Organics' is the third part of the *Naturphilosophie*, and in this work we start dialectically with the Notion, the Principle of

Universality, in its extreme impotence and self-alienation, in a state where it seems lost in a set of mutually indifferent individuals, whose unity is something sought for rather than actually realized, being manifest only in the tendency of bodies to group themselves about gravitational centers which themselves are outside of them. We then progress through the stage of Physics, in which there are all sorts of relations among specific natural bodies which force them into various strait-jacketed unities, such as the magnetic, the electrical and the chemical, but from which it is possible for them to spring back into their old looseness of association and merely mechanical interdependence. In chemical action we are, Hegel tells us, on the verge of life, for the acid and the base have practically an organic relation to one another, which only falls short of true organism because for them separate existence comes as easy as combination, and because it is only an external intrusion which brings them together or forces them apart. If acid and base went in quest of one another like male and female, and sustained their union against interferences and reproduced it in novel cases, they would be organic instead of being the half-way house to organism that they are. Their defective unity is also a falling short of teleology: their unity is not the self-directed, self-maintaining thing we find in organic unity, and, while not causal and mechanical, depends too much on external chances to be classed as truly purposive. In the living organism, however, everything in the individual and its relation to the environment, and in the successive generations of individuals, reflects the same generic unity, which in all internal or external change only brings forth itself. The organic sphere is then divided by Hegel into the Geological, the Botanical and the Zoological spheres, which we must separately consider.

As regards the geological part of 'Organics', one is inclined to feel that Hegel included it only because he needed *three* dialectical phases in his treatment of Life. He says that Geological Nature is only the ground, the basis of life, that it ought to be Life, individuality, subjectivity, but that it does not bring its different members into unity: it is even described as a crystal, a skeleton, which can be regarded as dead because its members still seem to subsist formally on their own, while the geological process falls outside of them (*PN,* § 337, *Z*). A careful reading of the paragraphs and *Zusätze* dealing with Geological Nature shows, however, that Hegel does attribute a sort of rudimentary life to the earth, the Universal Individual, and that he makes it break forth into manifestations which, though brief, can only by described as organic. We are told that this crystal of Life, the inanimate organism of the earth, which has its notion in the sidereal connections outside of it, but possesses its own peculiar process as a presupposed past, is the *immediate subject* of the meteorological process, through which this subject, as *in itself* the totality of life, no longer becomes merely an individual shape but is fructified into vitality. Land and especially sea, as thus the real possibility of life, perpetually erupt at every point into punctiform, transitory life-forms, lichens, infusoria, and, in the sea, countless hosts of phosphorescent points of life. But it is just because *generatio aequivoca* has this objective organism outside it, that it is restricted to this punctiform organization, does not develop internally into a specifically articulated organism, or reproduce itself *ex ovo* (*PN,* § 341).

There are accordingly flashes of life in the earth and the elements surrounding the earth which do not develop into lasting individuality, or the even longer lasting life of the species, but represent something more than chemico-mechanical materiality, having perhaps a spontaneity of origin and a brief self-maintenance of which inert matter is incapable. It is surely a great pity that *generatio aequivoca*, so unequivocally real to Aristotle and Hegel, has since been rendered so dubious: it should on general principles have been true. The doctrine that *omne vivum ex vivo* smells inexpugnably of the analytic Understanding. Modern science has, however, not failed to find forms of quasi-life which bridge the gulf between the organic and inorganic. Viruses are now the front-line candidates for this position, and I do not doubt that there will be others. But instead of being used to show chemistry and mechanism transcending themselves, I have no doubt that they will be used by some as but one more 'proof' of the mechanico-chemical nature of life.

From the Geological Organism we proceed to the Plant-Organism. Here it would appear that we have one of the many cases where the dialectic is best read backwards, where what comes later is necessary for the understanding of what comes earlier. We can only understand what Hegel has to say about plants by going on to what he has to say about animals. The animal organism is the organism fully actualized, the plant merely approaches animality in an inadequate manner, and falls short of it in ways that are only significant by contrast with animality. In the animal the moment of Universality is present in two fashions: on the one hand in the fluid organization of parts and functions, which have no fixed, separate character, on the other hand in a sort of rudimentary subjectivity, involving centralized feeling rather than cognition, into which all its specific differentiations are taken back. The animal not only is universal, but after a fashion feels its own universality. In the plant, on the other hand, there is no such detached, centralized universality: in so far as plants have a centralized self, this lies not in themselves, but in the light towards which they tend, and on which they depend for the full development of their functions. In the animal, likewise, Universality specifies itself quite differently in different parts and organs, and though all these are systematically coordinated, and can in some situations take on each others' functions, they are not simply interchangeable. In the plant, however, the total organic nature is not differently present, or not very differently present, in its various parts, and hence each part can readily become independent, or can take on the functions of another part. The animal, likewise, having its own complexly specified pattern stands in strong contrast to the mechanico-chemical environment, and can accordingly move about in it: the plant, with its much lesser internal differentiation, and consequently less differentiation from the unorganized environment, cannot prise itself loose from the latter. The animal, strongly relieved from its environmental background, has its characteristic temperature which it maintains whether it is hot or cold outside: there is, Hegel thinks, no analogue of animal warmth in the plant. The animal nourishes itself by consuming parts of the environment at times suitable to itself: the rooted plant cannot help taking in nourishment as long as the nourishment is there. Animals have voice, which, like their self-feeling and bodily heat, reflects their rounded individuality, whereas plants only produce sounds mechanically after the fashion of inorganic things. Animals, finally, reproduce their kind only by the coming together of sexually differentiated members of the same genus, whereas plants have a large number of less dramatic ways of preserving and multiply instantiating generic identity.

Hegel's comparison of plants and animals is an expanded version of Aristotle, but the essential point made is that instances of animality have an individuality lacking in instances of plant-nature. What does Hegel mean by this 'individuality' which he also connects with Subjectivity and Being-for-self? He plainly means that being a certain sort of animal is in some sense being both differentiated and rounded off: an instance of animality does not combine with other instances into an equally valid instantial unity, or break up into other similar subordinate units, but is a complete unit, and the genus present in it is a universal operating through a number of such complete units. This sort of individuality is not the mere *Einzelnheit* of instantiation which is one of the moments of the *Begriff*: it is an *Einzelnheit* of instantiation which after a fashion measures up to the universal it instantiates, nothing more or less being thus adequate or necessary, whereas the universal thus instantiated has built into it the need for just such a matching instance. The individuality represented by animal life is a stage of the Idea greatly underemphasized in Hegel's Logic, much as he also underemphasizes the *finite* conscious person or Ego, of which it is the dialectical predecessor. But plainly it is in the well-rounded individual organism, or finite personal mind, unified, differentiated and inexhaustibly rich, that the Absolute Idea has its most concrete fulfilment, rather than in any part or aspect which is a mere part or aspect, or in some universal sum total which is only a bad infinite, never complete. Plato ended by making his prime universals the ideas of natural units and particularly of living things, and it is in this sort of view that Hegel, with equally unclear emphasis, really ends.

Of the detail of Hegel's treatment of plants and animals I shall not say much because a great deal of it is too empirically intricate for a scientifically ill-versed person like myself. A great deal of Hegel's philosophical botany comes from Goethe's essays on the Morphology and Metamorphosis of Plants, where much is said as to the way in which plants fail to subordinate their parts to any total pattern, or as to the way in which one sort of part can, in appropriate circumstances, be transformed into another. A root is a sort of branch and vice versa, calices are sorts of leaves and so are fruits and their rinds, petals are sorts of calices, pistils and stamens sorts of petals, and so on. Hegel found his own dialectical principles in Goethe's ingenious biological speculations. Plant-life for him further organizes itself under the three heads distinguished in the Logic, the heading of internal organization, of the relation of the plant to the external environment and of reproductive genus-maintenance. In internal organization we see the plant dirempting itself into distinct parts, tissues and fluids, growing into a mature specimen and adding a finial to its structure in the bud which is also the rudiment of a new individual. In the relation to the environment we see the plant responding to fire (i.e. light and heat), air and water, and assimilating them to its imperfect selfhood. And finally in the genus-process we have an imperfect realization of sexual specification which only makes the sex-characters distinct without realizing them in distinct individuals, and which, being imperfect, is also superfluous, so that the plant can quite well reproduce its kind without going through the elaborate rituals of sexual union.

Hegel's treatment of animals has been anticipated in his contrasted treatment of plants, but the details of Structure, Assimilation and Genus-process are vastly more complex. We have a detailed discussion of Sensibility, Irritability and Reproduction as the three basic dimensions of animal existence, of which only the third is realized in plants. What plants undergo or do is all a part of their perpetual self-reinstatement in one or more individuals, whereas animals have somewhat of a comprehensive sense of what they are undergoing, and also a comprehensive total reaction to the latter. These three dimensions of animality realize themselves in three bodily systems: the nervous system, cerebro-spinal and sympathetic, is primarily sensitive; the muscular and circulatory systems are primarily 'irritable', reactive, whereas the digestive system with its glands and intestines is primarily 'reproductive' (in a sense which includes self-maintenance). The animal organism also organizes itself into individually structured members, head, hand, eye, liver, etc. in each of which there is a sensitive, a reactive and a sexual aspect. I was unaware that, e.g., the liver might be an erogenous zone, but on reflection cannot see why it should not be, even if I feel nothing sexual about it. All these structured systems and parts are involved in an endless, immensely complex process, each making use of the others and consuming them in its own interest.

The immanent structural process of the animal organism is nothing without the Assimilation to which Hegel gives the meaning of *any* sort of mastery over the environment, or of the imprinting of self on the latter. Such imprinting of self occurs in the various processes of sense-perception: though one speaks as if, in sense-perception, the environment imprinted itself on the animal, the case is rather that the animal assimilates the environment to itself, transforms the latter into an inner, qualitative affection of its own. Such need-assimilation does not involve conscious goal-foresight: it is therefore denominated 'instinctive', implanted, being wrongly ascribed to an external intelligence. This leads Hegel on to an elaborate study of defecation, which for him represents the organism getting rid of its own aggression against the environment, rather than merely of the environmental material that it has been unable to use. In defecation we do not merely eject waste material, but our own digestive juices also: we are like writers tearing up drafts after their manuscript has been typed. Defecation leads on bewilderingly to the constructive instinct, which is really a high-grade form of defecation in which the organism adds a little gratuitous beauty or utility to what it has defecated, e.g. the spider and its web, the bee and its honey, etc. The reproduction of living offspring is constructive defecation carried to a yet higher level. One can only regret that Freud knew nothing of these scatological passages. He would find that he was not the first to find depths and mysteries in defecation.

If defecation of various sorts is the highest form of organic assimilation, death is the highest achievement of the organism's generative activities. Its sexual activities end, in principle, in its own death: it has produced another embodiment of its genus, and can now rest in peace. Its necessary aggressions against the environment, and particularly against individuals of other species, likewise end, in principle, in death by violence, whereas its mere continuance in existence leads either to the milder disharmony of disease, or to the final discord of natural death. This stress on death in the last section of the *Naturphilosophie* anticipates Heidegger as well as Freud. Only that, for Hegel, the death of the individual points the way to the immortal life of mind, and not to the less differentiated quiet of inorganic existence.

A few remarks must suffice on Hegel's treatment of sex, the struggle for existence, and disease and death. Hegel sees in sex the Genus asserting itself over the Species and the Individual: members of the same kind, irremediably specified as to sex, seek to overcome both the one-sidedness of their sexuality and their individuality, by copulating with another member of the opposite sex. The outcome, unfortunately, is not the pure Genus, which Tristan and Isolde might have achieved in their *Liebestod*, but another individual of the same species, endowed with all the sexual and individual specificity which condemns him to further copulative proceedings, and so on to the 'bad infinite' of the generative treadmill. It is only in thought that the genus can exist in its purity, and procreation is therefore an endlessly inadequate substitute for thought. But Animality as Such not only specifies itself sexually within one genus, but in the interspecific differences of the various species of animals. Here Hegel believes, without any Darwinian background, in an endless necessary struggle for existence among animal species, and in a necessary development of those organs with which each attacks each or defends itself against each (as well, of course, as of the organs with which it attacks the environment). The purely logical differentiation of the Zoological Idea into various specific forms of animal necessarily goes with a mutual preying of organism on organism, which, taken together with the necessary deficiencies of the environment, makes animal existence violent, insecure, anxious and unhappy (*PN*, § 368, Remark). It is apparently only at the human level, in the mutual recognition of developed social intercourse, that this primal violence and unhappiness can be overcome. Hegel had not seen the great Central African parks where mutual tolerance rather than predation seems the rule of existence, and where only the members of the cat-family seem concerned to break this rule.

The struggle for existence among species transfers itself, however, to the interior of the animal organism, where various organisms or functions become entrenched in their self-sufficiency and become enemies to each other and to the whole: disease, therefore, is deeply natural, and is not some accidental falling-short of the Zoological Idea. But the Zoological Idea also displays itself in the *fever* which, by reducing all vital functions to fluidity, humbles the *hubris* of the diseased function or organ, and makes a cure possible. The Zoological Idea is, however, powerless against the individual's slow hardening in habit, which is the essence of growing old: growing less and less responsive to the environment, living more and more in the past, the organism becomes more and more like a dead thing. But, as we have said, "the goal of Nature is to destroy itself and to break through its husk of immediate sensuous existence, to consume itself like the phoenix in order to come forth from this externality rejuvenated as spirit" (*PN*, § 376, *Z*).

I have now come to the end of Hegel's treatment of Life and the Life-Sciences. I would comment, not only on the richness of its detail, and of its application of logical principle to that detail, but also on its general methodology. Hegel is neither a rationalist nor an empiricist: the basic forms of Life are logical differentiations of a single categorial conception, but he does not disdain to show us how these basic forms are further specified with infinitely rich, empirical contingency. Hegel is also neither a mechanist nor a vitalist. Life for him is led up to by processes which fall short of Life itself and is at all times dependent on such processes, it is itself a series of stages which lead up to and depend on one another, and it itself leads up to Spirit which, while transcending it, never breaks with it nor loses its dependence on it. Hegel's conceptual presuppositions are arguably those of a truly rational biology, not those of a biology intent of making nonsense of its own subject-matter and ultimately of itself.

Boston University

GEORGE DI GIOVANNI

MORE COMMENTS ON THE PLACE OF THE ORGANIC IN HEGEL'S PHILOSOPHY OF NATURE

I shall come shortly to Findlay's paper.[1] First, for reasons that will become apparent later, I want to broaden somewhat the scope of his discussion. It must be remembered that the 'organic', before appearing in the *Philosophy of Nature* as a form of the natural world, is already present in the *Logic* as one of the determinations of thought;[2] and the question I now want to consider is whether, after Hegel's treatment of it as logical category, it can still have legitimate meaning in the context of the *Philosophy of Nature*. Speaking briefly and lucidly about Hegel's *Logic* always presents, of course, virtually insurmountable difficulties. But I think that I can make my point with reasonable clarity using Kant as a point of reference. His Transcendental Logic had been built on a distinction between the concept of an object in general, such as would be obtained by reflecting upon the requirements of intelligibility that thought makes upon its object, and the concept of an object of experience.[3] The two were by no means identical, for an object of experience was *ex hypothesi* sensible; and although it too, in order to make genuine science possible, would have had to satisfy the general conditions of intelligibility, in its case these also had to be shown to be *facts* of experience. As long as one proceeded along reflective lines to determine what, in principle, constituted an intelligible object, one had no guarantees that such an object would in fact be realized in experiences, or that these, as Hume had in fact denied, were in fact a fitting subject of science. The big problem that thus faced the critical Kant was precisely to demonstrate that the concept of an object in general (as reflectively determined) was also a valid determination of the immediate objects of experience.

Whether Kant in fact resolved the problem, or whether it was even possible for him to resolve it on his principles, is a question that does not concern us here. But it is instructive to note how Hegel managed to avoid altogether the need of demonstrating the legitimacy of a transition from reflection to actual experience. What had led Kant to differentiate between object in general and object of experience had been precisely the immediacy that characterized experiences and that he thought detracted from their intelligibility. Hegel realized, instead, that the same immediacy is in fact a necessary condition of all objectivity; and while one could distinguish between types of objects,

R. S. Cohen and M. W. Wartofsky (eds.), Hegel and the Sciences, 101–107.

some being more intelligible than others, the difference between them was to be measured not by the extent they were free from immediacy, but by the degree to which they explicitly incorporated it as an integral moment of their being *per se*.[4] In other words, immediacy is for Hegel just as much a moment of thought-life as it is of sensible experience: it should be possible, therefore, to demonstrate from within thought itself, and on reflective terms, that a constant transition from the immediate to the reflective and *vice versa* is both possible and necessary. Once this point has been speculatively established, and the need to justify the existential relevance of thought is thus in principle absolved, one is free to consider whatever particular problem the interplay between immediacy and reflection poses in a given sphere of experience in terms that are appropriate to such a sphere. The resolution of the problem no longer calls for the impossible task of 'applying' reflective categories to the immediate content of experience.

Unless this point is understood, the form that Hegel's *Logic* assumes does not make any sense. Its goal is to develop the idea of 'concept' and this goal is to be realized by means of the highest possible reflection. It is clear, moreover, that for Hegel this reflection represents the most accomplished case of a free act; and that, in virtue of its freedom, it encompasses in principle all things. Like Aristotle's soul, Hegel's self-thinking thought is potentially a whole world. But precisely for this reason, the *Logic* opens with the category of Being, an indeterminate somewhat which is more the intimation of an object than anything actually recognizable. It is the form the thought itself must first assume in front of its own reflective glance inasmuch as, since in the reflection the distinction between itself as subject and itself as object becomes strictly relative, what its limits as thought actually are becomes a thoroughly open question. Thought must discover what it itself is: it re-enacts, albeit in a highly abstract and reflectively controlled conceptual medium, the same interplay between subject and object at work at any level of experience. It is difficult not to refer once more back to Kant. He had been right, after all, in arguing against Descartes that the *Cogito* does not offer to philosophical reflection an especially indubitable source of certainty,[5] for even thought, in order to know itself, would have to objectify itself. But it then would have to treat itself as 'other' just like any other object. Kant, however, had taken this peculiarity of conscious life to mean that logical reflection is strictly formal, and that in order to be relevant to the rest of experience, its determinations needed the complement of an extraneous, empirical material. Hegel interpreted it, instead, in a directly opposite manner. For him it meant that thought, even at its most reflective level, already exhibits the structure of all experiences; and that it is in a position, therefore, to indicate their speculative value without having to be extrapolated beyond its reflective limits.

In relation to experience in general, therefore, Hegel's *Logic* represents the highest form of rationality possible within it. One can also say that the *Logic* demonstrates the possibility of experience: not in the sense, however, that the determinations of an object it reflectively establishes are such that one can then legitimately use them to categorize the content of experience; but in the sense that, as the *Logic* develops the idea of thought, it also completes and epitomizes experience. The distinction Kant maintained, therefore, between categories of the understanding and ideas of reason becomes meaningless in the context of Hegel's *Logic*. None of the categories of the latter can be exhibited in sensible experience, for they are all idealizations. They all are, nonetheless, existentially relevant, for they express possibilities which are in fact implicit in experience. The point might seem over-ingenious; and it might be argued that Hegel ignored altogether the critical concern, which had been uppermost in Kant's mind, of denying that one can determine on the strength of thought alone what actually exists. A preoccupation of this sort, however, has meaning only in relation to a dogmatic frame of mind: only if one expects objects of reflection to have other than purely ideal existence. It is misplaced when directed to Hegel's *Logic*, for the only existential claim the latter makes concerns thought itself. There is only one body, after all, which responds to the determinations of the *Logic*, and that is nature itself inasmuch as, as transformed by the medium of reflection, it is expressly conceived as the context within which reason can emerge.

But with this last remark I am back to my original question, and also to Professor Findlay's paper. I too admire the lucidity and accuracy of Hegel's categories, especially those that have to do with 'life'. I am not ready to extend the praise, however, to include the pseudo-scientific reconstruction of the cosmos along organic lines that Hegel offers in his *Philosophy of Nature.* The categories of 'life' as developed in the *Logic* are indeed a paragon of rationality that would make the corresponding 'Baconian' conceptualizations appear to be the product of a mind "intent on making nonsense (in the words of Professor Findlay) of its own subject and ultimately of itself." But then Hegel's categories ought to be perfectly rational, for they are *ex professo* the product of pure reflection. I submit that these categories alone would be the only kernel of reason left if one were to abstract in the *Philosophy of Nature* from the mass of sundry items of empirical information and misinformation that Hegel collects under the general heading of 'life'. And what I have just said about the categories of 'life' in the *Philosophy of Nature* would apply as well to the rest of it. I find it difficult, in other words, to consider the place that the 'organic' might have in Hegel's *Philosophy of Nature*, for I do not accept the latter as a legitimate piece of philosophical thinking. I do not mean to say, of course, that after Hegel's *Logic* there is no room left for a speculative, even philosophical, study of nature.[6] My point is that, since the *Logic* has vindicated the speculative value of all experiences, even the most immediate among them, after it we should be ready to explore the sense that emerges out of our transactions with actual nature in a manner and with conceptual means that respect the immediacy of such transactions. This is not what Hegel does in his *Philosophy of Nature*. At worst, he simply uses logical categories in order to classify data gathered elsewhere. At best, he manages to throw upon nature a human profile: I mean to say, he interprets it as if it were an extension of the human body, attributing relative value to its many appearances according as they approximate the perfection of existence achieved in thought. An exercise of this sort might have a proper place and some value. It does not help us learn, however, about nature itself (something, incidentally, that the 'Baconian' concepts which Professor Findlay decries might well do); and while it might prove instructive about the ways of the spirit, even in this respect it does not add anything to what the *Logic* has already taught.

Rudolf Haym said of the *Phenomenology of Spirit* that it was at once Transcendental Psychology and History of Culture, and that the mixture of the two made for a completely imaginary account of the rise of man's spirit.[7] In a similar vein, one could say of Hegel's *Philosophy of Nature* that it represents an illegitimate extrapolation of logical reflection. But it must be noted that Kant too, no less than Hegel, had tried to determine *a priori* the properties of nature; and that his Transcendental Logic required, for reasons inherent to its very conception, that its categories be extended to encompass an empirical material. Kant needed a Metaphysics of Nature for basically the same reason he needed a Transcendental Deduction of the categories – viz., in order to demonstrate the *de facto* relevance of his logic to the process of experience. Hegel's *Logic*, however, as I have pointed out, circumvented this need of Kant. The use, therefore, to which Hegel puts the categories of the *Logic* in the *Philosophy of Nature* is specious not only *per se*, but also from the point of view of the *Logic* itself. Of course, saying that Hegel was wrong when he needed not be and that he was not in the clear with respect to his own conception of the system, might not appear at all flattering to his apologists. But it is an important point to make if one wants to adjudicate the nature and the value of Hegel's contribution to German Idealism. From a strictly historical point of view, we are today especially equipped to understand how Hegel got to his final position. Recent studies on his development during the Jena years throw a very interesting light on the process by which he came to formulate by stages the idea and the outline of his later System.[8] We know, for instance – if I may summarize in a few words what is in fact a very complex situation – that at Jena Hegel began by distinguishing between logic and metaphysics, and had assigned to the former the function of justifying dialectically the starting point of the latter by demonstrating the insufficiency of alternative positions. He gradually collapsed the two together, attributing positive speculative value to the dialectical method he had first used in the logic for purely critical purposes. The result was the creation of a new speculative logic which, while consisting essentially of a reflection of thought upon itself, provided nonetheless a positive doctrine of being. We also know that the form this logic finally assumed was very much influenced by the reflections, in which Hegel was engaged during the same years, on matters which were later to be incorporated as part of the System under the headings of *Philosophy of Nature* and *Philosophy of Spirit*. Some of the more concrete notions that Hegel had included in the metaphysics were relegated eventually to the two other parts of the System, while new categories were introduced in order to provide a complete series of logical counterparts to the notions in the *Philosophy of Nature* and the *Philosophy of Spirit*. The process of evolution extended well beyond the Jena years. At the end, however, the content of the *Logic* dovetailed pretty well with the *Philosophy of Nature* and the *Philosophy of Spirit*. At that point, Hegel might well have thought to have finally a logical blueprint on the basis of which the whole of nature and history could be interpreted as the outward manifestation of absolute thought. And so in fact they were in the *Enzyklopädie* of 1817.

But was this really the case? Was the possibility of a System the true meaning of Hegel's newly discovered logic? It is instructive to reflect on the official reason that Hegel gave in 1804, at a time when he had in fact already hit upon the idea of 'thought' which was to control his mature logic, for the need of a transition from the then *Metaphysics* to the *Philosophy of Nature.*[9] The alleged reason is that, by the end of the *Metaphysics*, spirit has been actualized only in the sense that the idea is finally at hand, which expresses the movement of 'remaining oneself while becoming the other' in which its life consists. The idea is itself the highest expression of this movement; and Hegel can say, therefore, that as spirit actualizes itself in the *Metaphysics*, it becomes Idea. But spirit does not yet recognize itself. It is not yet *spirit for itself*, for it still has to realize its own idea by exhibiting it in the immediacy of nature and eventually (I presume) in the vicissitudes of history. Hence the transition to the *Philosophy of Nature*, which Hegel justifies as the first step in the process of realization.

Now, Hegel's alleged reason is a highly disturbing one. Philosophically, it suggests a strange model on his part regarding the nature and the limits of reflection. Unless one seriously expects one's awareness of oneself, and the process that has led up to it, to become itself a 'person' (somewhat in the manner of a Trinitarian procession), it is difficult to see how else such awareness could be realized except as idea, viz., in the medium of abstract conceptualization. Anything else would involve suspending the stance of speculative thought, and becoming engaged instead in the concerns of actual existence. Historically, Hegel's words evoke issues from past debates between Fichte and Schelling that he was in fact in a position to silence once and for all. I am referring to the controversy regarding the possibility of a philosophy of nature that Schelling upheld and Fichte denied. Schelling had argued that his speculations on nature were needed in order to provide a more concrete counterpart to the otherwise purely abstract reflections of the *Wissenschaftslehre*; and Fichte had retorted by saying that it was wrong-headed, perhaps even dangerous, to try to recognize the process of thought in the appearances of nature, and that Schelling's attempts, in any event, lacked seriousness for they too were, and could not help being, the products of reflection.[10] Now, Hegel could have silenced both Fichte and Schelling by pointing out to each that his *Logic* was indeed only an idealization, yet, because of the continuity of ideal reflection and natural becoming, it provided the only purely reflective understanding of nature possible.

In brief, exactly when Hegel thought he had found a way of realizing the System which had been, in one way or other, the ideal of all his predecessors, he had in fact achieved an understanding of reason that opened it to all experiences, and made irrelevant the very notion of a systematic reconstruction of reality. As a response to an apology for Hegel's *Philosophy of Nature*, this amounts to saying that Hegel's merit is not to have succeeded where modern empirical sciences have failed, but on the contrary, to have provided the conceptual means for understanding the nature of such sciences, and the speculative value of the knowledge they provide.

McGill University
Montreal

NOTES

1 This paper was originally prepared, thirteen years ago, as a comment on Professor Findlay's 'The Place of the Organic in Hegel's *Philosophy of Nature*'. I have revised it somewhat and updated it for publication.

2 That is, in the *Logic* of 1812–16, in the third part of the third book, as the first form that Idea assumes.

3 This distinction is not usually adverted to by commentators. It is nonetheless crucial to Kant. Without it, for instance, he could not legitimately claim to be able to *think* of the thing-in-itself as affecting the mind, yet disclaim to *understand* the nature of the relation involved. As applied to the thing-in-itself and the mind, the cause-effect relation remains unschematized. It is used in the abstract sense of a determination of an object in general. Cf., for instance, *Critique of Pure Reason*, A290/B346–47.

4 Immediacy appears at every stage of the *Logic*, but has a different meaning in each depending on the specific form thought assumes. For the way immediacy re-asserts itself at the end of the *Logic*, cf. *Wissenschaft der Logik*, vol. II (Hamburg, Meiner, 1963), pp. 499 ff. I have tried to illustrate the place of immediacy in Hegel's *Logic* using some specific texts as my basis in two papers: 'Reflection and Contradiction: A Commentary on some Passages of Hegel's *Logic*', *Hegel-Studien* 8 (1973), 131–61; 'The Category of Contingency in the Hegelian Logic', in *Art and Logic in Hegel's Philosophy*, eds. W. E. Steinkraus and K. L. Schmitz (Atlantic Highlands, New Jersey, Humanities Press; Brighton, Sussex, Harvester Press, 1980), pp. 179–200.

5 *Critique of Pure Reason*, A347/B405–A348/B406; A354–55; B423, note.

6 This was Croce's position. Cf. *Logica come scienza del concetto puro* (Bari, Laterza, 1958; first ed. 1909), pp. 13 ff., 21–24.

7 *Hegel und seine Zeit* [1857] (Hildesheim, Georg Olms, 1962), pp. 240 ff. I am in no way endorsing Haym's estimation of the *Phenomenology of Spirit*.

8 I have in mind especially, Otto Pöggeler, 'Fragment aus einer Hegelschen Logik, mit einem Nachwort zur Entwicklungsgeschichte von Hegels Logik', *Hegel-Studien* 2 (1963), 47–70; 'Hegels Jenaer Systemkonzeption', *Philosophisches Jahrbuch* **71** (1963–64), 286–318; Heinz Kimmerle, 'Zur Entwicklung des Hegelschen Denkens in Jena', *Hegel-Studien* 4 (1969), 33–47; Johann H. Trede, 'Hegels frühe Logik (1801–1803/4)', *Hegel-Studien* 7 (1970), 123–68.

9 *Logik, Metaphysik, Naturphilosophie*, Fragment einer Reinschrift (1804/05), *Hegel – Gesammelte Werke*, vol. VII (Hamburg, Meiner, 1971), pp. 176–78.

10 For the controversy, see the following letters: 1800, Nov. 15 – Fichte to Schelling; Nov. 19 – Schelling to Fichte; Dec. 27 – Fichte to Schelling; 1801, May 29 – Schelling to Fichte; May 31 – Fichte to Schelling; Oct. 3 – Schelling to Fichte; 1802, Jan. 15 – Fichte to Schelling. *F. W. J. Schelling, Briefe und Dokumente*, ed. H. Fuhrmans (Bonn, Bouvier, 1973), vol. II, Zusatzband.

MILIČ ČAPEK

HEGEL AND THE ORGANIC VIEW OF NATURE

As the title of this paper may easily yield to a misunderstanding, my first obligation is to dispel it. I plan to deal not with Hegel's organic view of nature (for reasons which I am going to state) but with Hegel *and* the organic view of nature, which is something entirely different. The main reason I prefer not to dwell *primarily* on Hegel's own philosophy of nature, including his own views of *organic* nature, is that it makes specific claims, many of which are obviously and fantastically wrong, not only in the light of our present science, but even in the light of the science of Hegel's own time. On this point, I feel that Findlay's view that Hegel's science was "the science of his own time" is rather too charitable;[1] for the tragedy, or rather tragicomedy, of Hegel's philosophy of nature is that it was far *behind* the science of his own time. Hegel's errors fall into two main groups: (a) the plain and arrogant denials of those scientific discoveries which were generally accepted by the scientific community of his own time and without which classical nineteenth-century science was unthinkable; and (b) the peculiar, artificial, and often fantastic interpretations of some of the facts which even Hegel could not deny. I shall deal first with group (a).

To this group belong Hegel's strange views concerning physics, chemistry, and biology. In physics he remained persistently opposed to Newton, from his Latin dissertation, *De orbitis planetarum*, in which he wrongly predicted that no celestial body can exist between Mars and Jupiter, until the last edition of his *Encyclopedia of Sciences*. According to Hegel, three times in history an apple was disastrous to mankind: first, in causing the fall of Adam; second, as the apple of discord, which precipitated the Trojan war; and finally, when it fell on Newton's head and thus ruined the philosophy of nature.[2] Certainly this joke, intended to ridicule Newton, does far greater damage to its author; it illustrates graphically Benedetto Croce's sardonic judgement that there is no better caricature of Hegel's philosophy of nature than what Hegel unwittingly wrote himself. What is truly astonishing is that Hegel, according to his own admission, spent no less than twenty-five years in studying Kepler and Newton (*PN*, § 270, p. 67), and apparently not without some success; he reproduces correctly certain formulae such as that for the freely falling body, which he calls Galileo's "immortal discovery," but which

R. S. Cohen and M. W. Wartofsky (eds.), Hegel and the Sciences, 109–121.

he understands only in part and not without occasionally committing an incredible elementary algebraic blunder. (For instance, when he claims that $t: s/t = s: t^2$!) (*PN*, § 267, p. 60). What is even more puzzling is his metaphysical justification of this formula. The passage, which defies translation, is characteristic of Hegel's approach to the problems of philosophy of nature in general. He is obviously not satisfied with Galileo's mathematical derivation and subsequent confirmation of the formula $s = \frac{1}{2}gt^2$ (s=distance, t=time, g=acceleration of gravity), and adds his own interpretation to disclose its 'deeper' philosophical meaning. Here it is:

> It is quantity [to wit: t] coming out of itself, positing itself in the second dimension and thus expanding itself [*sich vermehrend*], but only according to its own specific nature; in this expansion sets itself as its limit to its own expansion, and thus in the process of becoming an Other [*Anderswerden*] it only relates itself to itself.

And Hegel adds confidently: "This is the proof of the law of descent of a falling body derived from the Notion of the thing."[3] Only one thing is clear in this obscure passage: Hegel's so-called proof is nothing but a vague and questionable application of his triadic method – first the thesis, then its negation, and, finally, the synthesis as a negation of the previous negation. There are hundreds of similarly obscure 'proofs' in Hegel's writings on nature.

It is not difficult to see that under such circumstances the whole meaning of Newtonian mechanics completely escaped him. Thus he denies the validity of Newton's deduction of Kepler's law from the law of gravitation. On the contrary, he claims that the law of gravitation can be derived from Kepler's third law and offers another of his 'proofs', based on an elementary confusion of symbols and the failure to understand the very meaning of mathematical proof.[4] But does he really believe in the law of gravitation itself when he says that the motion of the heavenly bodies is not a result of their being pulled this way or the other, but is a 'free' motion? "They go on, as the ancients used to say, like the blessed gods" (*PN*, § 269, *Remark, Z*, p. 65). As Eddington observed, "this sounds particularly foolish even for a philosopher."[5] Perhaps what Hegel was trying to say is that mechanics and astronomy describe what is a mere external aspect of a process whose intrinsic structure is akin to human will and to mind in general. This would be consonant with Hegel's view of nature as the self-externalization of the Absolute Idea; hence his opposition to Newtonian mechanism and his leaning toward an organic view of nature. But his own organic philosophy of nature is, at least in this particular context, of a very crude, hylozoistic kind; of the passage above it may be said without unfairness that it is on the same level as that of Thales talking about the soul of the lodestone. In truth, Hegel's view of magnetism differs from that of Thales only by its triadic jargon: the magnet with its pair of poles and a neutral field in between is a sensible manifestation of syllogism (*PN*, § 312, pp. 162–163)! No wonder that, as Hegel's disciple and editor, Karl Ludwig Michelet conceded, Hegel's philosophy of nature was met by scientists with a shrugging of shoulders and compassionate smiles. No wonder that Harald Høffding called it "une partie honteuse" of Hegel's philosophy,[7] and that Emile Meyerson was rather polite when he confessed that the resulting impression it leaves in the mind of the reader is that of "profound bewilderment" (*profond ahurissement*):

> Nothing which would resemble the science of today or the science of his contemporaries . . . not even any science in any era such as the Aristotelian physics or the chemistry of alchemists. It is as if, while expecting to see human faces, we were presented with a series of absurdly grimacing monsters. Sometimes the reader begins to doubt whether he understood properly and re-reads the text several times to be sure that the phenomenon of which the author speaks is really that which is known to science – to such degree does the interpretation given by the author differ from everything which science knows or has known. The examples abound and on this point one is embarrassed which one to choose.[8]

Benedetto Croce, who cannot be suspected of any antipathy against Hegelianism, but who neverthless was not blind to the absurdities of Hegel's philosophy of nature, gives several times the whole list of Hegel's incredibly strange views concerning physical nature. Indeed that list shows that Meyerson was correct when he wrote that examples of Hegel's theoretical monstrosities abound. Of Croce's list I select only a few: gravitation is regarded as the concept of material corporeality realized in the form of idea; light is material ideality, the abstract self of matter and infinite externality (*Ausser-sich-seyn*); the sun is more abstract and, as such, less perfect than the planets which are more concrete and thus more perfect; the senses of the animals are objectified in the physical world – sight as the sun, taste as water, and smell as air.[9] We could continue indefinitely; let these suffice.

Hegel's views on physics were plagued by his failure to understand Newton; his views on chemistry suffered by ignoring Lavoisier and Dalton. What may be regarded as a mitigating circumstance is that chemistry was then a recent science; its founders, Lavoisier and Dalton, were older contemporaries of Hegel. Furthermore, their ideas were not generally accepted; thus some of their contemporaries, like Priestley and Cavendish, still adhered to the phlogiston theory despite Lavoisier's researches, which discredited it. Yet it is odd for one who so confidently pontificated about sciences and scientists not to be aware of the chemical revolution of his own time. It is even more odd to see an outstanding nineteenth-century philosopher adhering to the old Aristotelian doctrine of the four elements (*PN*, § 281).[10]

As for Hegel's biological views, we have already mentioned his odd theory of the senses. Obsessed by the dialectical triad, he tried hard to prove that there are really only *three* senses. In a similar way he reduced the number of continents to three: Europe, Asia, and Africa; for America is a mere appendix of European civilization, and Australia is eliminated. But the most surprising feature of Hegel's biological views is the denial of biological evolution. It is true that he lived before Darwin; but Cuvier and Lamarck were his contemporaries. Paleontology, founded by Cuvier, was dismissed by him as a 'nebulous' science; and the fossils, according to him, have no evolutionary significance since they are not the relics of extinct organisms, but mere freaks of nature. The idea of evolution of plants and animals is explicitly rejected in § 249 of Hegel's *The Encyclopedia of Sciences.* Nature is devoid of history; history begins with historians. When Hegel speaks of the hierarchy of nature, he means by it the system of gradations without any evolutionary significance, the static *scala naturae* of the pre-evolutionary thought.[11]

The only possible conclusion, then, is that if there is any value in Hegel's philosophy of nature, it must not be looked for in his *specific* claims, which are mostly false, often absurd, at best irrelevant. It should be looked for in his more general, metaphysical ideas; among them the idea of becoming occupies first place. Yet, what was just said about Hegel's attitude toward evolution indicates that the status of becoming in Hegel's overall view of reality is uncertain and ambiguous. This question is of fundamental importance in our present context. On the way in which it is answered will depend our answer to another important question: What is the relation of Hegel's philosophy to the modern organic view of nature? Since the modern organic view of nature can hardly ignore evolution and the dynamic and historical aspect of reality, it is obvious how important for us is the status of becoming in Hegel's thought.

It is frustrating that the question concerning the status of becoming in Hegel's thought cannot be answered in a definite and unambiguous way. The reason is that the ambiguity is inherent in Hegel's thought, as various texts clearly indicate. On the one hand, in his own words, there was not a single idea of Heraclitus which was not incorporated into his own system (*HP* 2, p. 279). On the other hand, Hegel's leaning toward a dynamic view of reality was counterbalanced by his strong Eleatic emphasis on the timeless character of the Absolute Idea. The latter tendency was certainly strengthened by the

Kantian heritage, more specifically by the Kantian doctrine of the phenomenality of time, which is characteristic of all post-Kantian idealisms from Fichte to Bradley. The unresolved tension between Heraclitus and Parmenides pervades the system of Hegel and led to two opposite interpretations of his concept of dialectics.

Is 'dialectical process' an objective process whose phases are *genuinely* (and not merely phenomenally) successive, as the most common interpretation of Hegel holds? Or should we prefer McTaggart's interpretation, according to which the Absolute Idea is timelessly realized, and the dialectical movement, which consists of the succession of increasingly more and more adequate approximations, is merely in our minds?[12] Both interpretations can claim the support of authentic texts and a definite answer can be obtained only by selecting some of them and ignoring others. What makes any definite solution of this difficulty impossible is that Hegel uses the term 'dialectical triad' in two different senses and illustrates it in two different ways. Sometimes the triadic pattern 'thesis – antithesis – synthesis' designates three successive phases of the temporal, historical process. This is especially true of Hegel's philosophy of history. On the other hand, in other instances, the triadic pattern is clearly static, and thesis, antithesis, and synthesis appear as three complementary and *coexisting* aspects of one single, though complex, fact which itself is devoid of succession. It is clearly meaningless to say that in the triadic nature of magnetism – to use one of his odd illustrations – there is any successive relation between the north pole, the south pole, and the neutral field between them. In his philosophy of life, or, as he called it, 'organics', Hegel seems to deal with the successive phases of the development of life when he speaks of the Earth as the Geological Organism, generating by the process of spontaneous generation the Vegetable Organism and the Animal Organism. This looks like a successive triad, but it has merely an evolutionary cast, since we know on the authority of Hegel himself, quoted above, that he explicitly rejected the idea of the evolution of species. Findlay nevertheless maintains that

> if any philosopher is a philosopher of evolution, that philosopher is Hegel . . . Had the Darwinian and later data been available, he would almost certainly have acknowledged the historical trends in nature that he admits in the realm of Spirit . . .[13]

I am not so sure of it; in the first place, there were evolutionists prior to Darwin – Lamarck during Hegel's own life, Buffon and Diderot even before him; to believe that the idea of evolution came into being with Charles Darwin is a mere 'ethnic' prejudice. Furthermore, it is doubtful that Hegel was opposed to evolution because of the lack of scientific evidence; he was certainly formulating pseudo-empirical conclusions on the basis of much slimmer evidence than that available at the time of Lamarck in favor of the variability of species. Finally, judging by his cavalier treatment of the well-established Newtonian physics, I doubt that any additional evidence in favor of evolution would have impressed him; his opposition to it stemmed from the Eleatic strain in his thought, that is, from the characteristically idealistic contempt for time and change. Fortunately, he was not consistent in this respect and his own Heraclitism, which is so conspicuously absent in his philosophy of nature, showed itself fully in his philosophy of history.

On this second dynamic aspect of Hegel's thought I intend to dwell in the remaining part of my paper. Emphasis on it is sadly lacking in Hegel's philosophy of nature in general and in his thoughts on organic nature in particular. Yet, without becoming there can be no organic view of nature; only nature unfolded in time can meaningfully be called *living* nature – *die lebende Natur* – to use Hegel's own term.

II. Becoming appears as the third term of the first and most fundamental dialectical triad, that is, as the synthesis of Being and Non-Being. It is another instance of Hegel's dialectical triad, which cannot be understood as a 'dialectical process'; certainly neither Being nor Non-Being are anterior in a *temporal* sense to Becoming. Should then the priority of Being and Non-Being with respect to their synthesis be understood in a logical sense? If so, should we regard Hegel's attempt to deduce Becoming as successful? As early as ten years after Hegel's death it was pointed out by one of his critics that Becoming cannot be obtained by a synthesis of two equally static terms. This objection was formulated by Trendelenburg and was also raised by Victor Cousin, by Paul Janet, and much more recently by Emile Meyerson in his untranslated work *De l'explication dans les sciences* (1925).[14] Meyerson observed that the predicament in which Hegel found himself was not basically different from that of Parmenides; but while Parmenides realized that between two equally static, becomingless terms, no intermediate and non-static term can be smuggled, Hegel tried to do just that. But his deduction of becoming was fallacious; the only way that he was able to smuggle it in was to borrow it surreptitiously from experience, that is, from something altogether heterogeneous to the couple of static and abstract concepts. In the light of the history of ideas it was thus not accidental that Eleatism reappeared with neo-Hegelian thinkers, in particular in the thought of McTaggart.

Harald Høffding raised basically the same objection, even though superficially it looks different since it is formulated in the language of formal logic.[15] He pointed out that if each synthesis is a negation of the previous negation, we should obtain the original term – Being instead of Becoming. This is the well-known law of double negation, which is closely related to the law of the excluded middle. It is certainly valid within the two-valued logic. But is Hegel's thought intelligible within the framework of the traditional two-valued logic? Does it not transcend it as the thought of Heraclitus did before him and Bergson's dynamism after him? This, it seems to me, is hinted by Findlay when he wrote that

> Hegel is not wrong in pointing out that our thought hates borderlines or transitional situations, that it is averse to anything that would now be called a three-value logic, and that it seeks to break up its subject-matter into mutually exclusive aspects of phases, so that the conceptual position of anything is immediately clear.[16]

All three philosophers mentioned – Heraclitus, Hegel, and Bergson – tried to rediscover the continuity underlying the artificially isolated aspects which our traditional logic – called *Verstand* by Hegel – created on the surface of dynamic reality.

This sheds an altogether new light on the first dialectical triad. Hegel calls becoming the first concrete concept; this means that the two antithetical concepts, Being and Non-Being, are mere abstractions. Or in Croce's words, "becoming is not a concept added to or derived from two first concepts, taken separately, but is a *unique concept* which outside of itself has only two abstractions, two unreal phantoms, Being and Non-Being, mutually separated and as such united not by their struggle, but by their common vacuity."[17]

'The struggle of the opposites' has clearly a Heraclitean ring, and as such it shares a certain obscurity characterizing the highly metaphorical language of the 'dark philosopher'. I am convinced that becoming is a too fundamental and all-pervasive fact to be expressed adequately by the metaphors borrowed from our limited sensory experience, whether visual, kinesthetic, or auditory. The images of 'Fire' and 'Flow' or 'Tension' or 'Harmony' or 'Struggle', no matter how suggestive, are necessarily misleading precisely because of their specificity. The full meaning of becoming transcends each of its illustrations or instantiations. Heraclitus was aware of this, at least implicitly when he spoke of "ever-living Fire," omnipresent and everlasting and *distinct from the visible fire* which is mortal and sporadic; it is the latter, which according to Fragment 25 "lives the death of air and air lives the death of fire."[18] The distinction between universal Becoming and its particular and visible manifestation is clearly indicated. Heraclitus used also the metaphor of the flowing stream – another metaphor which never disappeared from the language of process philosophy. It reappeared two thousand years later in the vocabulary of Newton at the beginning of his *Principia*, and thanks to William James's term "stream of thought," it became a familiar term in contemporary introspective psychology. Yet, when Heraclitus spoke of Becoming on a more general and more abstract level, he consciously and pointedly indicated the inadequacy of any concrete imagery. This is clear when we compare Fragment 41 with Fragment 81. The first one is famous and very suggestive by its concreteness: "You cannot step twice into the same river; for fresh waters are ever flowing upon you." The second fragment is much less frequently quoted, probably because of its apparent obscurity: "We step and do not step into the same river; we are and we are not." The first part of this fragment defies any consistent visualization, since it consists of two conflicting and incompatible images. The clue to this contradiction is in the second part, when Heraclitus moves up to a higher level of generality in saying: "We are and we are not." It would be difficult to anticipate more fully Hegel's view that becoming synthesizes being and non-being.

This point, I think, is crucial. As I said above, the imagery of Heraclitus, by the opposition of two incompatible images, suggests the presence of contradiction in the very nature of becoming; and that it is so, is shown in the second part where the opposition of two *concepts* – instead of two images – suggests it: "We are and we are not." More accurately, *we neither are nor are not since we become* and Becoming transcends both mere Being and Non-Being. Using Bergson's language, we may say that both Being and Non-Being are merely two static snapshots of the continuity of Becoming. Thus I must correct myself: the contradiction does not inhere in the nature of becoming, but in our language; it arises out of the opposition between two exclusive and artificially separated aspects. But why does our thought have the tendency to assume such exclusive and mutually antagonistic aspects? More specifically, why this tendency to resolve Becoming into two aspects – that of Being and that of Non-Being? Two answers have been given in the history of Western thought and they are incompatible. The first was given by Plato in *Timaeus*: Becoming is something less than the true Being and more than a sheer nought; it is an eternal oscillation between these two. Its status is inferior to that of Being, yet it still possesses a shadowy sort of semi-reality; it is not flatly negated, as it was by Parmenides. The major part of the Western philosophical tradition sided with Plato; from Aristotle, who contrasted the immobility of Supreme Being with the realm of "generation and corruption," up to those interpreters – or rather misinterpreters – of contemporary physics who claim that the relativity theory eliminated becoming from the physical world.

The second answer is just the opposite: becoming is ultimate and it is more real than Being, which is nothing but an abstraction, a static snapshot of becoming artificially arrested by a fictitious, instantaneous cut made by our perception and thought. This answer underlies the dark metaphors of Heraclitus, and, apart from some ambiguities, is present in Hegel's own thought, at least in its dynamic aspects. In our temporal awareness we intuit Becoming in its full concreteness and we experience vividly the extreme difficulty of conceptualizing its structure. In trying to do so, we begin to understand the meaning of the obscure metaphors of Heraclitus and of some peculiarities of Hegel's language. In particular, we find at least a partial justification of his claim that becoming is the unity of opposites, the synthesis of Being and Non-Being. Ever since William James wrote his incomparable chapter on the perception of time, we know that the minimum temporal awareness "from the outset is a synthetic datum, not a simple one; and to sensible perception its elements are inseparable, although attention looking back may easily decompose the experience and distinguish its beginning from its end."[19] It is a perpetual fading of the vivid present into the penumbra of the past, and, as Bergson showed, it is illusory to draw a sharp line of demarcation between them and to indicate the precise point-like instant when perception ceases to be actual and is converted into the virtuality of recollection. This indivisible continuity of the past and the present or, as Lovejoy said, this perpetual fading of *cogito* into *memini,*[20] is a vivid exemplification of the unity of opposites, of the continuity of becoming. What Hegel called 'being' corresponds to the actuality and vivacity of the present, while 'non-being' designates the quality of pastness; these two aspects must not be taken separately since they are experienced as an undivided, synthetic whole, an indivisible, though complex temporal Gestalt. In this sense we truly "are and are not" as Heraclitus said.

Even Hegel's apparently mysterious term 'negation of negation' becomes intelligible in the example above. Every present moment is in a sense a negation of its immediate ancestor and it will be in its turn negated by the immediately subsequent moment. In truth, William James, who in his early years was sharply critical of Hegel and, more specifically, of Hegel's monistic idealism, in his last work, *A Pluralistic Universe*, while still critical of Hegel's "vicious intellectualism" and "intolerable ambiguity, verbosity, and unscrupulousness",[21] made a sympathetic comment on the Hegelian idea of "being its own other",[22] which characterized the way perceptual data interpenetrate each other. No psychological atomism, no doctrine of external relations can do justice to the mutual immanence of parts which is characteristic of immediate experience, sensory as well as introspective. This was systematically stressed by Gestalt psychology, in opposition to the atomistic and additive approach of associationistic psychology.

The emphasis on the unitary, Gestalt-like and dynamic character of organic life is probably the most valuable feature of Hegel's philosophy of nature. From the mechanistic standpoint, 'life' is a mere collective name for all living organisms; there is no supraindividual life in the sense of a concrete universal present in its different individual manifestations; in the mechanistic scheme, widely prevalent today in English-speaking countries and in Russia, life is nothing but a peculiarly complex case of lifelessness. In any organic view of nature just the opposite is true: life is the primary category and what we call 'lifeless' or 'inorganic' is merely a very rudimentary form of life or proto-life. The mechanistic view of life was a logical consequence of the mechanistic view of the universe; it remained justified as long as no serious flaws in the classical Newton-Laplacean view appeared. But anybody even superficially acquainted with the physics of the twentieth century, whose attention is not diverted to irrelevant problems of methodology, is well aware that the mechanistic view of the universe belongs to the past. The mechanistic view of life has been deprived of the basic premises which made it so plausible in the relatively recent past. But a detailed discussion of this would lead us far beyond the scope of this article.[23]

Hegel's philosophy of nature, together with the whole German Romantic movement of which Goethe and Schelling were the outstanding representatives, was the first revolt against the classical Newtonian physics. The main reason for its failure was that it was *premature* and was to a considerable degree merely a revolt of imagination and feeling against the depressing mechanistic view of reality. There were valuable insights in it, and Whitehead, in the fifth chapter of his *Science and the Modern World* (1925) did full justice to the English counterpart of the Romantic reaction, as he called it. But these valuable insights were fragmentary, and, in Hegel's philosophy of nature at least, wrapped in confusing jargon and pervaded by hasty generalizations, superficial and vague analogies, and even by weird and irresponsible guesses. When the course of time swept away the *dèbris*, little was left of Hegel's philosophy of nature; but what was left resembled, if I may use this metaphor, a few golden grains found after the mass of sand is washed away. There is no space left for me to show how Hegel's central idea of the relatedness of nature was fruitfully applied by his former disciple, Johannes Bernard Stallo, in his prophetic criticism of the mechanical theory of nature.

Stallo's criticism of Newton's rotating bucket experiment anticipated both Mach's criticism and Whitehead's much later criticism of the fallacy of simple location, as I tried to show some time ago.[24] It is true that Stallo, when he wrote his *Concepts and Theories of Modern Physics* (1882), was not a Hegelian any longer. He referred to his early Hegelianism as "the metaphysical malady which seems to be one of the unavoidable disorders of intellectual infancy."[25] Yet, an attentive reader of Stallo's intermediate writings as well as of his final work would find that the most fruitful of his insights – the inadequacy of the mechanization and fragmentation of nature – can be traced to Hegel's intuitive anticipation of the organic view of nature.

Boston University

NOTES

[1] John N. Findlay, *Hegel: A Re-examination* (New York, Humanities Press, 1958), p. 268.

[2] Cf. *Ce qui est vivant et ce qui est mort de la philosophie de Hegel* (Paris, Giard and Brière, 1910), 136. [Translation of *Ciò che è vivo e ciò che è morto della filosofia de Hegel* (Bari, Laterza, 1907), p. 161. Translated into English by D. Ainslie as *What is Dead and What is Alive in the Philosophy of Hegel* (London, Macmillan, 1915), p. 167.]

[3] My own translation does not differ substantially from that of A. V. Miller (*PN*, § 267, *Remark*, p. 59) and that of Michael Petry, (*Hegel's Philosophy of Nature* (London, Allen and Unwin, 1970), I, 255.

[4] Here is Hegel's 'proof' (*PN*, § 270, p. 66):

> The abstract derivation is simply this: in Kepler's third law the constant is A^3/T^2. If this is put in the form A^2A/T^2, and we call A/T^2 with Newton, universal gravitation, then we have his expression for the action of this so-called gravitation, in the inverse ratio of the square of the distance.

Hegel's blunder consists, first, in forgetting that T in Kepler's formula means time, more specifically, the period of revolution, whose square Hegel incredibly confuses with the square of the distance; second and equally incredibly he overlooks the fact that A in Kepler's formula means the distance and not the mass as in Newton's law. It is true that Newton's law of gravitation *can* be proved, if we assume jointly the third law of Kepler *and* the formula discovered by Huygens for centripetal acceleration, v^2/r (which Hegel does not mention): it is this fact to which Hegel refers when he says vaguely that "it is admitted by mathematicians themselves that the Newtonian formulae may be deduced from Kepler's laws." Hegel's third mistake was to believe that a certain formula can be proved when it is a *part* of another formula: he obviously confuses 'it contains' with 'it implies'. Thus one can understand Whitehead's remark about Hegel: "I have never been able to read Hegel: I initiated my attempt by studying some remarks of his on mathematics which struck me as complete non-sense." (Cf. 'Autobiographical Notes' in *The Philosophy of Alfred North Whitehead* (Evanston and Chicago, Northwestern University, 1941), p. 7.)

5 A. S. Eddington, *The Nature of the Physical World* (Cambridge, Cambridge University Press, 1933), p. 147. Eddington adds, rather jokingly, that Hegel, in a sense, unsuspected by him, was right since according to the relativistic theory of gravitation, there is no distinction between the natural ('free') and constrained motion (pp. 147–53).

6 J. N. Findlay in his Foreword to Miller's translation of the *Philosophy of Nature* (p. xx) says that "the phenomena of magnetism were for Hegel as for Schelling, of inexhaustible import, and it would be a pardonable exaggeration to see in the whole of Hegel's philosophy an interpretation of magnetism."

7 Harald Høffding, *A History of Modern Philosophy* (New York, Dover, 1955), Vol. 2, p. 183.

8 Emile Meyerson, *De l'explication dans les sciences* (Paris, Payot, 1921), Vol. 2, p. 23.

9 *PN,* § 269, § 276, § 270, § 316; Croce, *op. cit.*, Ch. IX.

10 Hegel regretfully admits that

> the conception of four elements which has been general since the time of Empedocles is nowadays rejected as a childish belief, because, forsooth, the Elements are composite. No physicist or chemist, in fact no educated person, is any longer permitted to mention the four Elements anywhere (*PN,* § 281, *Z*, p. 106).

11 Loren Eiseley, *Darwin's Century. Evolution and the Men Who Discovered it* (Garden City, N.Y., Doubleday, 1961), Ch. I.

12 John M. E. McTaggart, *Studies in the Hegelian Dialectics* [1922], 2nd. ed. (New York, Russell and Russell, 1964), Ch. V, 157–69.

13 J. N. Findlay, *op. cit.*, p. 272.

14 F. A. Trendelenburg, *Logische Untersuchungen* (Berlin, Bethge, 1840), Vol. 1, p. 25; V. Cousin, *Fragments et souvenirs* (Paris, Didier, 1857), p. 183; Paul Janet, *Etudes sur la dialectique dans Platon et Hegel* (Paris, Ladrange, 1861), p. 352; Meyerson, *op. cit.*, Vol. 2, p. 62.

15 Høffding, *op.cit.*, Vol. 2, p. 182.

16 Findlay, *op.cit.*, p. 158.

17 Croce, *op.cit.*, p. 22.

18 John Burnet's numbering in his *Early Greek Philosophy*. On the distinction between the visible flame and the 'metaphysical Fire' cf. F. M. Cornford, *From Religion to Philosophy. A Study in the Origins of Western Speculation* (New York, Harper, 1957), p. 188.

19 W. James, *The Principles of Psychology*, 2 Vols. (New York, Holt, 1890), Vol. 1, p. 610.

20 A. O. Lovejoy, *The Revolt Against Dualism* (Lasalle, Ill., Open Court, 1955), p. 381. On this problem cf. my article '*Memini ergo fui*?' in *Memorias del XIII congresso international de filosofía*, 10 vols. (Mexico, Universidad Nacional Autónoma de Mexico, 1963–1966), Vol. 5, pp. 415–26.

21 W. James, *A Pluralistic Universe* (London, 1947), p. 107.

22 *Ibid.*, p. 109:

> If, however, the conceptual treatment of the flow of reality should prove for any good reason to be inadequate . . . then an independent empirical look into the constitution of reality's pulses might possibly show that some of them *are* their own

others, and indeed are so in the self-same sense in which the absolute is maintained to be so by Hegel.

Whitehead's view on the combination of self-identity and self-diversity of each actual occasion has a truly Hegelian ring. Cf. *Process and Reality* (New York, Macmillan, 1930), p. 38.

[23] Cf. my books, *The Philosophical Impact of Contemporary Physics*, 2nd ed. (Princeton, Van Nostrand, 1969) and *Bergson and Modern Physics. A Re-Interpretation and Re-Evaluation*, Boston Studies in the Philosophy of Science, vol. 7 (Dordrecht, Boston, D. Reidel, 1971).

[24] Cf. my articles, 'Two Critics of Newton Prior to Mach: Boscovich and Stallo' in *Actes du XII congrès international de l'histoire des sciences*, Paris, 1968 (Paris, Blanchard, 1971), Vol. IV, pp. 35–37; and 'Johann Bernard Stallo' in *Dictionary of Scientific Biography*, ed. by C. C. Gillispie (New York, Scribner, 1975), Vol. XII, pp. 607–10.

[25] *The Concepts and Theories of Modern Physics*, ed. Percy W. Bridgman (Cambridge, Mass., Harvard University Press, 1960), p. 6.

DIETRICH VON ENGELHARDT

HEGEL'S PHILOSOPHICAL UNDERSTANDING OF ILLNESS

I

In the course of time the understanding of illness has changed profoundly; however, many 'moments' have been preserved throughout this change. Even within a historical period the understanding of illness has by no means been uniform. Especially varied is the spectrum around 1800, in the decades of German Idealism and Romanticism. Hegel's philosophy of nature and of man is a speculative metaphysics of nature. In this period a metaphysical conception of nature had first been developed in systematic form by Schelling. The Romantics had allowed themselves to be essentially influenced by this system; their highly varied concepts stand somewhere between empiricism and metaphysics. Not infrequently the Romantics were natural scientist and philosopher or physician and philosopher in one person. Platonic, neoplatonic, Spinozist and Leibnizian currents also influenced the Romantics directly, just as Schelling himself was affected by these. Hegel also based himself in part on Schelling, but at the same time he differs from him as well as from the Romantics. Schelling furnished the essential precondition of the Romantic as well as the Hegelian view of nature; speculative as well as Romantic aspects can be found in his work. Empirical medicine, its investigation of man, his health and his illness, is clearly to be distinguished from these tendencies.

Hegel's conceptions about illness and the world of medicine could find no interest until the present.[1] In this study, we begin with Hegel's position between empirical and Romantic medicine. Then we clarify Hegel's philosophical understanding of illness in terms of several fundamental 'moments' of his concept of illness, and the stages of illness and of therapy.

II

A systematic presentation of the concept of illness around 1800 will thus have to take into consideration quite varied points of view. These differences also determined the self-understanding of the Romantics and the idealist philosophers. In his understanding of illness, as well as fundamentally in his philosophy of nature, Hegel, the dialectical philosopher, takes issue above all

R. S. Cohen and M. W. Wartofsky (eds.), Hegel and the Sciences, 123–141.

with both views toward research into nature and medicine, empirical and romantic. Especially complex is his relation to Schelling. The differences among the innumerable Romantic proposals on the one hand, and the difference between these and Hegel's philosophy of nature on the other, remained unnoticed by the physicians of his time. Hegel's ideas about illness were basically disregarded,[2] his philosophy of nature sharing the general fate of the Romantic philosophy of nature.

Already during Hegel's lifetime a pronounced resistance to metaphysical conceptions of illness was to be found in medicine. According to Hufeland (1800), for medicine during these years of crisis there was only "the middle road between fruitless speculation and blind empiricism"[3]; an enduring and practice-oriented doctrine of illness will be especially careful not to "impute to nature some sort of speculative system".[4] In 1805, Hecker warns of the "slippery ice", of the "thorns, cliffs and rushing forest streams of the most recent philosophies", which "climb at once to the highest viewpoint of all, while lacking any path to get up there."[5] On the contrary, it was the task of empirical medicine to find a path between the "mania for systems, hair-splitting dogmatism and blind empiricism."[6] Especially in an age of anarchy – in political as well as scientific respects – it was essential to have solid foundations; the imperative of the hour was a "rational scepticism."[7] Even if such scepticism also had an eclectic character, the interest of the patient was to be preferred to any system. In the year 1816, a physician, using the pseudonym 'Candidus', published a critique of philosophy of nature, under the title 'Not an Accusation, But a Complaint'.[8] According to this Candidus, philosophy of nature of a Romantic-speculative character led to absurdities; to be sure it was "famed only in Germany, in France and England it was either unknown or held in contempt."[9] One could only wish that all the "fantasies and errors of young and old which philosophy of nature had caused in Germany would be forgiven and as soon as possible forgotten."[10] But not only are erroneous concepts of man and of illness due to this philosophy of nature, it also makes the individual human beings who follow it unhappy; the death of Reil is oppressive, as is his melancholy, his despair, his dreary views of man and the world. Hufeland devoted an Afterword to Candidus's critique. Above all, medicine must defend itself against this, "because it is just the medical world where this view can most easily gain admission, and the physical world is the enchanted forest through which the spirit is led to moral eclipse."[11] Around 1830, the preceding decades were condemned by the anatomist Hartmann because during these times philosophy of nature "subjugated medicine and transformed it to accord with its basic views";[12] fortunately this time had now been superseded forever.

Hegel repeatedly separated himself from the empiricist and Romantic conceptions of his time. Just as he was against the natural scientists, so he also finds himself in a dual opposition to medicine: on the one hand against the Romantic designs for medicine, and on the other against the fact-centered form of contemporary empirical medicine. Still with this, Hegel considers a combination of empirical work and speculation as a *thinking* and *comprehending* [*begreifend*] examination of nature to be in principle both possible and necessary (*PN* § 246, p. 6). According to him philosophy of nature is not an alternative to natural science, nor is it to be defended or rejected as a preparatory stage or a menace to natural science. However, Hegel sees the medicine of his time surrendering to the effect of the *Zeitgeist*, making everything shallow, just like all the other sciences. In a letter of April 11, 1824 he thanks Windischmann for sending him his latest work, "Über etwas, das der Heilkunst Noth thut [About something that is needed by medicine]", which, according to Windischmann, should fall like "a bomb into the vile junk and apothecary rubbish of medicine."[13] In his evaluation of this work Hegel remarks:

> Your attack is disordered at its root, and even if those who are sunk in this morass can no longer hear, still on the other hand, for those whose confidence cannot attain force in the face of the universal climate, these words will prove fruitful for their power and their sentiments, and the courage of knowing will grow in them. To begin with, you have turned to medicine, and the presentation will be most convincing when it presents the particular distress and need of a special field. However, other fields deserve equal attention, especially theology, from which all confidence and intrinsic worth issue for the others. But the state of this science almost justifies that of the others, for the sacred realm is not entrusted to them, and if the priests (and among these I also include the philosophers and in its way also the government) allow the people to fall into such superficiality, then medicine too will be forced to conceive of their illnesses and suffering in such an external manner, for then it finds no point of support for an intellectually deep activity that both enters inward and acts from within.[14]

According to Hegel, mind and nature are torn apart by the thought of the time; but what is required is their combination, to be sure, a specific combination of concept and reality, which, according to him, the Romantics (who have turned against the dominant medical science) have by no means achieved.[15] Their conceptions are familiar to him. Hegel criticizes the formalism in all the Romantic attempts, the external analogizing, the involution and uncritical application of the concept of metamorphosis.[16] Hegel also speaks of a "wild forest stream that threatens to sow confusion in reason and science";[17] a philosophic critique must decisively oppose these tendencies, in order that the attitude toward philosophy of nature, which had fallen from astonishment into indifference could be changed back again to respect and recognition. The external manner which had already been at work in Schelling's philosophy had been carried still further by Oken, Troxler and other Romantics, who had lapsed "completely into an empty formalism" (*PN* § 359, *Zusatz*, p. 388); their method of comprehending nature and the living organisms was "crude and thoroughly unphilosophical" in spirit (*PN* § 359, Remark, p. 386), and this because of a thoroughgoing confusion of logical categories and the categories of natural science:

> . . . this ignorance and contempt for the Notion which, in general, gives rise to the facile formalism that, in place of the determinations of the Notion, makes use of sensuous materials such as chemical substances, as well as relationships belonging to the sphere of inorganic Nature, such as the north and south poles of the magnet, or even the difference between magnetism itself and electricity, a formalism which, in its apprehension and exposition of the natural universe, attaches externally to its various spheres and differences a ready-made schema compounded of such material (*PN* § 359, Remark, p. 387).

Conceptuality and phenomenality, according to Hegel, are developed with special sharpness among the Romantics, because of this incapacity to distinguish the dimensions between idea and sensibility and to relate them immanently to each other: "the crudest empiricism embellished with a formalism of substances and poles and with analogies empty of reason and drunken lightning-stroke insights."[18]

Thus Hegel's critique of Romantic medicine aims in two directions: its handling of phenomena, its empirical work, is found just as lacking as its philosophic procedure. For example, in his judgment Hegel rejects Göden's Romantic conception of illness:

> Fine philosophers these, who fancy that in essence they have the truth, and that if only they keep on saying 'essence', then this makes it the inner truth of the matter! I have not the slightest respect for their talk of essence, for it is nothing but an abstract reflection. To explicate essence, however, is to make it appear as a real existence (*PN* § 371, *Zusatz*, p. 430).

The essential concepts and convictions of Romanticism cannot be found in Hegel. The widely held esteem in which the Romantics hold Dr. John Brown is rejected by Hegel; however, Brown's achievement will remain, for he had gone beyond the view of the "merely particular and specific both in diseases and remedies" and had committed himself to recognize "the *universal* in them as the essential element". However, to see in Brownianism "a complete system of medicine" (*PN* § 373, Remark, p. 437) and to try to explain illness with the categories of sthenia and asthenia [strength and weakness] alone, cannot convince Hegel: "A theory of medicine built on these arid determinations of the Understanding is complete in a half dozen propositions; it is no wonder that it spread rapidly and found many adherents" (*PN* § 359, Remark, p. 386).

Nor will Hegel accept the historization of illnesses, according to which the taxonomic system of illness is at the same time supposed to be the law of their appearance in history.[20] Hegel also does not follow the natural history approach to illness, which induces Oken to designate "pathology" as the "physiology of the animal kingdom", and illnesses as the "life-processes of animals",[21] and which leads Hoffmann to the conception that rickets is a sinking-down to the condition of mollusks, arthritis a regression to animal regeneration by means of the joints, and epilepsy a coming-to-life again in human beings of the natural stage of the Oscillatoria [algae].[22] Hegel also does not agree with the religious view of illness as set forth by a Heinroth, a Windischmann, a Ringseis.[23] With all his interest in illness and its significance in nature and in the realm of the spirit, the philosopher Hegel cannot agree with the emphatic judgments of the Romantics about illness and the pathological. Instead he sees the Romantic preoccupation with illness as dangerous to the equanimity of soul and mind of human beings; above all from such confrontations with mental illnesses, psychic damage may result. He is very eager, Hegel writes to Windischmann on May 27, 1810, to see his work on magic; he himself does not dare to draw near to "this obscure side or mode of mental nature or of natural mind", for "health and a merry, and indeed ruggedly merry, disposition" was needed for this work more than for any other. Without doubt, the deplorable state of his feelings, of which Windischmann had reported, was to be blamed on this work:

> . . . this descending into dark regions, where nothing shows itself to be firm, definite and secure, where shimmering lights flicker everywhere, but beside chasms, still more obscured by the brightness, misled by the surroundings that cast false reflections rather than illuminate – where the beginning of every path breaks off again or dwindles away into the indeterminable, is lost and tears us away from our destiny and our direction.[24]

From his own experience he knew that state, where one had steeped oneself in the "chaos of appearances" and had not yet arrived at "clarity and a detailed view of the whole"; he himself had "suffered several years from this hypochondria to the point of debilitation." Probably every human being had "such a turning point in his life, the dark point of contraction of his being, through the narrows of which he is squeezed to be strengthened and confirmed in his self-assurance. One must surmount this state, and if the "security of ordinary every-day life" could no longer give one fulfillment, seek the "security of an inwardly nobler existence."[25] And so Hegel follows the classical rejection of the Romantic idealization of weakness and sickliness. To these phenomena, he gives a deeper significance in his speculative philosophy of nature than he does to a positivistic view of nature – but to elevate them to the norm or bring them near to the ideal, from these he was prevented by his commitment to a well-considered and unpretentious conception of reality.

III. THE IDEA OF ILLNESS

In Hegel's philosophy, illness is integrated into a world of nature and of mind that is articulated within itself and at the same time coherent. The doctrine of health and the doctrine of illness are immanent 'moments' in the series of phenomena ranging from inanimate matter to the highest forms of spirit: art, religion and philosophy. The organic follows upon mechanics, physics and chemistry; it is the most concrete and most differentiated part of nature – here nature ends, here begins the world of mind, and here illness has its place. Illness receives its significance in just the same degree as the death of the living individual has its significance, too, for the interconnection of nature and spirit. Illness, for Hegel, is an essential characteristic of the organic, the organic is *original* illness, it bears within it the "*germ of death*" (*PN* § 375, p. 441) – in illness human beings can find "the manner of their death" (*PN* § 374, p. 440). Illness is a "transitory disharmony" (*PN* § 367, *Zusatz*, p. 411); the "general inadequacy" (*PN* § 374, p. 440) of life is natural death. Through the phenomena of individual illness, and the death of the individual, and in terms of these, the species maintains itself in the philosophic perspective, and spirit originates. With the discord between individual integrity and totality on the one hand and the various bodily functions and dimensions of consciousness isolating themselves on the other, the individual makes real within itself the relation between individual and genus: "the individual is, both in its own self and in its opposition to itself, genus; it is itself alone the genus, and has it within itself" (*PN* § 371, *Zusatz*, p. 428).

As the individual human being fundamentally cannot transfer the spiritual universality possible for him into his bodily individuality, cannot let it appear in sensual reality, he must die, and ever again becomes ill. Spirit is eternal. In illness and in death, the aliveness and mental character of human kind manifest themselves; pain is the privilege of the higher developmental stages of nature. All processes of illness ultimately pass over into a process that can no longer be recovery:

The organism can recover from disease; but disease is in its very nature, and herein lies the necessity of death, i.e. of this dissolution in which the series of processes becomes the empty process which does not return to itself (*PN* § 375, *Zusatz*, pp. 441–442).

Illness documents the power of nature over the individual; in the sexual relationship, man relates to nature in another person, in taking nourishment he relates to external nature; but in becoming ill he has to carry out this confrontation within himself. Sexuality, as the foundation of the species and illness are philosophically related phenomena. [In German, *gatten* means to mate or to copulate, and *Gattung* means species – Ed.] In the perspective of Hegelian philosophy, nature and mind must necessarily fall asunder for a positive interpretation of reality, man must become estranged from his material environment and also from his own body.

For Hegel illness is a disturbance of the organism, a falling apart of the capacities and body systems which are maintained in a unity in a state of health. At the same time, Hegel rejects Brown's proposal to understand illness as the consequence of too strong or too weak a stimulus – fundamentally illness is a discrepancy between the individual as a living ego [*Ich*] and his bodily reality:

Disease does not consist in an irritation (*Reiz*) being too great or too small for the susceptibility of the organism, rather is its *Notion* a *disproportion* of its being and its self (*PN* § 371, *Zusatz*, p. 428).

A stone is destroyed by a strong external action; but this destruction is not illness, because a stone "is not the negative of itself which overlaps (*übergreift*) its opposite" (*PN* § 371, *Zusatz*, p. 429); it cannot reconstitue itself, cannot become healthy again. Illness is to be distinguished from other states in which a lack is perceived, for instance that of desire (or appetite). Here the feeling of the negative is directed toward something alien, not toward one's own existence:

Appetite, too, the feeling of a lack, is to its own self the negative, relates itself to itself as a negative: is itself and is also in relation to itself as a being feeling a lack; but with this difference, that in appetite this lack is something external, that is, the self is not turned against its structure (*Gestalt*) as such, whereas in disease the negative thing is the structure itself (*PN* § 371, *Zusatz*, p. 429).

In an equally fundamental way the psychic illnesses are thought of as a falling apart of consciousness; aside from these illnesses, Hegel emphasizes contagious diseases and epidemics, in which the level of the individual is exceeded. The three kinds of illness, under which the multiplicity of the phenomena of illness can be ordered, are to be understood as a transition of the logical categories of the universal to the particular and the individual.

Just as the Romantic view of nature, so too Hegel's philosophy of nature has been reproached with a fundamental hostility to empirical research; this reproach must be examined for every natural philosopher individually, and better still in terms of specific passages. According to Hegel, a philosophy of nature and of man cannot forego natural science and empirical medicine, but must base itself on empirical knowledge; it presupposes it (*PN* § 246, Remark, p. 6). As part of this contempt for empiricism, an alleged basic rejection of the principle of causality is also listed. But Hegel by no means disregards etiology; besides the somatic conditions of disease, the philosopher explores the psychological, social and political-cultural determinants of falling ill. But the significance of illness is not exhausted within the perspective of concrete causation. Dimensions of reality are manifested in illness which are not only also present in the state of health, but even possess outstanding significance for life. Illnesses show the structure of reality in especially sharp relief, more so than does health. As a form of negativity, illness has a constitutive function in the transition from one domain of reality to the next – a function always assigned to the negative in Hegel's philosophy. In the health state those fixations occurring in illness are constantly being overcome, for health too is related to those dimensions which assume independence in illness.

Hegel's philosophy of illness is neither nosology nor concrete therapeutic doctrine; it does not describe the self-perceptions of the patient, the manifestations of illness in his self-consciousness, in his experience of the world and in his relations to other human beings, nor does it treat the social mooring of illness, or the relation of the physician to the patient. But these 'moments' do not confront the philosophic perspective as something irreconcilable. According to Hegel, philosophy comprehends being and essence and not "preconditions which present themselves in psychological, anthropological and other forms" (*WL* [M] p. 761). Hegel wants to give a deduction of nature and of man which is at once conceptual and real; to this belong the illnesses, their genesis and how they are overcome. Empirical investigation and theory, phenomena and concepts, all are accorded their rights to an equal degree. How open the conceptualization is to the phenomena can be made clear by the weight carried by the multiplicity of both the concrete and chance in Hegel's philosophy.[26] Hegel by no means claims to deduce all forms of illness, their causes and their courses, the possibilities of treatment, therapeutic methods, and the limits of therapy. According to him that cannot be the task of philosophy.

Typology of Illness

According to Hegel the disproportion of subjectivity and the somatic-psychic world of the individual appears in three general kinds of illness: contagious diseases and epidemics, physical illnesses and psychic illnesses.

Even though illnesses are always bound to an individual human being, not infrequently their true character is only revealed within a superindividual perspective. Here Hegel is thinking of contagious diseases and epidemics, of syphilis, typhus and yellow fever. In these diseases, the harmful aspect appears to be set by the organism and by external nature; both aspects belong to these illnesses, they are falling ill of living nature as such, which also includes the individual organism. As general elementary disturbances, these illnesses manifest themselves "chiefly in the skin, the lymph, and the bones" (*PN* § 371, *Zusatz*, p. 431). In their general nature, they are bound to supra-individual vicissitudes of history; they are "not only climatic but also historical, since they appear at certain periods of history, and then disappear" (*PN* § 371, *Zusatz*, p. 431). In the same way they can also arise when a human being is removed to a climate to which he is not accustomed.

In the more individual physical illnesses, the organism is overpowered by inorganic nature

> when one of its systems or organs, *stimulated* into conflict with the inorganic power (*Potenz*), establishes itself in isolation and persists in its particular activity against the activity of the whole, the fluidity and all-pervading process of which is thus obstructed (*PN* § 371, p. 428).

The ideal nature of organic life is destroyed:

> Whereas in health, all the vital functions are held in this ideality, in disease the blood, e.g., is heated, inflamed; and then it is active on its own account (*PN* § 371, *Zusatz*, p. 429).

Gallstones are to be understood as excessive activity of the gall bladder, illnesses of the stomach as an isolation of the stomach, which ceases to be a 'moment' subordinated to the whole, but instead elevates itself to become the central activity on its own. According to Hegel these illnesses have two modes of manifesting themselves: acute and chronic. In acute illnesses, the disease of one region of the body spreads to the whole organism, while in chronic illnesses the disease cannot become illness of the whole organism; the essential vital functions remain unbroken: "In such diseases, appetite and digestion remain quite unimpaired, and the sexual instinct retains its strength" (*PN* § 371, *Zusatz*, p. 432). Cirrhosis of the liver and consumption are typical examples of chronic illnesses for Hegel.

The third kind of illness relates to the subject in its subjectivity – these are the "*diseases of the soul* (*Seele*)" (*PN* § 371, *Zusatz*, p. 432). Just like the physical illnesses, these can be so acute that death takes place. In the psychic illnesses, according to Hegel, the psychic dimensions absolutize themselves against the unity of the mind and destroy the identity of the individual, which is now no longer capable of remaining itself through all its different feelings and affects:

so, too, in the *psychical life* illness results if the merely *psychical* side of the organism, freeing itself from the power of the *mental* or *spiritual* consciousness, usurps the latter's function and mind or spirit, in losing control over the psychical element belonging to it, no longer retains its self-mastery but itself sinks to the form of psychical life and in doing so surrenders that relation to the actual which to the sound mind is essential and objective, that is, the relation resulting from the reduction to a *moment* of what is posited as external (*PM* § 406, *Zusatz*, p. 106).

Sorrow and fright can be the immediate precondition for the onset of psychic illness, but also general conditions – above all "religious and political exaltation". During the war of the Cévennes, such complexes appeared;[27] another famous example was Joan of Arc in whom "patriotic enthusiasm . . . and . . . a kind of magnetic state" were combined (*PM* § 406, *Zusatz*, p. 107). The climate, and also the time of day or year could also have damaging effect, and could find expression "in morbid states (including insanity) and at periods when the self-conscious life suffers depression" (*PM* § 392, p. 37). Romantic conceptions about a cosmic dependence of life are explicitly rejected by Hegel: "The history of the world is not bound up with revolutions in the solar system, any more than the destinies of individuals with the position of the planets" (*PM* § 392, pp. 36–37).

Sickness and health, in their origins and their development, are not determined by the movement of the planets (*PM* § 392, *Zusatz*, p. 38); Hegel does not support any sidereal concept of illness. Animals and plants are bound to such a common life joined with nature; modern man, however, has freed himself from these bonds. Psychic illnesses and those of the body according to Hegel, are linked concretely with each other and dependent on each other, since in the dissociation of consciousness, the "corporeity which is as necessary for the empirical existence of mind as it is for that of soul, is divided between these two separated sides and accordingly is divided within itself and therefore sick" (*PM* § 406, *Zusatz*, p. 106).

The division of illnesses in three basic kinds finds its confirmation in reality, according to Hegel; but this distinction is also justified (onto-) logically. The philosophic systematization would be legitimized on the one side by a dialectic of the transition from concept [*Begriff*] to phenomenon and phenomenon to concept, which should be capable of being carried out subsequently in empirical work; but on the other side this mediation of notions and phenomena has its foundation in logic. From the principle of this double perspective – empirical research and logic – comes the decisive methodological difference between Hegel's speculative understanding of illness and the Romantic conceptions. According to Hegel, the first kind of illness – epidemics and contagious diseases – are to be traced back to "a *general* determinateness lying in the non-organic nature as such" (*PN* § 371, *Zusatz*, p. 430). The second kind is caused by "*particular*, external harmful influences" (*PN* § 371, *Zusatz*, p. 432); in these illnesses above all, specific parts of the body are affected. The third kind of illness, finally, originates "in the universal subject" (*PN* § 371, *Zusatz*, p. 432). In the three kinds of illness, for Hegel, a transition from the universal to the individual, by way of the particular, is carried out, a transition that is developed in the logic and receives its justification there. However, at the same time Hegel's conviction of the possibility of a logical foundation of reality is combined with insight into the limits of such a foundation; Hegel considers a deduction of all forms of illness to be impossible.

Stages of the Course of Illness

Hegel's philosophical doctrine of illness does not stop at a general concept of illness and a phenomenal differentiation of three kinds of illness. The developmental character of illness is fully presented. In the process of becoming ill and recovering, the individual runs successively through the states of sensibility, irritability and reproduction – in these Hegel sees the three basic functions of form and formation which during health act together harmoniously, but separate in illness; the succession of these constitute being ill, an essential part of which is the phenomenon of fever:

The characteristic *manifestation* of disease is, therefore, that the identity of the entire organic process displays itself as the *successive* course of the vital movement through its distinct moments: sensibility, irritability, and reproduction, i.e. as *fever*, which, however, as process of the *totality* in opposition to the *isolated* (*vereinzelte*) activity, is just as much the effort towards, and the beginning of, *cure* (*PN* § 372, pp. 432–433).

Accordingly illness develops through three stages; in these, illness also unfolds its significance for the self-perception of the subject. In the first stage the illness is only present *in itself* [*an sich*], the beginning of the physical disturbance does not cause one to 'feel ill' – the individual is unaware of it; in the second stage the illness is perceived consciously, and as a part of the organism, making itself independent, it turns against the organism's subjectivity. The organism, as Hegel says, is "irritated beyond its capacity to respond" so that "one particular part, a single system, gains a foothold in opposition to the self" (*PN* § 371, *Zusatz*, p. 433). The third stage, following upon the sensibility of the first and the irritability of the second, is that of reproduction. With an antiquated expression, Hegel designates this stage as 'coction' [a stage of digestion prior to elimination of morbid matter – Ed.]; here it "consists in the affection of the one system becoming an affection of the whole organism" (*PN* § 372, *Zusatz*, p. 434). But with that, the possibility of a cure is given at the same time. Fever is a symptom of this state, this complete embarrassment of the organism by an illness and at the same time a sign of overcoming it; fever is the "pure life of the diseased organism", it is the 'fluidization' of the shattered basic functions of the living being, it stands between illness and health – it is the high point of the illness, and it is also the beginning of recovery: "Therefore, even if fever is, on the one hand, a morbid state and a disease, yet on the other hand, it is the way in which the organism cures itself" (*PN* § 372, *Zusatz*, p. 434).

Fever is also development, and this too passes through the three basic organic functions. It appears to begin with

shivering, heaviness of the head, headache, twinges in the spine, twitching of the skin, and shuddering. In this activity of the nervous system, the muscles are left free and consequently their own irritability functions as an uncontrolled trembling and powerlessness. Heaviness of the bones sets in, tiredness of the limbs, withdrawal of the blood from the skin, a sensation of cold.... The organism dissolves all its parts within itself in the simplicity of the nerve, and feels itself withdrawing into the simple substance [of its being] (*PN* § 372, *Zusatz*, p. 435).

Then this development will be a transformation into "*heat, negativity*, where the blood is now the dominating factor" (*PN* § 372, *Zusatz*, p. 435). Simultaneously this is now the time of delirium. Added to this is sweating and secretion of perspiration. According to Hegel, this secretion cannot be regarded as the expelling of diseased matter, which would only have to be "emptied out with a spoon" (*PN* § 372, *Zusatz*, p. 435) in order to overcome or shorten the illness. Rather, in this act, which is an act of production, of formation, the organism regains its healthy identity and wins back the capacity of self-formation:

> The significance of this product is, that in it, the isolation, the single system (*das Einzelne*), the determinateness, ceases, for the organism has produced itself as a whole, in general, has digested itself; sweat, to use an expression of the ancient physicians, is *cooked morbid matter* – an excellent notion. Sweat is the *critical secretion*; in it, the organism attains to an excretion of itself, through which it eliminates its abnormality and rids itself of its morbid activity. The *crisis* is the organism which has gained the mastery over itself, which reproduces itself, and exercises this power by sweating (*PN* § 372, *Zusatz*, p. 435).

Therapy

In illness the organic succumbs to the inorganic, the unity of consciousness succumbs to a single psychic dimension; organs and organic functions, drives and fields of consciousness, make themselves independent. This perspective gives therapy its orientation – it must remove and transcend [*aufheben*] the particularization. This happens in three ways. Either the organism is intensified and strengthened in its activity, as due to the medication it "must direct itself outwards" or through therapeutic means "the activity of the conflict is weakened" (*PN* § 373, *Zusatz*, p. 439). Finally there is magnetic hypnotherapy and other psychic forms of treatment. For Hegel therapeutic remedies are inorganic substances; by confronting them the body can more readily pull itself together than when it confronts inorganic forces or subentities within itself that have made themselves independent:

> The medicine provokes the organism to put an end to the *particular* irritation in which the formal activity of the *whole* is fixed and to restore the fluidity of the particular organ or system within the whole (*PN* § 373, p. 436).

Illness can also be called a "hypochondria of the organism" – the organism must be pulled away from the state of sinking down into itself and rejecting the external world, and set back again "into the general activity of assimilation" (*PN* § 373, *Zusatz*, p. 438). The organism is led back to itself from outside, it is *bewitched*, transformed by magic. Under the influence of the healing remedy, it is restored to itself, gains strength in the confrontation with this remedy, and now can also overcome the process of illness within itself. The relation between a specific medication and a specific illness, to be sure, was still scarcely understood, according to Hegel; as yet "experience alone is supposed to decide this" (*PN* § 373, *Zusatz*, p. 439). Here a future research task is sketched for empirical medicine, which, for Hegel, cannot fall within the province of philosophy. A fundamentally different action of medication is to weaken the organism: "Its purpose, then, is to depress the activity of the organism so that in eliminating from it all activity, that of the diseased system, too, is eliminated" (*PN* § 373, *Zusatz*, p. 439).

Fasting, diet, but also bleeding and severe chilling for inflammation, should be effected in this direction. If until now medicine has enjoyed greater success in treating acute rather than chronic illnesses, for Hegel the reason is that in acute illnesses the entire organism suffers, while the specific fixations of chronic illnesses can only be overcome with difficulty, and this difficulty becomes greater the more "this organ or system is affected" (*PN* § 371, *Zusatz*, p. 432). Finally psychic diseases are to be treated by magnetic hypnotherapy and other psychic or physical therapeutic procedures. Just as the medication from outside overcomes the isolation of the diseased body region again, so the hypnotizing physician overcomes the independence that certain fields of consciousness have established for themselves. The consciousness of the patient is 'fluidified', in the artificial sleep the disturbed self [*Ich*] is to reconstitute itself, is brought back "to the feeling of its inner universality" (*PN* § 373, *Zusatz*, p. 440). The sleep of hypnosis is related to natural sleep, and it can be replaced by the latter as well:

> But instead of this magnetically induced sleep, a healthy sleep, too, can produce this turning-point in an illness, i.e. the organism can then spontaneously gather itself together into its substantiality (*PN* § 373, *Zusatz*, p. 440).

'Animal magnetism', its empirical background and theoretical consequences, as well as the controversies around this approach, are known to Hegel; there is no unconditional approval in his work.[28] Psychic therapy is frequently combined with physical methods. But the physical aspect of the cure cannot be understood philosophically: "The medical remedies employed are, on the contrary, for the most part empirical and are therefore uncertain in their action" (*PM* § 408, *Zusatz*, p. 136). However, the psychic is decisive. Hegel considers an essential presupposition for every psychic therapy to be the insight that psychic disease is not a complete loss of rational mental power, but rather "only a contradiction in a still subsisting reason" (*PM* § 408, p. 124). This fundamental principle also remains valid for therapy of physical illnesses; here, too, one cannot proceed from a total loss of health, but only from a contradiction within health. Treating a sick human being from this perspective, according to Hegel, is to be designated as *humane* [*menschliche*] treatment:

This humane treatment, no less benevolent than reasonable . . . presupposes the patient's rationality, and in that assumption has the sound basis for dealing with him on this side – just as in the case of bodily disease the physician bases his treatment on the vitality which as such still contains health (*PM* § 408, p. 124).[29]

IV

To take into consideration other dimensions of the understanding of illness besides the perspective of natural science seems to have been a continuing human interest Aside from the sociology, psychology and anthropology of illness, and investigation of the relationship between art and illness, a philosophical conception also finds recognition. In the history of the confrontation of philosophy with the nature of man, of health, of illness and death, Hegel occupies a special place. Hegel's speculative concept of illness differs from the Romantic conceptions of his time; it is also distinct from the interpretation of illness in Schelling's philosophy. Hegel's conceptions are not to be understood as an alternative to empirical medicine; he is convinced of the possibility of linking speculation and empirical work in a unity and in recognition of the difference between them. For Hegel, illness is understood as a phenomenon of organic life, which can occur only on this level, and only within this level can insight be gained into it. The multiplicity of illnesses is ordered in three kinds of illness, a taxonomy that is to satisfy logical principles as well as find confirmation in reality. Universal and individual physical illnesses as well as psychic illnesses are the processes of organs and functions making themselves independent, whereas in the healthy state these are subordinated to the unity of the body and the identity of consciousness. The process of falling ill and recovering health is a falling apart of the essential vital functions, which, during the stages of this process, successively predominate and then, in health, return to their harmonious combination. Recovery is not always possible; illnesses anticipate death, organic life must die. Therapy corresponds to the original arising and the phenomenal character of illness; in it the drive toward independence of the diseased body systems and psychic domains is again eliminated and transcended [*aufgehoben*]. Hegel's philosophy of illness is determined by the interrelation between nature and mind; in therapy the viewpoint of humaneness is emphasized; treatment has a moral aspect.

Ruprecht-Karl-University
Heidelberg

NOTES

[1] After completion of work on this manuscript there appeared W. Jacob's brief description 'Der Krankheitsbegriff in der Dialektik von Natur und Geist bei Hegel' in *Hegel-Studien*, Beiheft **11** (1974), pp. 165–73; and also T. J. Bole: 'John Brown, Hegel and Speculative Concepts in Medicine', *Texas Reports on Biology and Medicine* **32** (1974), 287–97. Previously to that only W. L. Leibbrand had presented Hegel's conception of illness in quite detailed form (see below Notes 15, Leibbrand (1956), 187–90, 263–70). In the comprehensive works on the history and systematics of the concept of illness, Hegel has not been included: H. Ribbert, *Die Lehren vom Wesen der Krankheiten in ihren geschichtlichen Entwicklung* (Bonn, Cohen, 1899); T. Meyer-Steineg, 'Der Gang der Krankheitslehre in ihren wichtigsten Phasen', *Deutsche Medizinische Wochenschrift* **50** (1924), 311f., 347f., 380f., 412; H. Müller, 'Über den Krankheitsbegriff im Wandel der Zeiten, mit besonderer Berücksichtigung neuerer Anschauungen und Folgerungen', Diss. Berlin 1939; A. Dietrich, 'Die Entwicklung des Krankheitsbegriffes', *Hippokrates* **12** (1941), 122–26; E. Berghoff, *Entwicklungsgeschichte des Krankheitsbegriffes*, 2nd ed. (Vienna, Maudrich), 1947 (Here alone a quotation from the *Philosophy of the Subjective Mind* is to be found.); W. Riese, *The Conception of Disease, Its History, Its Versions and Its Nature* (New York, Philosophical Library, 1953); L. J. Rather, 'Zur Philosophie des Begriffes Krankheit', *Deutsche Medizinische Wochenschrift* **83** (1958), 2012–18; also: 'Towards a Philosophical Study of the Idea of Disease', in *The Historical Development of Physiological Thought*, ed. C. C. M. Brooks and P. F. Cranefield (New York, Hafner, 1959), pp. 351–73; O. Temkin, 'The Scientific Approach to Disease; Specific Entity and Individual Sickness', in *Scientific Change*, ed. A. C. Crombie, (London, Heinemann, 1963), pp. 629–58; P. Diepgen, G. B. Gruber and H. Schadewaldt, 'Der Krankheitsbegriff, seine Geschichte und Problematik', *Handbuch der Allgemeinen Pathologie*, (Berlin, Heidelberg, New York, Springer, 1969), vol. I, pp. 1–50; K. E. Rothschuh, *Zwei Beiträge zur Allgemeinen Krankheitslehre* (Stuttgart, 1973).

[2] H. A. Göden published a totally negative account in 1819 in his essay 'Über Hegel's Begriff vom Wesen der Krankheit und der Heilung' in *Okens Isis*, vol. 2 (1819), col. 1127–38; in 1829 K. H. Scheidler reviewed the 'Grundzüge der Hegelschen Philosophie in Beziehung auf die Medizin [The Basic Principles of the Hegelian Philosophy in Relation to Medicine]" within the treatise 'Über das Verhältnis der Philosophie überhaupt and der Psychologie insbesondere zur Medicin', *Minerva Medica*, **1** (1829), 211–48; he regretted the widespread disinterest in philosophy among physicians; this disinterest has brought about "that not even the system of Hegel, recently epoch-making in the focal point of North German scientific culture, and already applied in many ways to theology, jurisprudence, history in general, as well as to the history of philosophy especially and to aesthetics, has not been considered or noted by them." (p. 245f.)

[3] C. W. Hufeland, *System der practischen Heilkunde* (Frankfurt and Leibzig, 1800), vol. 1, p. VIII.

[4] *Ibid*., 1802, vol. 2, p. VIf.

[5] A. F. Hecker, *Kunst, die Krankheiten der Menschen zu heilen* (Erfurt, Henning, 1804; 4th ed., 1813), p. V.

[6] *Ibid*., p. VI.

[7] *Ibid*., p. 298.

[8] Candidus, 'Nicht Anklage, sondern Klage', *Journal der practischen Heilkunde,* **33** (July 1816), 110–16.

[9] *Ibid.*, p. 113.

[10] *Ibid.*, p. 116.

[11] Hufeland, 'Nachwort zu Candidus', *op. cit.*, p. 119.

[12] P. C. Hartmann, *Theorie der Krankheit* (Vienna, Gerold, 1823), p. 60.

[13] C. J. Windischmann to Hegel, 13 October 1823 in *Briefe von und an Hegel*. 3rd ed. Ed. by J. Hoffmeister. 4 vols. (*Sämtliche Werke*. Neue kritische Ausgabe, 29, 1959), Vol. III, p. 34.

[14] Hegel to Windischmann, 11 April 1824, *ibid.*, p. 40.

[15] For Romantic medicine see the more general studies: Fr. v. Müller, *Spekulation und Mystik in der Heilkunde. Ein Überblick über die leitenden Ideen der Medizin im letzten Jahrhundert* (Munich, Lindauer, 1914); P. Diepgen, *Deutsche Medizin vor hundert Jahren. Ein Beitrag zur Geschichte der Romantik* (Freiburg and Leipzig, Spever and Kaerner, 1923); W. Fischer, *Die Krankheitsanschauungen der Romantik* (Rostock, Hinstorff, 1926); E. Hirschfeld, 'Romantische Medizin', *Kyklos* 3 (1930), 1–89; F. H. Garrison, 'The Romantic Episode in the History of German Medicine' (1931) in *Contributions to the History of Medicine* (New York and London, Hafner, 1966), pp. 115–40 W. Pagel, *Virchow und die Grundlagen der Medizin des XIX. Jahrhunderts* (Jena, Fischer, 1931); P. Diepgen, 'Alte und neue Romantik in der Medizin', *Klinische Wochenschrift* **11** (1932), 28–34; W. Pagel, 'The Speculative Basis of Modern Pathology. Jahn, Virchow and the Philosophy of Pathology', *Bulletin of the History of Medicine* **18** (1945), 1–43; K. Schober, 'Die Vorstellungen der Ärzte der Romantik von der Wirkung der Heilmittel', diss. med. Mainz 1950; G. Rosen, 'Romantic Medicine: a Problem in Historical Periodization', *Bulletin of the History of Medicine,* **25** (1951), 149–58; W. Leibbrand, *Die spekulative Medizin der Romantik* (Hamburg, Claassen, 1956); E. Heischkel, 'Pharmakologie in der Goethezeit', *Sudhoffs Archiv* **42** (1958), 302–11; K. E. Rothschuh, 'Ansteckende Ideen in der Wissenschaftsgeschichte, gezeigt in der Entstehung und Ausbreitung der romantischen Physiologie', *Deutsche Medizinische Wochenschrift* **86** (1961), 396–402; O. Temkin, 'Basic Science, Medicine and the Romantic Era', *Bulletin of the History of Medicine* **37** (1963), 97–129; G. B. Risse, 'Kant, Schelling, and the Early Search for a Philosophical 'Science' of Medicine in Germany', *Journal of the History of Medicine and Allied Sciences* **27** (1972), 145–58; H. Sohni, *Die Medizin der Frühromantik* (Freiburger Forschungen zur Medizingeschichte, N. F. vol 2. Freiburg, Hans Ferdinand Schultz, 1973). – There is a series of specific concepts and groupings within Romanticism, as well as about the concept of illness in literature; thus, for example: W. Milch, 'Zum Problem der Krankheit in der Dichtung der deutschen Romantik', *Sudhoffs Archiv* **23** (1930), 213–35; K. E. Rothschuh, 'Joseph Görres und die romantische Physiologie', *Medizinische Monatsschrift* 5 (1951), 128–31; E. Seidler, 'Entwicklung naturwissenschaftlichen Denkens in der Medizin zur Zeit der Heidelberger Romantik', *Sudhoffs Archiv* **47** (1963), 43–58; H. H. Lauer, 'Krankheit und 'Heilung' bei Windischmann', *Sudhoffs Archiv* **47** (1963), 59–72; W. Artelt, *Der Mesmerismus in Berlin* (Mainz, 1965); H. Schipperges, 'Krankheit als geistiges Phänomen bei Novalis', *Der Horizont* 8 (1965), 116–29.

[16] For Hegel's relation to Romanticism, see: O. Pöggeler, *Hegels Kritik der Romantik* (Bonn, Bouvier, 1956).

[17] Hegel, *Maximen des Journals der deutschen Literature*, 1806. *Sämtliche Werke*

(*Jubiläumsausgabe*) **1**, 4th ed. (Stuttgart-Bad Cannstatt, 1965), p. 545.

[18] Hegel, 'Aphorismen aus der Jenenser Zeit', in *Dokumente zu Hegels Entwicklung*, ed. J. Hoffmeister (Stuttgart, Frommann, 1936), p. 355.

[19] J. N. Neubauer, 'Dr. John Brown (1745–88) and Early German Romanticism' *Journal of the History of Ideas* 28 (1967), 367–82; G. B. Risse, 'The History of John Brown's Medical System in Germany during the Years 1790–1806', diss. Chicago 1971.

[20] Such a view was reported to Hegel, for example, by C. D. v. Buttel on Dec. 16, 1829: "So among other matters I can report with joy how one of my friends in Jever, Dr. Med. Pinner, continually ascertains results, which I am convinced can be incorporated by philosophy almost without further effort Among other matters he has found in the history of medicine, that the periodically dominant and changing forms of illness are simply based on the trilogy of vital functions, and as in the present day the decaying (of reproductivity), under the emergence of carbon, imparts to the constitution of illness a consistent character, as previously (only a few years ago) the inflammational (irritability), under the predominance of oxygen, and still earlier the nervous (sensibility) under the potentiation of nitrogen was the *dominant* constitution" (Hegel, *Briefe von und an Hegel*. 3rd ed. Ed. by J. Hoffmeister. 4 vols. (*Sämtliche Werke*. Neue kritische Ausgabe **29**, 1959), Vol. III, p. 290f. The historical character of the illnesses was maintained around 1800 repeatedly and in different ways. Hufeland sees an evolution of illnesses in the history of mankind, which arises solely "through mechanical causes by way of violent external interventions," to nervous diseases "not only in the cities and among the higher social ranks, but this nervous character is also noticeable in the countryside, and what was unheard of in Antiquity, there are now peasants who are hypochrondriacs, and peasant women who suffer from vapors just as much as ladies in the town." (Chr. W. Hufeland, *Geschichte der Gesundheit nebst einer physischen Karakteristik des jetzigen Zeitalters* (Berlin 1812), p. 13 and 19 f.).

[21] L. Oken, *Lehrbuch der Naturphilosophie*, 2nd ed. (Jena, Frommann, 1831), § 3063, p. 388.

[22] K. R. Hoffmann, *Vergleichende Idealpathologie* (Stuttgart, Balz, 1834). Other important representatives of this tendency are: K. W. Stark, *Pathologische Fragmente* (Weimar, Landes-Industrie-Comptoir, 1824–25); *Allgemeine Pathologie oder allgemeine Naturlehre der Krankheit* (Leipzig, Breitopf and Härtel, 1834–44) who is regarded as its founder; R. W. Volz, *Medizinische Zustände und Forschungen im Reiche der Krankheiten* (Pforzheim, Dennig, 1839) and F. Jahn, *Die Naturheilkraft* (Eisenach 1831); for the natural history school see W. Karst, 'Zur Geschichte der natürlichen Krankheitssysteme', *Abhandlungen der Geschichte der Medizin und der Naturwissenschaften* 37 (1941).

[23] J. C. A. Heinroth (*Lehrbuch der Störungen des Seelenlebens oder Seelenstörungen und ihrer Behandlung*, Pts. 1–2 (Leipzig, Vogel, 1818); *Lehrbuch der Seelengesundheitslehre*, vols. 1–2 (Leipzig, Vogel, 1824); C. J. Windischmann, *Über etwas, das der Heilkunst Noth thut. Ein Versuch zur Vereinigung dieser Kunst mit der christlichen Philosophie* (Leipzig, Gnobloch, 1824); J. N. v. Ringseis, *System der Medizin* (Regensburg, Manz, 1841); for this point of view see W. v. Siebenthal, 'Krankheit als Folge der Sünde', *Heilkunde und Geisteswelt*, vol. 2 (Hannover 1950). It is this conception which is involved in the fundamental difference between them, of which Hegel speaks repeatedly in his correspondence with Windischmann, and which is also mentioned by Windischmann in his replies.

[24] Hegel to Windischmann, Letter of 27 May 1810 in *Briefe von und an Hegel*. 3rd ed.

Ed. by J. Hoffmeister. 4 vols. (*Sämtliche Werke*. Neue kritische Ausgabe **29**, 1959), Vol. I, p. 314.

[25] *Ibid.*

[26] See D. Henrich, 'Hegels Theorie über den Zufall', *Kantstudien* **50** (1958/59), 131–148; reprinted in Henrich, *Hegel im Kontext* (Frankfurt, Suhrkamp, 1971).

[27] [The Cévennes, a mountain range forming the southern and eastern fringe of the central plateau of southern France, was the locale of an early eighteenth century uprising of Huguenot Protestant peasants – the Camisards – against the Church and the King. The Camisards carried on organized military resistance for some years after 1702, following nearly two decades of violent repression of their movement after the revocation of the Edict of Nantes (1685). Their religion was marked by visions, ecstasy, miracles and other spiritual manifestations, as well as by prophetic and mystical witnessing, and the leadership of young children. – Ed.]

[28] Hegel expressly points to the observations of P. G. van Ghert who studied with him and who formed a positive judgement of Hegel's philosophy; a 'Tagebuch einer magnetischen Behandlung' (Diary of a magnetic treatment) (Holland 1814) by van Ghert appeared in German translation in the *Archiv für den thierischen Magnetismus* **2** (1817) St. 1, pp. 3–24, 55–188, St. 2, pp. 3–51; also a 'Sammlung merkwürdiger Erscheinungen des thierischen Magnetismus' (Collection of Remarkable Phenomena of Animal Magnetism) (Holland 1815) appeared in translation in the same journal, vol. 3 (1818) St. 3, pp. 1–97. Hegel calls van Ghert a "reliable physician who is at the same fertile in ideas and educated in the most recent philosophy" (see *PM* § 406, *Zusatz*, p. 118).

[29] In this relation Hegel recalls to mind Pinel and his commitment to a humane, moral treatment of mental illness (see *PM* § 408, p. 124; *Zusatz*, p. 137). Pinel himself is convinced that "the moral treatment of insanity is one of the most important parts of observational medical science and one that has till now received little attention." His residence in the Hospital Bicêtre, his communication with the insane patients, his observation of the "regular order" of their behavior and well as of "changeable and often bizarre scenes" led him to the insight, how frequently it would be possible in these "cases of misled reason to restore its use either by means of mild remedies or an energetic but wise and humane discipline". P. Pinel, *Philosophisch-Medizinische Abhandlung über Geistesverirrungen oder Manie*. A. de Fr. (1801) tr. by M. Wagner (Vienna 1801), p. 112. [This is a translation from the original French work: *Traite médico-philosophique sur l'aliénation mentale, ou la manie* . . . (Paris, 1801). The English version is *A Treatise on Insanity*, trans. by D. D. Davis (London, 1806; reprinted New York, Hafner, 1962). – Ed.]

RÜDIGER BUBNER

ON HEGEL'S SIGNIFICANCE FOR THE SOCIAL SCIENCES

After the breakdown of the Hegelian encyclopedic system, the individual parts suffered different fates. One set of questions seems to have survived better than the others: Hegel's social philosophy has continued to incite controversies in the social sciences, the humanities and philosophy from the nineteenth century to the present. German sociology with its so-called 'Positivism-dispute' has been the most recent scene of debate.[1] This controversy between Popper's philosophy of science and a Neo-Marxist defence of the dialectic reiterates motives and prejudices with a long-standing tradition. However, a true appraisal of Hegel's significance for the social sciences must renounce the established chain of prejudices on both sides. In my opinion the fecundity of Hegel's philosophy for fundamental questions of history and society first becomes clear in the light of a full understanding of the famous thesis in the Preface to the *Philosophy of Right*: "Philosophy is its own epoch comprehended in thought".[2]

The recourse to Hegel himself does not arise out of a purely historical interest in textual exegesis. To the contrary, the attempt to reconstruct Hegel's original dialectic independently of traditional controversies reveals its significance and fruitfulness for dealing with problems of our day. Consequently, my paper traces the contemporary discussion back into the nineteenth century and ultimately to its source in Hegel. The first section reviews pertinent aspects of the recent Positivism-dispute (I). I then sketch out the tradition of Hegel's influence on the methodology of the social sciences. These historical remarks should serve to clarify some commonplace ideas (II – The impatient reader may skip that section). An analysis of Hegel's view of the inescapable tension between philosophy and its epoch completes the argumentation (III). Finally, I draw some general conclusions on the role of dialectics for social thought (IV).

I

Professor Popper places Hegel, as we know, among the most dangerous enemies of an Open Society.[3] He is just as ready to cast him out of the ideal state of classical liberalism as once Plato was to cast the Sophists and the

R. S. Cohen and M. W. Wartofsky (eds.), Hegel and the Sciences, 143–159.

poets out of his Republic.[4] Instead of recognizing the individual's freedom of decision and moral responsibility, Hegel, according to Popper, believes in objective laws of history which permit predictions about social developments. This prophetic *historicism* is criticized for excluding rational and critical discussion which is constitutive of all scientific progress, in the natural as well as in the social sciences. Popper's political objection to Hegel is thus based on a methodological argument, and the plea for democracy represents the obvious political consequence of the fundamental insights of the *Logic of Scientific Discovery*.

And yet, Popper often concedes that his methodological approach of "trial and error" has much in common with *dialectical* procedure.[5] He argues as follows. Growth of knowledge originates from problems which emerge in science or life outside the realm of philosophy. These problems present themselves as difficulties or contradictions between our prior knowledge and the facts. The epistemological method suggested by Popper to remedy this situation has as its goal the permanent elimination of future contradictions. Human knowledge can only advance by seriously taking such contradictions into account and by critically dissolving them through a search for a better theory. This approach bans historicist claims of all-embracing laws of history.

Popper's position was challenged by dialecticians. During the last two decades in Germany there has been an animated dispute between sociologists and philosophers of Hegelian and Popperian persuasions. It commenced with an attack by the Neo-Marxists of the Frankfurt School on Popper's concept of methodology in the social sciences. Although Popper's critics have employed Hegelian arguments, they use these arguments in the spirit of *Marxist social theory* and critique of ideology. They assert that contradictions do not simply arise in the course of scientific research but are objectively given in the character of current capitalist society. Hence, the prime concern is not to eliminate the contradictions but to understand the historical necessity of their existence.

The appropriate method for this task is not to be found in critical discussion among researchers, because this discussion is carried out in the socially detached sphere of institutionalized science. According to the Neo-Marxists, Popper's methodological concept harbors the danger of ideological obfuscation, since it necessarily can not grasp all of the relevant contradictions. The critical discussion of contradictions must consequently be expanded from the methodological realm to a material critique of socio-historical reality itself. The political model of liberalism favored by Popper then falls victim to such a critique because it represents the illusion of individual freedom in the midst of actual bondage and inequality. The Neo-Marxist position aims not at true knowledge of existing society, but at knowledge of the falsity of existing society viewed against the background of a projection of another truer, though not yet actual, society.

In response to this position Popper's followers simply repeated the old arguments against the dialectic, detecting again what they call an irrational myth of totality which Popper had discarded as historicism. The positions hardened to a kind of stalemate and at the level of this controversy no satisfactory solution has been reached. This dispute, however, represents only the most recent phase of the discussion of Hegel's significance for the social sciences. It will be useful to recall this tradition of Hegel's influence in order to remove some persistent misunderstandings and to evaluate the enduring significance of Hegel's legacy.

II

Philosophy comprehends its own time in thought – this might be considered the motto of the left-wing movement of *Young Hegelians.*[6] Arnold Ruge[7] as well as the early Marx[8] understood it as a radical encouragement to *criticism*. The Young Hegelians established a strict opposition between philosophy and reality. They prescribed for philosophy unconditional confrontation with the political, social and economic conditions of its time. Since philosophy finds itself *ab origine* in opposition to reality and since its own claim to reason seems to be well founded, the duty of the critic is clear: to expound the prevailing irrationality theoretically and to attack its supporters practically.

The first critique of this sort was directed against Hegel himself, who, according to the Young Hegelians, constrained the critical power of thought in a speculative system that existed only for itself and possessed no further relationship to its time. Hegel was accused of an "accommodation" to the political situation; his philosophy of right, which had formulated the task of philosophy with regard to its epoch, was regarded as the most extreme ideological glorification of the Prussian state.

In his famous 'Introduction to the Critique of the Hegelian Philosophy of Right'[9] *Marx* developed a program which included both a critique of the ideological function of pure philosophy and the foundation of revolutionary practice. If philosophy is to comprehend its time, then it cannot merely capture it in thought and sublimate it theoretically while the wretched reality of the age continues to exist under this ideological cover. The full sense of

Hegel's dictum implies more than Hegel himself recognized. The "task of philosophy for its time" means overcoming isolated theory and practically changing reality.

In contrast to such a critical application of the dialectic Dilthey's *hermeneutical theory* of the humanities (*Geisteswissenschaften*) represents a more conservative line of Hegelian influence on the social sciences.[10] This theory seeks to found the method of the humanities on Hegel's concept of the objective mind (*Geist*), in the attempt to circumvent the hubris of an absolute system of encyclopedic philosophy. Dilthey's central category, 'life' (*Leben*), designates the inner connection of historical phenomena, and can be considered a terminological translation of Hegel's concept of objective mind.[11]

According to Dilthey, the unitary stream of life not only inexhaustibly manifests itself in innumerable concrete phenomena, but also enables us to understand these phenomena. If the phenomena of the objective mind are expressions of the all-embracing stream of life, then these phenomena are immediately accessible to the understanding, for he who understands, participates in that very stream of life. The object of the humanities is historical-social reality, and if it is to be grasped it must be ontologically distinct from nature. It must neither be artificially objectivized nor alienated through analogy with the method of the natural sciences.[12] The appropriate method of the humanities is the identification of the given object as a manifestation of life. Thus an act of *anamnesis* occurs in which the subject recognizes himself, as it were, in the object. The comprehensive stream of life to which both the subject and the object essentially belong mediates between the two sides of one whole. Dilthey calls this process '*Verstehen*'.[13]

As a natural consequence of his conception of the humanities, it was impossible for Dilthey to accept Hegel's idea of an absolute mind, under which art, religion and philosophy are subsumed. The realm of absolute mind should be abolished as a metaphysical dogma. Dilthey complains of Hegel's approach: "The plan of construction of the '*Geisteswissenschaften*' is reversed from the outset. Historical understanding is sacrificed to the metaphysical schema".[14] In fact, Dilthey is guided by a thoroughly un-Hegelian epistemological intention which he conceives analogously to Kant as a "critique of historical reason". His central concern is to clarify the possibilities for knowledge of historically limited, finite reason; in his view this means rectifying Kant's transcendental abstractions by focusing on the living individuals' concrete acts of reason, thereby at the same time avoiding the speculative extravagance of Hegel's absolute system.

Thus Dilthey had to challenge the *Neo-Kantians* as well, who strictly in the Kantian spirit developed a merely methodological distinction between the natural sciences and the humanities. The first of these was Wilhelm Windelband.[15] He declared that the ontological distinction between nature and mind was obsolete since the sciences of nature and mind are both to be conceived as sciences of experience. The distinction between them lies only in the sphere of methodology, that is in the logic of the different ways of experiencing reality.[16]

Windelband asserted: "The principle of division is the formal character of the goals of their knowledge. The one seeks general laws, the other particular historical facts". He coined the designations 'nomothetic'[17] and 'idiographic' for the two methodologically distinct branches of science. The nomothetic sciences strive for general laws, under which all particulars can be subsumed. The idiographic sciences try to grasp and describe individual significant events, persons or phenomena.[18]

For the purpose of characterizing particulars as meaningful or significant, Windelband employs a concept introduced by Hermann Lotze, and which later, through Heinrich Rickert and Max Weber, decisively shaped the methodology of the social sciences – the concept of 'value' (*Wert*). The term was obviously borrowed from economics. Interestingly, however, Lotze first applied this notion to problems of aesthetics in order to explain what a judgment about beauty signified. Lotze wanted to express in realistic sounding terms what German idealism, not least that of Hegel, conveyed with the concept of "idea", namely the reality of reason in contrast to the merely empirical reality of multifarious facts.[19]

Heinrich Rickert adopted Windelband's methodological distinction and developed it through use of the concept of value so that it assumed the most influential position in the philosophy of science at the beginning of our century.[20] He bases his approach on the logical distinction between the general and the particular. Since, as he argues, the natural sciences strive only for general laws, they must systematically neglect the specific qualities of individual occurrences and they can never adequately come to terms with the elusive concreteness of empirical phenomena. Precisely this ungrasped concreteness remains as the object of the historical or cultural sciences. Hence his characterization of the object of the cultural sciences is derived negatively from the shortcomings of their counterpart.

However, Rickert also offers a positive definition by means of a so-called 'theoretical reference to value' (*Wertbeziehung*). The 'reference to value' does not involve practical value-judgments but is theoretical and value-free. It selects out of the abundance of details the specific individual fact which is related to a value and therefore is of general significance in the historical-social realm. As is well known, Rickert's notion of value-reference was adopted by *Weber* as a useful tool for sociological understanding. This aspect of the social sciences continues to remain controversial.[21]

Rickert's treatment of historical-social reality as the object of an individualizing and value-referring science does not suggest Hegel's speculative idealism as much as Dilthey's. He spoke of *Kulturwissenschaften* rather than *Geisteswissenschaften*. However, he did admit they and their subject-matter essentially rested on Hegel's provocative and ambiguous principle stated in the introduction to the *Philosophy of Right* that what is rational is real and what is real is rational.[22]

There are yet two further examples of Hegel's influence on the social sciences which should be considered: the work of *Hans Freyer* and *Karl Mannheim*. Freyer could easily appeal to Hegel when he charged that the sociology of his day, which was methodologically oriented towards the *Kulturwissenschaften*, was too removed from life. He maintained that a sociology which is primarily concerned with methodological questions tends to forget its actual duty vis-à-vis history. He reminded the sociologists of the legacy of Hegel's philosophy and Marx's socio-economics. Sociology should deal with the reality of changing society and not with the constructs of *Kulturwissenschaften*.

Sociology must therefore consciously recognize its ties to its time and seek to maintain this connection. Only after recognizing the essential dependence of theory on its time can sociology become aware of its genuine object. Freyer's formulation of the task conforms completely to the Hegelian dictum that the epoch must be comprehended in thought.[23] "A sociology is the scientific self-understanding of a given society. In the nature of its problems and in the inner form of its thought, a sociology is inescapably determined by its time."[24]

This program of sociological realism differs from Karl Mannheim's sociology of knowledge, although it too has roots in Hegelian thought. Mannheim begins as well by postulating the crucial bondage of theory to its time. But instead of deducing the present duties of sociology from this postulate, he formulates a thesis which, he believes, rests securely on the methodological basis of the *Kulturwissenschaften*. He raises the historical limitation of all manifestations of mind to the universal principle that all thought is bound to its time and place. The research of the sociology of knowledge takes individual, historical phenomena which Hegel described as manifestations of objective mind and then sets them into a deterministic relationship with the given social conditions which Hegel's speculative idealism had overlooked.[25] The "relative synthesis", the link between every idea and its historical point of origin, becomes the modest, un-speculative truth which a sociology of knowledge retrieves from the principle of the comprehension of an era in thought.[26]

Mannheim mediates between Hegelian philosophy and Marx's objections. Marx classified Hegel as an ideologist insofar as he neglects the political relationships of his time by concentrating on pure thought. Marx saw the need to criticize this secret endorsement of the *status quo*. However, Mannheim neutralizes the practical, revolutionary import of the concept of ideology, since he treats it as a general, theoretical category of knowledge which a value-free, scientific sociology can then employ as a tool.

Of course, he must take into account the significance of the concept of ideology for the position of the sociology of knowledge itself; he must consider how his own theories are bound to their historical situation. For this purpose he falls back on the notion of a *pseudo-class of socially detached intellectuals* (*freischwebende Intelligenz*) who are not bound by their particular circumstances. In this way Mannheim is able to rescue the methodological consistency of the sociology of knowledge while maintaining the concept of ideology. Moreover, he can defend his answer to Max Weber's call for a link between value-free research and the political *engagement* of the researcher:[27] The pseudo-class of intellectuals serves as a rational guide for decisions in current politics by virtue of its better overview of the social whole.

The Marxists very early reiterated the reproach of ideology against this neutral re-interpretation of the concept of ideology in the sociology of knowledge. They suspected an unconscious political partisanship behind the retreat to theoretical self-certainty.[28] And even Popper, who stands in opposition to the Marxist camp, objected to the easy answers of the sociology of knowledge on the grounds of his logic of scientific discovery.[29] The process of critical discussion among scientists necessarily presupposes the subjection of the researcher to his situation. The assumption of the sociology of knowledge – that scientists belong to a special group of socially detached intellectuals – seems to eliminate the necessary social aspect of scientific method. It therefore represents an unfruitful abstraction, indeed an illusion.

We have thus returned to our starting point – the controversy between Popper and the dialecticians[30] and the as yet unclarified role of Hegel's approach. I am afraid this short survey of the forms in which Hegelian thought guided the development of the social sciences has been, more than one might have wished, a German affair. If the excursus appears all too provincial, I think it does nevertheless reflect the course of his influence. Although Hegel did exert a rather intense influence on English thought at the end of the previous century, he was expelled from the Anglo-Saxon world like Beelzebub himself by the great English philosophers of this century, Russell and Moore. The main object of their criticisms was, perhaps, the reception of Hegel by Bradley and McTaggart. Nevertheless, Hegel suffered as well under their attacks. Afterwards he remained unfamiliar to the positivists and the philosophers of science.

If they know something about him at all, it is often that he is not worth knowing and they even maintain that lucid thinking, clear ideas and exact methods must be immunized against the evils of Hegelianism. I consider this an unjust treatment of Hegel which unfortunately overlooks his potential contribution to the social sciences. I do not believe that the current value of Hegelian thought for the social sciences lies in a renewal of the category of objective mind, which Dilthey and Neo-Kantianism tried to render serviceable.[31] Nor do I think that the lesson which the contemporary social sciences may learn from dialectics consists only in the critical confrontation of theory and reality, initially proposed by the Young Hegelians and recently repeated by the Neo-Marxist school. The following offers an alternative evaluation of the tension between philosophy and its age which can provide a framework for the interrelationship of social theory and practice.

III

The introductory chapter of Hegel's first publication, the *Differenz des Fichteschen und Schellingschen Systems der Philosophie* (1801)[32] reflects on the relationship of philosophy to its historical situation. Hegel sees his point of departure as a stage which "has such a great number of philosophical systems behind it as its past", that philosophy must inevitably develop a historical consciousness. From this standpoint he makes the straightforward general observation that every philosophical conception and every system originates in a specific time, "arises out of its age".[33] He does not rest content with showing an external parallelism of philosophical systems and trends, compared with dates and epochs in the style of customary histories of philosophy. On the contrary, he defines the relationship more deeply, not as a mere doxographic model, but as the *duty of philosophy* itself.

One may ask why such value is placed on the relationship of philosophy to its age, since, after all, emphasis on the historical epoch seems extrinsic. However, the problem is embedded within the notion of philosophy itself. One speaks of philosophy and means something definite and unitary. But in reality there are only philosophies, differing systems, alternating tendencies. Philosophy as such is not a visible entity; one sees only attempts to realize the essence of philosophy, attempts which have followed upon each other throughout history.

So considered, the central problem of philosophy's self-understanding becomes how to grasp the *unity* of the intended idea of philosophy in conjunction with the *multiplicity of its historical manifestations.* Philosophy cannot begin at all before it has solved the paradox of the single idea and its many realizations. The paradox is that of philosophy itself and does not reside somewhere outside of philosophy. He who begins to philosophize and wants to be clear about what he is doing encounters the difficulty that that which he intends exists only in multifarious historical forms even though he does not mean these various forms when he thinks of philosophy. What then does he mean?

The question cannot be referred to the *propaedeutic* preliminaries of philosophy. It is impossible to decide the question before beginning to philosophize, in order thus to be able to proceed undisturbed with philosophy itself; for, in order to be able to say what precedes philosophy, it is necessary to know first what philosophy itself is. Otherwise there is no criteria to draw the desired boundary. But what philosophy itself is – that was exactly the question at issue. It is obvious that one cannot begin to philosophize without knowing what philosophy itself is and that one can only arrive at a definition of philosophy through the history of philosophy. Hence, the task of philosophy is primarily to come to terms with its own history. The elucidation of that history represents the first step in philosophizing and, at the same time, philosophy thereby provides its own secure foundation.

Within the fully developed Hegelian system, the *Phenomenology of Mind* takes up this task of a philosophical clarification of the historical constellations of thought. As the first part of that system the *Phenomenology* serves as a preparatory stage[34] for everything that follows. Today we no longer believe in *the* philosophy as emphatically as did Hegel. Nevertheless, the unity of philosophical thinking still represents a philosophical problem of undeniable and fundamental import. One tends to ignore the search for this unity in favour of an unarticulated aggregate of independent disciplines and highly sophisticated special techniques. And yet, the question remains: What holds this aggregate together?

Of course, this question can be avoided if one attempts – like the positivist philosophers – to equate a particular theoretical discipline (i.e. the logic of scientific research) with philosophy *per se*. Those who believed in the logic of scientific research denied the name of philosophy to anything not corresponding to their own tradition. However, this kind of dogmatism has never been able to survive for the simple reason that its self-assuredness is dependent upon the denial of its factual relativity in the historical context of a plurality of existing philosophical orientations. In the long run, the absolutist claim of a philosophical dogmatism necessarily becomes historically relativized. Its intentional blindness to its historical position is bound to give way in the course of philosophical debate. The recent history of the philosophy of science from the orthodoxy of the Vienna Circle to the pragmatism of the community of researchers and finally to the quasi-hermeneutic historicism of paradigm-change *à la* Kuhn offers an instructive example of the significance of this process.

Philosophy cannot abstract from its own historicity. This does not mean, however, that philosophy is nothing more than an intellectual expression of given circumstances that change from one epoch to the next. Rather, the awareness of historical conditions is the only way to escape the dominating influence of those conditions. Philosophy must continuously insist on its *theoretical autonomy*, but such an autonomy cannot be secured by assuming a position high above the relativity of the *hic et nunc*. A philosophy that is not jolted out of such a dogmatic slumber is bound to fall prey to the unexpressed presuppositions of what the actual tendencies in the social and intellectual world prescribe. It will unconsciously bear the mark of what goes without saying in the existing paradigm of theoretical activities. Only by reflecting upon these historical conditions can philosophy defend its legitimate claim to truth, a claim which reaches beyond historical relativity.

Paradoxically, then, this transhistorical claim is better supported by taking the undeniable historicity into account than by neglecting it. Cynical fatalism that believes in nothing but historical change is not the lesson to be derived from Hegel's perspective on philosophy through history. What can be learned is that the claim to truth is no inherited privilege, but can only be raised in a productive critical exchange with those variable conditions that are given in a broader historical context and that would seem to hinder autonomous philosophizing. Thus, the main presupposition of the autonomy of philosophy is the recognition of its limitations.

When effectively clarified, the historical conditions lose their strictly determining character. This does not imply that reflection liberates philosophy once and for all from its limitations. For, however great the penetrative power of philosophical reflection may be, what is in fact reflected upon is never a matter of free choice. The object is simply given and this givenness must be taken into account by every serious interest in philosophical autonomy. Despite continuous reflection upon a potentially infinite series of conditions, we shall never be able once and for all to break out beyond the boundaries they present.

Up to this point, I have discussed the historical factors and the dominating character of an epoch as the given conditions which initiate the work of philosophical reflection. However, this is but one aspect of the problems inherent in the dictum from which we set out: "Philosophy is its epoch comprehended in thought." If it is to be expected that a philosophy constituted by reflection upon its epoch can attain deeper insights than are otherwise available, one must still ask: What, then, does one come to know of an epoch once it has been *comprehended in philosophical thought*?

Clearly, this cannot be something already expressed in the common-sense interpretation of everyday life. Every epoch develops a certain self-understanding through which it gives itself a specific identity in relation to the past and to the future. An epoch is more than the sum of natural, economic, political and cultural data. It is the result of a more or less vague collective *process of interpretation.* At this level, nothing new is to be expected from philosophy. A philosophy that limits its activity to this level and which is thus reduced to a conceptual justification of the commonly held character of an epoch can be of no great theoretical value. Philosophical thinking has to bring forth insights other than those already held by current opinion. On the other hand, philosophical insights have to be clearly related to the epoch in question. They must formulate an aspect of an epoch that cannot be expressed in terms of the common self-understanding of that epoch.

The recognition of how the dominating factors of an epoch affect the philosophy emerging in it allows for a differentiation inaccessible to everyday understanding. Philosophical reflection is able to *distinguish surface-phenomena from underlying structures*, whereas the common understanding normally mixes them together in the melting-pot of popular opinion. More precisely, the differentiation between surface-phenomena and fundamental structures is non-sensical, as long as one moves within the sphere of the self-understanding of a given epoch. That which appears to be a predominant and essential trait at a certain historical moment may later be exposed as having been a mere vogue. Conversely, historical factors that are *prima facie* held to be of less importance may, in the course of philosophical reflection, be found to have a primary significance. The *immediate* view is not necessarily the most penetrating one.

The physiognomy of an epoch is in no sense evident and easy to grasp. Mere involvement in a historical situation is not a sufficient basis for a comprehension of that situation. The conceptual articulation of what is truly substantial in a manifold of historical phenomena presupposes more than contemporaneous participation in the general consciousness of an epoch. What is required is the view from a distance, whereby the observer is not ahistorically removed from the situation but is also not involved in its details. Only from such a standpoint is one able to uncover *what is essential* under that which is commonly held to be real. Only from such a standpoint is it possible to comprehend the essential traits, the long-range tendencies, in short, the actual make-up of a given society. Hegel entrusts this task of comprehension to philosophy. The question still remains open: Why philosophy?

Since philosophy is interested in its own paradoxical autonomy within the framework of an historical epoch, it is not interested in that epoch as such. Although its historical conditions are fetters which philosophy cannot cast off, they are only seen as *contingencies that must be relativized.* Of course the reduction of those factors to the status of mere contingency cannot dispose of history. It does, however, put an end to the exaggerated role ascribed to seemingly essential factors that temporarily eclipse other characteristics before they, too, yield to change. In other words, change, looked at as mere contingency, no longer has the power to influence the comprehension of an epoch, since it no longer obscures the fundamental structures that are of true historical relevance behind the veil of superficial phenomena.

The result of disinterest in the contingency of the moment over against the genuine philosophical interest in reason is that those factors in any given historical situation which resist the activity of rational elucidation prove to be the fundamental structures of an epoch. *What philosophy is unable to relativize to any greater degree, it must recognize* in the name of reason itself. This is precisely the meaning of Hegel's provocative motto of the actuality of the rational and the rationality of the actual.[35]

IV

I have not undertaken a rigorous philological exegesis of all the texts of Hegel that are relevant to our question. This could easily be carried out and would provide a coherent picture of the genesis of Hegel's idea of the task of philosophy *vis à vis* its epoch.[36] Rather, through a reconstruction of the argument underlying Hegel's claim for philosophy, I have attempted to demonstrate its plausibility.

We have finally arrived at the point where we can readdress the main question at issue: What is Hegel's *significance for the social sciences*? I am well aware of the difficulties of making Hegel's mode of philosophical thinking acceptable. A first step in this direction would be to cast doubt upon the self-assured objectivity of the methods of the social sciences. I have, with intentional indefiniteness, characterized the prevailing opinions in everyday consciousness, or the immediate picture of an epoch, as a reality that philosophy can and has to take into account. Science on the other hand, far from escaping this reality, is intricately bound to it. Science exerts a strong influence upon the formation of our everyday opinions, just as common opinion has an indirect effect upon the general orientation of scientific research. The process of 'scientification' of common, or public opinion becomes ever more pervasive, while, reciprocally, the obligation of science to be 'socially relevant' is increasingly emphasized. It would be simple-minded to believe that scientific research proceeds according to independent standards of rationality and methods of self-evident legitimacy.

Especially the social sciences have gained the authority of being a permanent and definite source of knowledge concerning all those questions everybody is interested in. No wonder, therefore, that sociologists, political scientists, economists, psychologists, psychoanalysts constantly throw themselves upon the market of public opinion. Privileged by their reputation as scientists, they are busy contributing to the picture every epoch constructs of itself. This symbiosis of the social sciences and common opinion is quite natural and should by no means be denounced in the name of some false purism. Instead one must turn, so to speak, to a higher court of appeal. Since the days of the *sophists*, philosophy has had to deal with this very problem. Philosophy has thereby learnt that it does not, in supposed contrast to the social sciences, enjoy a kind of splendid isolation. That was the impulse behind my effort to destroy the fiction of a *philosophia perennis* that purports to hover above the sphere of common life.

What in fact distinguishes philosophy from the social sciences although both are entangled in the self-understanding of an epoch is that the philosophical interest of reason is not primarily oriented towards the epoch and its immediately evident features. Philosophy does not so much aim at a knowledge of those economic, political and social factors relevant to the social sciences, but uncovers the structures of an epoch that, under given conditions, can be called rational. In other words, philosophy brings to light that which, independent of the historical moment, is worth considering on generally valid grounds. This philosophical claim to insight and comprehension can only be substantiated by the lasting significance of those structures which appear in a perspective that reaches beyond the contingency of historical change.

To summarize, then: even if one at first believes it is possible to circumvent the controversy between philosophy and the social sciences, one can hardly deny the importance of the philosophical accomplishment of reason that distinguishes the essential from the inessential within the framework of socio-historical reality. Furthermore, if one admits that the social sciences are thoroughly integrated in the self-understanding of an historical period, then the task remains of conceptually elucidating the character and lasting features of an epoch. Given the entanglement of the social sciences in what they should elucidate, philosophy has to play a critical role in the effort of discovering the essential structures of socio-historical reality. Even if philosophy can not appear on the scene as a *Deus ex machina*, she can help with the directing.

Tübingen University

NOTES

[1] T. W. Adorno, H. Albert, R. Dahrendorf, J. Habermas, H. Pilot, K. Popper: *Der Positivismusstreit in der deutschen Soziologie* (Neuwied; Berlin, Luchterhand, 1970). [*The Positivist Dispute in German Sociology*, tr. G. Adey and D. Frisby. London, Heinemann, 1976.] For the theories of the 'Frankfurt school' cf. one of the principal documents: M. Horkheimer, *Eclipse of Reason* (New York, Oxford University Press, 1947), and the more recent works by J. Habermas, *Knowledge and Human Interest* (Boston, Beacon Press, 1971); *Theory and Practice* (Boston, Beacon Press, 1973). See also P. Lorenzen, 'Szientismus vs. Dialektik, in: *Hermeneutik und Dialektik* (Festschrift for H.-G. Gadamer) ed. Bubner, Cramer, Wiehl (Tübingen, Mohr, 1970).

[2] "Philosophie ist ihre Zeit in Gedanken erfaßt" (*GPR*). A similar thesis can be found in *GP*, p. 85 and *WL* [L, 1951] **2**, p. 226.

[3] *The Open Society and its Enemies* [1945]. 4th ed. rev. 2 vols. (London, Routledge and Kegan Paul, 1962).

[4] Interestingly enough, at the same time as Popper wrote his book which branded Hegel as a precursor of Fascism, Herbert Marcuse sought to defend Hegel, from a Marxist standpoint, against exactly this charge, presenting him as a progressive theorist of freedom and critical reason in *Reason and Revolution* (New York, Oxford University Press, 1941).

[5] 'What is Dialectic?', *Mind, N. S.* 49 (1940); reprinted in *Conjectures and Refutations* (London, Routledge and Kegan Paul, 1976, pp. 312–335); *Open Society*, Vol. 2, p. 39; *Objective Knowledge* (Oxford, Clarendon Press, 1972) p. 126, 297. See my essay: 'Dialektische Elemente einer Forschungslogik', in: *Dialektik und Wissenschaft*, 2nd ed. (Frankfurt, Suhrkamp, 1974).

[6] Cf. Karl Löwith, *From Hegel to Nietzsche* (New York, Holt, Rinehart and Winston, 1964); D. McLellan, *The Young Hegelians and Karl Marx* (London, Macmillan, 1969).

[7] *Der Liberalismus und die Philosophie* (*Sämtliche Werke*. Vol. IV, Mannheim, 1847^2, p. 291); *Was wird aus der Religion? ibid.*, p. 250); *Über das Verhältnis von Philosophie, Politik und Religion* (*ibid.*, p. 273 ff.)

[8] Leading article of the *Rheinische Zeitung* of 14th July 1842; An exchange of letters from 1843 in the *Deutsch-französische Jahrbücher* (both in: K. M., *Frühe Schriften*, ed. Lieber (Stuttgart, Cotta, 1962, Vol. 1, p. 188, p. 450).

[9] Cf. L. D. Easton and K. H. Guddat (Eds.) *Writings of the Young Marx on Philosophy and Society* (Garden City, N.Y., Doubleday, 1967).

[10] See the interesting evaluation, from an analytic point of view, by G. v. Wright in *Explanation and Understanding* (Ithaca, Cornell University Press, 1971), chapters 1 and 4.

[11] *Der Aufbau der geschichtlichen Welt in den Geisteswissenschaften* (1910), *Gesammelte Schriften* 7 (Stuttgart, Göttingen 1958^2), pp. 148 ff. See the English translation by Rickman (Dilthey, *Selected Writings*, Cambridge, Cambridge University Press, 1976, Parts III and IV).

[12] Dilthey's target in this respect is J. S. Mill (*System of Logic*, Part VI, London, Parker, 1843) and H. T. Buckle (*History of Civilization in England*, Vol. I, Chap. 1, London, Parker, 1857–61).

[13] *Einleitung in die Geisteswissenschaften* (1883), *Gesammelte Schriften*, Vol. 1 (Leipzig Teubner, 1922), pp. 93 ff.

[14] *Gesammelte Schriften*, 4 (Leipzig, Berlin, Teubner, 1921), p. 249. – This volume contains Dilthey's 'Jugendgeschichte Hegels', which had the merit of stimulating the Hegel research of this century.

[15] *Geschichte und Naturwissenschaft* (Strasbourg, Heitz, 1894).

[16] Dilthey immediately protested against reducing the difference to the purely methodological, emphasizing instead the disparity between manifestations of the historical mind and facts of external nature (*Gesammelte Schriften*, Vol. 5, pp. 242 ff, 253).

[17] A Kantian concept. Cf. *Critique of Pure Reason*, A 424; *Critique of Judgment*, Introduction, 1st version (end), A 415.

[18] Because of his disavowal of 'historicistic' laws of history, Popper could only affirm Windelband's division. (Cf. *Der Positivismusstreit*, p. 118; *Open Society*, Vol. 2, pp. 264, 364; *The Poverty of Historicism* (London, Routledge and Kegan Paul, 1957) § 30). Historical sciences offer interpretations on the basis of a so-called 'logic of situations' which is contrasted with laws.

Philosophers of science such as E. Nagel and E. Zilsel deny that there is a clear differentiation and assert that one type blends into the other. (Nagel, 'Some Issues in the Logic of Historical Analysis', in: P. Gardiner (ed.), *Theories of History* (Glencoe, Ill., Free Press, 1959), p. 373 ff.) From the historian's point of view, Louis O. Mink makes a similar criticism ('The Autonomy of Historical Understanding', in: W. Dray (ed.), *Philosophical Analysis and History* (New York, Harper and Row, 1966), 165 ff).

As a matter of fact, Windelband himself referred to interference between historical description of individual occurrences and the use of general laws; thus, the objections prove nothing new.

[19] *Über den Begriff der Schönheit* (1845), in: H. Lotze, *Kleine Schriften* I (Leipzig, Hirzel, 1885), pp. 300 ff.; p. 333 f.; *Seele und Seelenleben* (1846, *ibid.*, Vol 2), p. 175.

[20] *Die Grenzen der naturwissenschaftlichen Begriffsbildung* (1896) (Tübingen and Leipzig, Mohr, 1902).

[21] Max Weber, *Die Objektivität sozialwissenschaftlicher und sozialpolitischer Erkenntnis* (Tübingen, Mohr, 1904); trans. by W. J. Goode as *The 'Objectivity' of Sociological and Socio-political Knowledge* (s. l., 1941). *Der Sinn der 'Wertfreiheit' der soziologischen und ökonomischen Wissenschaften*, 1917/18. Trans. by E. A. Shils and H. A. Finchas as *Methodology of the Social Sciences* (Glencoe, Ill., Free Press, 1949). Cf. W. Runciman, *A Critique of M. Weber's Philosophy of Social Sciences* (Cambridge, Cambridge University Press, 1972).

[22] Rickert, *Grenzen der Naturwissenschaftlichen Begriffsbildung* (Tübingen, Mohr, 1921), XVII, 139, 398. The proclamation of a Hegel-renaissance already appears in: *Kulturwissenschaften und Naturwissenschaften* (Freiburg, Mohr, 1899), p. 13; cf. W. Windelband, 'Erneuerung des Hegelianismus', *Sitzungsberichte der Heidelberger Akademie der Wissenschaften*, 1910. The *kulturwissenschaftliche* interpretation of Hegel is also clear in the work of Rickert's student, Emil Lask: *Fichtes Idealismus und die Geschichte* (1902) in: E. L. *Gesammelte Schriften* I (Tübingen, Mohr, 1923), p. 17 *passim* (esp. the Introduction: *Die Logik des Wertens in der Geschichtsphilosophie des deutschen Idealismus*); *Hegel und sein Verhältnis zur Weltanschauung der Aufklärung* (1905), *ibid.*, p. 340 ff.

[23] *Soziologie als Wirklichkeitswissenschaft* (Leipzig, Berlin Teubner, 1930), p. 211. (Reprinted Stuttgart, Teubner, 1964.)

[24] *Ibid.*, p. 5; cf. p. 11.

[25] *Ideologie und Utopie* (1929). (Frankfurt, Schulte-Bulmke, 1965) p. 62 ff. (*Ideology and Utopia*, trans. L. Wirth and E. Shils. London, K. Paul, Trench and Trubner, 1936.)

[26] *Ibid.*, p. 132 f.

[27] *Ibid.*, p. 142.

[28] M. Horkheimer, 'Ein neuer Ideologiebegriff?', *Archiv für die Geschichte des Sozialismus und der Arbeiterbewegung* **15**, 1930; T. W. Adorno, 'Das Bewußtsein der Wissenssoziologie', in: *Prismen* (Frankfurt, Suhrkamp, 1955); both reprinted in K. Lenk (ed.), *Ideologie*. 2nd ed. (Neuwied, Berlin, Luchterhand, 1964).

[29] *Open Society*, Chap. 13.

[30] Cf. the critical remarks of Popper and Adorno on the sociology of knowledge: *Der Positivismusstreit,* p. 111 ff.; p. 136 f.

[31] Compare, however, K. Popper, 'On the Theory of Objective Mind', in *Objective Knowledge.*

[32] *Werke* I, ed. H. Glockner (Stuttgart, Frommann, 1927–30).

[33] K. L. Reinhold's reflections on time and philosophy provided the occasion for this discussion: *Beiträge zur leichteren Übersicht des Zustandes der Philosophie beim Anfang des 19. Jahrhunderts* (Hamburg, Perthes, 1801–3); *Der Geist des Zeitalters als Geist der Philosophie*, in: *Neuer Teutscher Merkur* (ed. C. M. Wieland) 1801.

[34] *Phän* [H, 1952], Preface, p. 31; *Phen* [B] p. 95. Cf. my study, 'Problemgeschichte und systematischer Sinn der *Phänomenologie* Hegels' in *Dialektik und Wissenschaft* (Frankfurt, Suhrkamp, 1974).

[35] *PR*, Preface.
[36] Cf. J. Ritter, *Hegel und die französische Revolution* (Frankfurt, Suhrkamp, 1965); J. Habermas, '*Nachwort*' to: Hegel, *Politische Schriften* (Frankfurt, Suhrkamp, 1966); S. Avineri, *Hegel's Theory of the Modern State* (London, Cambridge University Press, 1972), Chap. VI.

MURRAY GREENE

HEGEL'S CONCEPTION OF PSYCHOLOGY

The present paper seeks to understand Hegel's notion of a philosophical or 'scientific' psychology, these terms being synonymous for him. First the paper notes Hegelian criticisms of various approaches to psychology, whose diverse failings, Hegel often says, derive from a proceeding through "external reflection," or the "reflective understanding."[1] External reflection, according to Hegel, is a philosophizing from the standpoint 'outside' the *Sache selbst*, and we may say that in Hegel's view, external reflection constitutes the main obstacle toward a philosophical psychology. With regard to such a psychology, external reflection takes on an especially problematic form in post-Cartesian philosophy as a philosophizing from the subjective consciousness. For when such a philosophizing turns to consciousness itself, should not its procedure be *eo ipso* immanent to the *Sache selbst*? In fact, says Hegel, it is not; rather its procedure in dealing with the soul or mind or consciousness is but a special form of external reflection. It is this post-Cartesian form of external reflection, particularly in its most challenging embodiment in the Kantian Critical philosophy, that Hegel's approach is designed to overcome. The attempt to elucidate the nature of the problem here forms the second part of our paper.

In the third part we try to show how Hegel purportedly overcomes the barrier of philosophizing from the subjective consciousness by placing consciousness itself within a deeper principle of subjectivity, the notion of 'subjective spirit.' Psychology is presented by Hegel as the third and culminating science of subjective spirit, following the sciences of anthropology and phenomenology. We thus have a triad of sciences corresponding to the three moments of subjective spirit in its notion. While each science has its own province and to a certain extent its own principle, the three sciences are internally linked. The science of anthropology demonstrates an 'emergence' of consciousness through the working of a principle prior to and deeper than consciousness. Through this same principle – namely, spirit as identity in opposition of substance and subject – the science of phenomenology demonstrates the overcoming of the subject-object dualism in consciousness by a necessary movement within consciousness itself. Only at this vantage point of the third moment of subjective spirit, only now that the subject-object dualism has been overcome, is a psychology possible that is not an

R. S. Cohen and M. W. Wartofsky (eds.), Hegel and the Sciences, 161–191.

external reflection but a philosophic science of "spirit as such." The fourth and last part of our paper will attempt to indicate the nature of the treatment in Hegel's philosophical psychology.

Our paper is based mainly on Hegel's doctrine of subjective spirit as presented in the third part of the *Encyclopedia of Philosophical Sciences.*[2] Since we view Hegel's effort especially in terms of resolving a particular problem, we shall deal with the psychology itself only in its general plan and idea. A more detailed exposition of specific features must await another occasion.

I

According to Hegel, a philosophical psychology had not emerged prior to his own presentation of subjective spirit in the *Encyclopedia*. Virtually dismissing all modern approaches as unscientific, Hegel tells us that the Aristotelian treatises on the soul remained up to his own day the sole works of merit in psychology, and that he hoped to rekindle the Aristotelian speculative approach by his own Philosophy of Spirit (*PM*, p. 3). Hegel's sweeping rejection of all post-Cartesian efforts in this field indicates that for him some principle was at issue, and not merely that the great moderns – Descartes, Spinoza, Locke, Hume, Kant – happened to be poor psychologists. Despite the advance of thought in modern times, despite the acuity of these thinkers, a false tack had somehow been taken that set psychology back to a position not even equal to its status in Aristotle. What was this false tack?

While modern philosophy's proceeding through the subjective consciousness marked a great advance, nevertheless, according to Hegel, this procedure had to result in an impasse as long as the method native to its principle of subjectivity remained external to that principle itself. As long as subjective philosophizing continued unconscious of its proper methodology, its principle that nothing is to be accepted as true which does not possess inward evidence in consciousness turns into the very barrier that separates consciousness from a knowledge of itself. But had not the very birth of the new philosophy meant a new stress on method, a new insistence on demonstration according to the inward evidence of consciousness?

According to Hegel, all the new methods that were put forward beginning with Descartes were modes of external reflection. The Cartesian 'new beginning', notes Hegel, was born in the abstract resolve of the subjective consciousness to accept no outer authority but "produce everything oneself and accept only one's own deed as what is true" (*Phen* [M], p. 50). But this approach, says Hegel, had not yet "equipped itself" even to grasp truly its "own deed." In proceeding straightway to criticize and test, it employed canons that were uncritical and untested – such as its standard of "clear and distinct," which, says Hegel, is a "psychological" (in the unphilosophic sense) rather than a "logical" criterion.[3] The fact that one's opinions come from one's self does not mean that truth has taken the place of error. Lacking the proper method of showing how its certainty becomes truth, Descartes' proceeding through the inner certainty of the subjective consciousness turns out to be but a form of external reflection.

The post-Cartesian methods of synthesis and analysis, or composition and resolution, Hegel maintains, are likewise modes of external reflection. Analysis dissolves the concretum of experience into lifeless abstractions. Synthesis presents the universal in definitions that have no "constraining necessity" for the content, and is a "purely subjective cognition which is external to the object" (*L* [W, 1892], p. 366). The method of demonstration taken from mathematics is unsuitable to philosophy in general and particularly inappropriate to a science of mind (*Phen* [M], p. 24ff). Spinoza's demonstration *more geometrico*, according to Hegel, is the "fundamental defect in the whole position" (*HP* 3, p. 283), and Spinoza tells us that substance is the unity of thought and extension "without demonstrating how he gets to this distinction, or how he traces it back to the unity of substance" (*L* [W, 1892], p. 275). Leibniz was perhaps the only pre-Kantian thinker who put forward a claim which, as we shall see, is also central to Hegel's own notion of subjective spirit: namely, that self-consciousness arises as an internal development of the preconscious subjectivity. But here too, says Hegel, the philosopher demonstrated no immanent necessity. While the Leibnizian monad evolves its representations "out of itself," says Hegel, "they arise in the monad like bubbles; they are indifferent and immediate over against one another and the same in relation to the monad itself" (*SL* [M], p. 396). In the so-called "method of construction" – the method that "construes its notions" – there is, Hegel grants, a "dim consciousness of the Idea, of the unity of the Notion and objectivity" insofar as, in demonstration, the subjectivity "shows itself a modifying and determining principle" (*L* [W, 1892], p. 371). But in construction the demonstration is "ruled by an external end." The proof is not truly a "*genesis* of the relationship that constitutes the content of the theorem," and therefore it too is "an *external* reflection" which proceeds "from without inwards," "infers from external circumstances the inner constitution of the relationship" (*SL* [M], p. 812).

While these Hegelian criticisms of external reflection apply to philosophic method generally, Hegel offers a number of criticisms directed specifically to certain approaches to psychology. The so-called "knowledge of man" that supposedly probes the secrets of the human heart, says Hegel, is a symptom of the modern preoccupation with the self in its particularity (*PM*, p. 1). Where it does not indulge in self-analysis, it tends to view men in general in terms of how they 'react' under different circumstances, and, as in the pragmatic view of history, points out how great historical figures were after all motivated by petty passions. This sort of *Menschenkenntnis*, says Hegel, does not deserve serious attention as science. Because it has no inkling of the Notion, it cannot attain to the universal and substantial in man; its effect is rather to reduce him to a bundle of motivations and reaction patterns.

At least the old pneumatology or rational psychology sought to view man in terms of universal determinations. This metaphysic of the soul proceeded by inquiring whether the soul is 'simple', 'selfsame', a 'substance'. Kant rightly rejected this kind of psychology, says Hegel. But Kant was right for the wrong reasons (*L* [W, 1892], p. 97). The defect of the old metaphysical psychology was not, as Kant maintained, that it argued paralogistically from empirical phenomena to pure thought categories. Rather, says Hegel, its categories themselves were of such a kind as "neither can nor do contain truth" (*L* [W, 1892], p. 66ff). The old psychology approached the soul as a thing, and Kant did well to rid philosophy of the soul-thing. But this metaphysic deserved to be cast out not because its categories 'soared beyond' the powers of reason but because it never attained to reason's speculative use. Far from being a static *ens*, says Hegel, the soul is absolute restlessness, pure activity, "the negating or ideality of every fixed category" of the understanding (*PM*, p. 3). The soul is not abstractly simple but self-differentiating in its simpleness; it is not a selfsame essence hiding behind its manifestations but only actual in its manifestations. Hence in inquiring whether the soul is simple or composite, immaterial or material, selfsame or not selfsame, pneumatology was not so much arguing paralogistically in Kant's sense but rather employing the rigid dichotomies of the reflective understanding in a way that can never comprehend spirit as displaying "contradiction in its extreme form," as the "absolute unity of opposites in the Notion" (*SL* [M], p. 776).

As against the empty abstractions of the old metaphysic, the empirical psychology that accompanied the rise of the sciences sought a 'solid footing' in experience. But in its relating itself passively to a "prefound existent confronting it" (*ein vorgefundenes ihm gegenüberstehendes Seiendes*) (*L* [W, 1892], pp. 364–365), empiricism bars itself from a genuine knowledge of spirit, whose essence as ideality is such that it does not let itself be thus 'found'.[4] In its analysis of the 'given', empiricism means to introduce nothing but its own act of separating. But its breakdown of the concrete object of experience only leaves it with abstract universal attributes, and thus brings in again a metaphysic of the "thing and its properties" (See *Phen* [M] p. 77ff). Yet empiricism remains all the while oblivious to its own metaphysical presuppositions and employs its thought categories in an "utterly thoughtless and uncritical" fashion. In separating the given into "simple ideas" and the mind into 'faculties' and 'forces,' empiricism renders the living totality an aggregate of "parts" and the mind a "skeleton-like mechanical collection" (*PM*, p. 189).

In reconstructing the concretum of experience and arriving at the higher connections of mind, empiricism proceeds essentially by showing a 'natural evolution' of what happens in "pictorial and phenomenal thinking" (*SL* [M], p. 588). But philosophy, says Hegel, is not meant to be a 'narration' of what happens but a cognition of 'what is true' in happenings. This sort of genetic method, whether showing an origin of ideas or of faculties, Hegel argues, carries a built-in limitation of knowledge to sensuous experience. For when there is a question of history rather than truth, then the "substrate of feelings and intuitions" from which understanding abstracts its universals need not be seen as sublated in conceptual thinking but "remains for representation in the same complete reality with which it first presented itself".[5] In the so-called laws of association of ideas as employed by empiricism, some principle is assumed, such as a "force of attraction" of like images. But such 'attraction,' Hegel points out, would just as well have to be a 'repulsion,' a "negative power of rubbing off the dissimilar elements against each other" (*PM*, p. 208).

Hegel also rejects any procedure through the 'facts of consciousness', whether employed by empiricism or some form of intuitionism, as in Jacobi or Fichte. If the criteria of truth are to be taken in "the immediacy, or self-evident way, in which a fact or body of truths is presented in consciousness," says Hegel, then "every alleged truth has no other basis than the subjective certitude and the assertion that we discover a certain fact in our consciousness." What I find in my consciousness, I can exaggerate into a "fact of the consciousness of all" and even pass off "for the very nature of consciousness" (*L* [W, 1892], p. 134). If indeed we are to examine the nature of consciousness, says Hegel, then we must be able to strip it of its particular and accidental elements and "by the toilsome operation of reflection" disclose "the universal in its entirety and purity."

If we are not then, like the old metaphysic of the soul, to speak of the mind in abstract thought categories; if we are not to be deceived by empirical psychology's spurious concreteness, how do we obtain the categories proper to the nature of mind? If we are not to employ forms of demonstration alien to the subject matter, or philosophize directly from the so-called facts of consciousness, where do we discover the method proper to mind in its notion? According to Hegel, the method that can grasp mind as at once "absolute restlessness" and abiding self-identity, that can demonstrate the development of spirit as an immanent self-unfolding, is the "speculative method." This method proceeds according to the "logical Idea" and derives its categories from the "form of the Notion," which is immanent to the *Sache selbst* (*SL* [M], p. 830; p. 826).

But what sort of enterprise will yield the "form of the Notion" and the "logical Idea"? To answer this question adequately would require a lengthy excursus into Hegel's overall conception of the philosophic sciences as a system. For the purposes of understanding Hegel's overcoming of the kind of post-Cartesian external reflection that bars the way to a philosophical psychology, perhaps our most direct route is through Hegel's criticism of the Kantian critical enterprise.

II

Hegel's criticism of those approaches which wield their thought categories in a "thoughtless and uncritical" manner has the ring of the Critical philosophy's critique of dogmatism. In Kant's view the dogmatists were engaged in external reflection inasmuch as the "dogmatic procedure of pure reason" advances straightway upon its subject matter "without previous criticism of its own powers" (*CPR*, B xxxv; B xxxvi). To make use again of a Hegelian expression, we can say that the dogmatists were unaware that "subjectivity shows itself a modifying principle" (see p. 163 above). But if any approach oblivious to the role of subjectivity must remain dogmatic, what is the proper critical approach to subjectivity itself? In Hegel's view, Kant's critique of subjective cognition as an 'organ' of knowledge is itself a form of external reflection. By first investigating cognition as an 'instrument' or 'medium' through which the truth reaches us, says Hegel, this form of critique already presupposes a distinction of ourselves from this knowledge.[6] The demand that thought examine its capacity for knowledge is a "fair enough" demand, says Hegel. But the problem is, how shall this examination proceed? We can test an ordinary instrument before we put it to use. But with what can we criticize cognition except another cognition (*L* [W, 1892], p. 17)? If thought is to investigate itself, says Hegel, the action of the forms of thought must be their own criticism. The forms of thought must "examine themselves," and the critical work of thought, "instead of being brought to bear upon the categories from without," must remain "immanent in their own action." Hegel's first application of this method of immanence – or, as he terms it, the "speculative" or "absolute" method – was in the *Phenomenology of Spirit* of 1807.

The *Phenomenology* of 1807 was not meant by Hegel – the mature Hegel at least (see below, p. 175) – to constitute an absolute beginning of philosophy or indeed of his own system as such. It is not a beginning without important presuppositions. In commencing with the 'natural consciousness,' for example, the notion of consciousness is in a sense presupposed. We might also say the commencement is historical, inasmuch as the very term 'natural' or 'empirical' consciousness is hardly conceivable prior to the empirical thinkers and Kant's own distinction of the empirical and the pure consciousness. Kant wanted to go behind the empirical consciousness to its *a priori* forms, but his presentation of the forms of the *a priori*, according to Hegel, was "solely based upon psychological and historical grounds" (*L*, [W, 1892], p. 84). Hegel in the *Phenomenology* purports at once to take the empirical consciousness on its own terms and allow it to unfold its pure forms in the concrete context of its movement of 'experience.' In this way the forms of thought are seen to "test and examine themselves." Each form of thought, as a shape of experience, succeeds its predecessor as its "determinate negation," while the philosophic observer need only "look on" (*Phen* [M], p. 54). In fact the philosopher does more than look on.[7] He assembles and organizes the shapes of consciousness so as to bring out their necessary movement from the natural to the philosophic consciousness (*Phen* [M], p. 55; p. 480). But insofar as the philosopher thereby demonstrates the necessity of his own emergence in the movement of consciousness, he may be said to overcome the standpoint of external reflection.

Through the pathway of experience demonstrated in the *Phenomenology* of 1807, consciousness has purportedly attained to the standpoint of pure science. In this sense the *Phenomenology* is an introduction to the system of philosophic sciences and in the first instance to the science of logic. What is most important for our purposes is to note Hegel's stress on the fact that the *Phenomenology*, by traversing the shapes of experience of consciousness, has made possible that 'liberation' from consciousness which is the precondition of all philosophic science.[8] The standpoint of consciousness as such, we can now say, is a standpoint of external reflection. For reasons both historical and systematic, it had to be overcome – but not by another external reflection. The great truths of transcendental philosophy, which were only expressed by the transcendental thinkers in forms of external reflection, have been concretely demonstrated in the *Phenomenology*: that self-consciousness is the truth of consciousness (*Phen* [M], p. 102), that ego is Notion (*SL* [M], p. 583); that the category is not merely subjective but rather thinking actuality (*Phen* [M], p. 142). The *Phenomenology* has hereby yielded the 'notion' of the science of logic (*SL* [M], pp. 68, 49). In the overcoming of every shape of experience where thought and Being are held separated, the *Phenomenology* has led to that science of "pure essentialities," the forms of the Notion, which are at once determinations of substance and subject, Being and thought. Henceforward in the system of philosophic sciences – in logic as well as the concrete sciences of nature and spirit – the movement of thought in demonstration is no longer in terms of "specific shapes of consciousness" but in "specific notions" and as the "organic self-grounded movement" of these notions (*Phen* [M], p. 491).

The speculative method, which is the overcoming of external reflection, obtains its first application in the *Phenomenology* of 1807 and its first exposition in logic (*SL* [M], p. 53). In the realm of consciousness the *Phenomenology* brought forward the simple logical principle "necessary to achieve scientific progress": that "the negative is just as much positive." In the succession of shapes of consciousness demonstrated in the *Phenomenology*, each "has for its result its own negation – and so passes into a higher form." Hereby the supremely important principle emerges that "what is self-contradictory does not resolve itself into a nullity" but is rather a "negation of a specific subject matter" and therefore a "specific negation" and "has a content." In this way, each succeeding determination is "higher and richer than its predecessor" inasmuch as it "contains it but also something more," and is "the unity of itself and its opposite." Where the *Phenomenology* showed this movement in the realm of consciousness, the science of logic shows it in "the simpleness of knowing." Logic demonstrates the movement of the Notion "in this pure ether of its life."

In the science of logic, all those thought categories wielded in a "thoughtless and uncritical" manner by external reflection obtain elucidation in their aspect as pure thought forms. Their intrinsic limits are exhibited so that they may receive their proper roles in the concrete sciences. The categories 'parts' and 'whole,' for example, are shown to express the "external and mechanical relation" appropriate to mechanism, while they are seen to be unsuitable to organism and even more inappropriate to mind or spirit.[9] The same holds true for 'thing' and 'property' (*SL* [M], p. 498), for 'quality' (*L* [W, 1892], p. 171), 'force' (*L* [W, 1892], p. 247ff) and other categories uncritically employed by external reflection in the realm of mind. At the same time such categories as identity and difference, substance and accident, appearance and essence are shown to be appropriate to spirit as reflecting into one another in their opposition, and not – as they were taken by *Verstand* – as fixed opposites.

Since the categories of the logic are shown to derive from the immanent movement of the Notion as inner self-differentiation, the speculative method combines analysis and synthesis. It does so, however, not in "bare juxtaposition or mere alternating employment" (*L* [W, 1892], p. 376) but by showing that difference is nothing other than the necessary opposition contained in all determinate thought. Hence the true *arthra*, which were demonstrated by the methods of analysis and synthesis in a movement of thought external to the demonstrandum, now emerge as a movement of thought immanent to the *Sache selbst*. This movement of the pure Notion or logical Idea is at once a self-externalization and inwardization, a 'forthgoing' (*Äussersichgehen*) that is simultaneously a 'withdrawal inwards' (*In-sich-gehen*) (*SL* [M], pp. 840–841). The movement in the logic, as in the system of philosophic sciences as a whole, is a circle, a coming back to self of the Notion. Each step away from the 'beginning' is also a 'getting back to it,' so that the "retrogressive grounding of the beginning" and its "progressive further determining of it," in truth "coincide and are the same" (*SL* [M], p. 841). This principle, demonstrated in the logic in the "simpleness of knowing," will prove to be important in the concrete sciences of subjective spirit. In the science of psychology, for example, such determinations as intuition, imagination, memory will not appear as 'parts,' 'faculties,' or 'forces' but as 'moments,' each of which is also a definition of mind in its living totality, just as in logic each form of the logical Idea "is only a closer determination and truer definition of the Absolute" (*L* [W, 1892], p. 162). In the doctrine of subjective spirit as a whole, we shall see spirit as 'natural soul' in the anthropology dividing itself into ego and object in the phenomenology and returning to itself in its higher unity and truth as reason in the psychology. It is this speculative method alone that is appropriate for a science of psychology, Hegel claims, since it "does not behave like external reflection" but "takes the determinate element from its own subject matter, since it is itself that subject matter's immanent principle and soul" (*SL* [M], p. 830).

But if the speculative method shows us the proper categories and method of demonstration for the sciences of mind, how specifically does it overcome the external reflection built into a philosophizing from the subjective consciousness and its subject-object antithesis? Here again we may begin with Kant, whose denial of the possibility of a pure science of psychology in a way sums up the 'false tack' taken by post-Cartesian philosophy even while its notion of the subjective consciousness marked the great advance of modern thought.

While, according to Kant, the original identity of self-consciousness is the highest principle of knowledge, the pure ego or self as it is 'in itself' must remain for us a 'transcendental *x*.' All our certain knowledge is a knowledge of "objects" and requires the possibility of a sensible intuition. Since I can have no such intuition of myself as pure ego, I can have no knowledge of the "identity of the subject" whereby it can be "given as object" (*CPR*, B 408).[10] Further, "I cannot know as an object," says Kant, "that which I must presuppose in order to know any object" (*CPR*, A 402). The pure ego, which is ever the subject in judgment, cannot make its own self the object of its judgment without an awkward circularity. For these and other reasons, Kant says, "the hope of succeeding by *a priori* methods" in a science of the soul must be "abandoned" and psychology "banished from the domain of metaphysics" (*CPR*, A 849-B 877). Kant relegates psychology to a position within anthropology, where it can play a role as "the pendant to the empirical doctrine of nature" (*CPR*, A 849-B 877). Such a psychology would deal, for example, with the empirical laws of association in the reproductive imagination (*CPR*, B 152). In his own writings and lectures on anthropology, education, and applied ethics, Kant offers such empirical psychological material. But by the critical philosophy's own canons, these treatments do not rank as theoretical science in the sense of apodictically certain knowledge, and do not belong to the science of the *a priori* principles of knowledge.[11]

According to Hegel, the Kantian notion of a 'transcendental *x*' is an external reflection that derives from a philosophizing from the standpoint of consciousness as an absolute. The Kantian philosophy, in Hegel's view, is a phenomenology, since it takes spirit only in its aspect as consciousness (*PM*, p. 156). For this reason it does not even do justice to the truth of consciousness. That Kant's restriction of knowledge to sense experience is untenable, Hegel claims to have demonstrated in the first three chapters of the *Phenomenology* of 1807. We shall pass this discussion by in order to focus on Hegel's critique of the supposed circularity of a knowledge of the 'transcendental *x*.'

It is ridiculous, says Hegel, to stigmatize as circular – as though it implied a fallacy – "this nature of self-consciousness, namely, that the 'I' thinks itself, that the 'I' cannot be thought without its being the 'I' that thinks" (*SL* [M], p. 777). It is not as though, says Hegel, we "must already make use" of the ego whenever we want to judge about it. The ego does not 'make use' of something like an instrument – even its own self – in judging. Ego *is* the subject-object relation (*PM*, p. 151ff), *is* judgment. And its actively performed judgments in the theoretical, practical, and other realms are precisely what render it concrete and knowable. For its every judgment of and relation to an other is at the same time a self-relation. It is in just this self-relation in all relating to other, says Hegel, that ego reveals itself as Notion, as "the absolute relation-to-self that, as a separating judgment (*Urteil*), makes itself its own object and is solely this process whereby it makes itself a circle" (*SL* [M], p. 77). In this statement, I believe, we have the key to Hegel's organization of the sciences of subjective spirit and the possibility of a philosophical psychology.

According to Aristotle, the soul attains truth in affirming or denying a predicate of a subject, that is to say, in judgment. But in Aristotle the problem of judgment can hardly be said to be viewed in terms of subjectivity and objectivity in the modern sense. One may measure the Kantian revolution in philosophy by Kant's definition of 'objective judgment.' Such a judgment, Kant says, "is nothing but the manner in which given modes of knowledge are brought to the objective unity of apperception. This is what is intended by the copula (*Verhältniswörtchen*) 'is' " (*CPR*, B 142). Hereby Kant makes 'objectivity' a function of the subjective consciousness: apperception. This, as we shall see more specifically later, may also be said to be the case for Hegel. In Hegel, however, it has a different sense. In the first place, according to Hegel, the copula 'is' is not to be understood only as a function of the subjective consciousness as apperception. "The copula 'is' springs from the nature of the Notion, to be self-identical even in parting with its own" (*in seiner Entaüsserung*) (*L* [W, 1892], p. 298). It is not only consciousness but 'the Notion' that distinguishes itself into subject and predicate; and it is the forms of this distinguishing that provide the logical categories. Hegel makes use of the German word *Urteil*. Ordinarily in judgment, says Hegel, we view subject and predicate as independent extremes which we somehow connect by the copula 'is.' However, "the etymological meaning of the judgment in German goes deeper, as it were declaring the unity of the Notion to be primary, and its distinction to be the original partition." And that, says Hegel, "is what the judgment really is" (*L* [W, 1892], p. 297).

But while the science of logic demonstrates the pure forms of the *Scheidung in sich* of the Notion, it is up to the concrete sciences of subjective spirit to show that *ego* is this very *Urteil.* In the doctrine of subjective spirit, Hegel demonstrates consciousness itself to be the 'original partition' whereby subjectivity comes to hold itself 'in relation' to its content as a self-subsistent object and thereby to speak of the latter in terms of the *Verhältniswörtchen* 'is'. In the science of anthropology, Hegel purportedly shows the 'emergence' of consciousness and the subject-object 'relation' as an *Ur-teil* of a primary unity, that of soul and body. In the course of the movement of consciousness, this split is healed and, as demonstrated in the science of phenomenology, spirit returns to itself and knows itself as reason. Hereby Hegel overcomes the Kantian argument against circularity – the argument that the 'I think' cannot make itself object of its own judgment. If Hegel is successful, we can discard the notion of a 'transcendental *x*' as a figment of an external reflection. Thus in Hegel's approach, the initial possibility of a philosophical psychology consists in showing the emergence of the subject-object dualism as an *Urteil* of the original soul-body unity.

III

Anthropology, the first of the three sciences of subjective spirit, deals with the natural foundation of man, spirit as first arising out of its 'immersion' in physical nature. A main theme of the anthropology is the soul-body relationship. The demonstration in the anthropology shows how the soul (*Seele*) as preconscious subjectivity comes to power in its corporeality (*Leiblichkeit*). Having permeated the latter and rendered it the outer 'sign' (*Zeichen*) of the soul's own spiritual inwardness, the psychical subjectivity 'lets go' (*entlassen*) the pre-objective content of sensation and feeling to become an objective outer world. This development, according to Hegel, is the genesis at once of the ego of consciousness and of its content as an independent externality to which the ego relates itself in its knowing.

In discussing the question of how a soul-body unity is possible, Hegel says the problem remains unanswerable for any standpoint on which the object as such already has the shape of a prefound existent standing opposite it (*PM*, p. 30). As long as the subjective consciousness is related to its outer object – or the immaterial to the material – as one particular to another, says Hegel, the mind-body relation must remain a mystery. But the posing of the question of a 'community' of two such particulars is an external reflection. In the Philosophy of Nature, Hegel has purportedly shown how natural materiality sublates itself to spirit. In the sentient animal soul, which feels itself a one at every point in its corporeality, and in the sublation of the singular animal life in the abiding genus, says Hegel, nature has overpassed itself as self-externality and the realm of the *partes extra partes.*[12] But in this spatial and temporal asunderness consists the very essence of materiality in its notion. Spirit, as individual natural soul, has emerged as the 'result' of nature and at the same time as nature's own 'truth' and 'absolute *prius*' (*PM*, p.8).

The individual soul in its identity with its corporeality, according to Hegel, is the 'microcosm' in which the macrocosm of universal nature is 'concentrated' and sublated in its asunderness (*PM*, p. 36). My body, says Hegel, is the "mean" (*Mitte*) whereby I "come together with external nature in general."[13] This is exhibited in Hegel's treatment of sentience (*Empfindung*). Hegel seeks to show that the five senses comprise a system of specified 'bodiness' (*Körperlichkeit*) which contains the essential moments of external body or materiality in its notion.[14] Thus sight reveals materiality in its moment of "pure ideality," light; touch, the moment of "solid reality."

The transient content of the 'sentient soul' attains a higher form in the 'feeling soul' (*fühlende Seele*), whose manifold as a feeling-life comes to it in the form of dream and presentiment. Hegel here deals with a realm of the psyche that later came to be explored by depth psychology. The psychical subjectivity is termed by Hegel a self-enclosed 'monad,' since its content does not yet come to it in the form of an objectively structured outer world. The 'immediate' knowing of this inner subjectivity is evidenced in forms of trance-like 'gazing-knowing' (*schauendes Wissen*) of an inner content not yet mediated by such objective categories as space and time, cause and effect. The phenomena of 'animal magnetism' are seen as displaying a psychical union (*Seeleneinheit*) of two subjectivities where the passive feeling selfhood of the patient is permeated by the dominating subjectivity of the hypnotist, as in a kind of foetus-mother relationship. Sicknesses of insanity are discussed by Hegel as forms of 'self-feeling' (*Selbstgefühl*) where a particular content of passion has taken on the spurious objectivity of a fixed idea, thus forming a self-sunderance and 'blocking' (*Hemmung*) of the psychical subjectivity in its fluid universality. In 'habit', which is the concluding moment of the feeling soul, the subjectivity sublates the particular corporeal determination and comes back to itself in its fluid universality. Hereby the soul wins release from its immediate corporeal affections and frees itself for occupation with other things. In reducing the corporeality to an 'instrument' of its 'subjective purpose', the soul has raised itself to 'subjective substance' and developed itself as the 'foundation' of consciousness.

The soul 'actualizes' itself as ideality of the body by so 'informing' (*einbilden*) it with its subjective inwardness that the body is but the soul's own outer 'sign'. In stance and attitude, in gesture, demeanor, and mobile play of features, the body has become the soul's own 'work of art.' The body stands not for itself but is the externality wherein the spiritual inner "feels itself and makes itself felt," the *Dasein* of the inner subjectivity. Having attained to its being-for-self as actual soul, the subjectivity 'separates' itself from its immediate being, sets itself opposite (*gegenüberstellen*) its own being as corporeality while sublating it and determining it to its own. In its very externality the soul is hereby 'inwardized' (*erinnert*) within self and infinite self-relation. This being-for-self of the free universality, says Hegel, is the 'awakening' of the soul to ego: to the "abstract universality insofar as it is for the abstract universality" (*PM*, p. 151). Subjectivity is no longer immersed (*versenkt*) in the immediacy of 'feeling' but is now 'thinking' and subject 'for itself.' More particularly, says Hegel, it is subject of the *Urteil* in which ego excludes its formerly monadic content as an 'object.' To this the subject now relates itself in such a way that in it, it is directly "reflected within itself," is consciousness (*PM*, p. 151).

Thus where modern philosophy from Descartes to Kant began on the level of consciousness, Hegel demonstrates consciousness as a 'result.' This means that a deeper principle has been exhibited than that of consciousness and its subject-object relation. Indeed it is a principle that first makes that relation possible. In the anthropology the body was 'object,' the soul 'subject'; the body was the 'in itself,' the soul the 'for itself'; the body was substantiality, the soul subjectivity. In its rendering the body an 'outer' of its own spiritual 'inner,' the subject has posited itself opposite itself and in this its very opposite is reflected back into itself. It is this that constitutes the subject an 'ego' and its content 'objective.' Thus we can say that for Hegel as for Fichte, ego is the act of positing itself. But to exhibit this, Fichte called upon an 'intellectual intuition' and needed the *Anstoss* from without, while Hegel demonstrates it in the form of an immanent development of the soul's relation to its corporeality as its natural substantiality. The demonstration of the emergence of consciousness as an *Ur-teil* (see above, p. 171), in which each side is implicitly reflected into itself in its relation to its opposite, provides the notion of consciousness:

> Ego, as this absolute negativity, is implicitly the identity in the otherness: the *ego* is itself that other and stretches over the object (as if that object were implicitly cancelled) – it is one side of the relationship and the whole relationship – the light, which manifests itself and something else too (*PM*, p. 153).

This identity in opposition of subject and object constitutes the principle of the science of consciousness: phenomenology.

In the doctrine of subjective spirit, says Hegel, phenomenology "stands midway between the sciences of natural spirit [anthropology] and spirit as such [psychology]" (*SL* [M], p. 781). Phenomenology, for Hegel, is the science of consciousness that makes possible science's liberation from consciousness. Consciousness, according to Hegel, is spirit in its 'appearance,' spirit as subjective 'certainty' (*Gewissheit*) of itself. It is in this onesided subjectivity that the transcendental philosophies essentially remained and what they could not overcome.[15] But the 'aim' of spirit as consciousness is "to make its appearance identical with its essence, to raise its self-certainty to truth" (*PM*, p. 157). Neither Kant nor Fichte could show this.

Subjectivity, it has been shown in the anthropology, has first raised itself to ego, to reflection. For this reflection, the content formerly belonging to the natural life of the soul has been 'let go' and is now an 'independent object' to which the ego 'relates' itself. The object is thus implicitly the subject's own substantial being as the negative of the subject and something lying beyond it. Hence, says Hegel, consciousness is a contradiction: it is at once the independence of both sides and their implicit identity. This contradiction provides the dialectic of consciousness in its movement to sublate its own diremption. For the subjective consciousness in this movement, alteration seems to come wholly from the side of the object. Nevertheless, ego as pure universality is 'thinking' (*Denken*), and the progressive logical determination of the object is "identical in subject and object" (*PM*, p. 156). Hence the demonstration in the phenomenology deals with the 'logical *Fortbestimmung*' at once of subject and object.

In a treatment of subjective spirit the question cannot but arise as to the relation of the phenomenology as the second science of subjective spirit and the *Phenomenology* of 1807. While we cannot offer a detailed comparison of the 'two phenomenologies', there are certain considerations we may briefly note.

The *Phenomenology* of 1807, we suggested, has a historical starting point and also a starting point in the 'natural consciousness,' whose nature is there presupposed. In the Heidelberg *Encyclopedia* of 1817, Hegel refers to the *Phenomenology* of 1807 as "the scientific history of consciousness," which was originally intended as the "first part" of philosophy "since it would precede pure science as yielding its notion.[16] But "at the same time," Hegel goes on to say, "consciousness and its history, like every other philosophic science, is not an absolute beginning but rather a link in the circle of philosophy." This role of phenomenology as 'link,' as we now have noted, consists in its comprising the science 'midway' in the doctrine of subjective spirit. As having this place within 'the circle of philosophy,' phenomenology is preceded by anthropology, which has brought forth the notion of consciousness and thereby demonstrated what the *Phenomenology* of 1807 began with as a presupposition.

In the *Encyclopedia* the aspect of phenomenology as a 'history of consciousness' appears to recede in favor of phenomenology in its systematic role in the doctrine of subjective spirit. The demonstration in the 1807 work was explicitly and thematically in terms of a notion of 'experience.' Yet the word 'experience' hardly occurs in the *Encyclopedia* treatment.[17] The apparent de-emphasis of the historical may account for the fact that the stoic, sceptic, and 'unhappy' forms of consciousness are omitted in the phenomenology of the *Encyclopedia*. Instead of an explicit treatment in terms of consciousness's 'experience,' the demonstration emphasizes the movement in terms of the thought categories already explicated in the logic as 'pure essentialities' of knowing. In these categories now expressed concretely as forms of consciousness's subjective certainty, the object is successively determined as a singularity that just 'is,' a 'thing' as container of 'properties' and 'matters,' an 'inner' and an 'outer,' 'force,' 'law,' and 'life.' These categories more or less recur where ego as self-consciousness moves to actualize its certainty of self in 'desire' (consumption of the object as singularity, thing), in 'struggle for recognition' (inner and outer, law, life), and the resulting emergence of the universal self-consciousness as reason.

In the treatment of the sense consciousness there occurs an important shift in the discussion of space and time. Without elaborating on the significance of the shift – whether it means a correction or perhaps betokens some difference in approach from the *Phenomenology* of 1807 – Hegel tells us in the *Encyclopedia*:

> the treatment of the 'here' and the 'now' which in the 1807 work appeared in connection with the sense-consciousness, "properly (*eigentlich*) belongs to intuition," which takes its object not as external to consciousness, as in phenomenology, but as self-external (*PM*, p. 156).

Intuition, as we shall note shortly, is reserved for psychology, and the reason why this is so will be an important indication of the nature of organization of the sciences of subjective spirit.

After these few remarks on the nature of the demonstration in the phenomenology as the second science of subjective spirit, we are finally in a position to view in broad outline Hegel's conception of a philosophical psychology.

IV

We have argued that, in its organization and conception, Hegel's doctrine of subjective spirit is designed to overcome the kind of external reflection inherent in post-Cartesian philosophizing from the standpoint of consciousness. For a philosophic science of psychology to be possible, Hegel declares, thought must be liberated from the subject-object antithesis of consciousness. Otherwise we remain in the Kantian position where certain knowledge can only be of phenomenal 'objects' external to the subjectivity that can only look outside itself in order to make apodictic judgments. The Kantian position is in effect restated by William James, when he says, "the psychologist's attitude towards cognition" must be a "thoroughgoing dualism" that "supposes two elements, mind knowing and thing known, and treats them as irreducible."[18] This "irreducible" dualism has been shown by Hegel to be in fact derivative, a dualism that first emerged with consciousness itself as an *Urteil* of the soul-body unity. On the basis of the implicit identity of the two sides, Hegel's science of phenomenology shows the sublation of the dualism and the restoration of the identity of subject and substance as reason. Freed from this dualism that claims absoluteness out of ignorance of its own origin, philosophy is in a position to investigate the concrete determinations of cognition as identity of subject and object. The standpoint of external reflection has been overcome, and in psychology the intelligence has consciousness itself as its object.

Psychology is thus for Hegel the third and culminating science of subjective spirit. In psychology, says Hegel, spirit "has determined itself to the truth of the soul and of consciousness." In this science we deal with "spirit as such," spirit as rational intelligence and will. As intelligence, spirit no longer stands "in relation" to its content as subject to object but begins with its "own proper being" and relates itself only "to its own proper determinations." Every content from the outset has the twofold determination that it "is" and is spirit's "own" (*die des Seienden und die des Seinigen*) (*PM*, p. 184). The course of the movement of theoretic intelligence consists in that what spirit 'finds' in itself as *seiend* it 'posits' as *das Seinige.*

Spirit's "own proper determinations" are intuition, representation, memory, desire, etc., which are its universal modes of activity as "spirit as such." Their demonstration will not proceed as a "natural evolution" in the individual of various faculties and powers but as a necessary unfolding of spirit's modes of activity according to its inherent notion as intelligence. Each new determination will mean a further step in spirit's overcoming of its own immediacy and its 'show' of being determined by a content 'found,' until spirit freely and explicitly determines itself through its self-creation as rational thought and will.

Although intelligence has emerged as the truth of soul and consciousness, and psychology is possible as a result of the demonstrations in anthropology and phenomenology, the science of psychology, Hegel points out, contains its own principle and must not be mixed with forms from consciousness or soul. In the anthropology, the psychical subjectivity could not yet say, 'I think' and was only implicitly for itself; the modes of its being-for-self were made explicit by the philosopher. In the phenomenology the subjectivity could think itself and put forward successive claims to a knowledge of the object as the nonself; nevertheless, for this 'appearing' knowing, alteration seemed to come about wholly on the side of the object, and it was the philosopher who pointed out that in the altering of the object, consciousness itself was altered. The special principle of psychology consists in that the relation of spirit to its determinations is a self-relation that is explicitly 'for' spirit itself. We have thus attained to a relation of pure transparency and immanence that is impossible to an external reflection.

But what now does this mean specifically in terms of the treatment in psychology? What does it mean to say that intuition, representation, memory, etc., are the 'proper' modes of activity of intelligence? Do not these forms of mental activity also belong to consciousness, and perhaps even in a measure to the soul? Is not 'attention,' for example, a determination of consciousness? Why then is attention only first treated by Hegel in psychology and not in phenomenology or anthropology? Why should intuition, with its forms of space and time, only first come up in the third and concluding science of subjective spirit? As a matter of fact, intuition has in a way been treated earlier – as the sensuous consciousness of the phenomenology, and even as the sentient soul of the anthropology. Why then the three levels, and what marks the difference in treatment?

With these questions we approach directly the notion of subjective spirit and its demonstration in the form of a triad of sciences. What in anthropology was 'sentience' and in phenomenology 'sense consciousness' becomes in psychology – as the intelligence itself in its 'immediacy' – 'intuition.' We see other forms recur. What in anthropology was the permeation of the corporeal *Dasein* by the soul as subjective purpose in 'habit,' becomes in psychology the permeation of the 'inner' *Dasein* – the stock of images and representations – by the intelligence in 'sign' and particularly language. What in anthropology was intersubjectivity on the psychical level as 'animal magnetism,' and in the phenomenology the master-slave relation and 'recognition,' becomes in psychology the ethical unity of free rational wills. In these and other instances, what determines the difference in treatment on the respective levels? In anthropology, as we saw, the determinations of the 'monadic' subjectivity were those of its natural corporeality and its self-enclosed feeling life. There was as yet no question, properly speaking, of an 'object.' In phenomenology, on the other hand, the ego is 'thinking' and determines its object according to thought categories in which it puts forward its claims to a knowledge of the object as an "other." But such an 'objective' knowledge was in truth but a one-sided subjective certainty. Hence the *Fortbestimmung* of object and subject in the phenomenology can only be called a 'constitution' in a purely logical sense. The *Fortbestimmung* consists in that the forms of *Gewissheit* are sublated to *Wahrheit*, while the 'appearing' ego is sublated to ego in its 'essence.' Only once the logical status of object has been clarified and established as identical with that of subject is psychology possible as a science that is not an external reflection. Only now can psychology demonstrate the constitution of the object – not in terms of a onesided subjective certainty of an 'other' but in the object's explicit identity with the subject as but the latter's own modes of activity. The first form of this explicit identity, according to Hegel, is not sense consciousness, for which the object is an other, but 'intuition.' For this reason, intuition is treated for the first time in psychology.

Intuition, according to Hegel, is intelligence's 'finding' itself determined and for this reason a mode of intelligence in its self-externality. At the same time the content intuited has the form of self-externality as determined in the universal modes of externality as such: namely, space and time. Space and time are indeed forms of intuition, as Kant maintained. But in the first place, says Hegel, they are not simply picked up by an external reflection on the part of the philosopher, but demonstrated from the very notion of intuition as intelligence's own form of self-externality. Secondly, they are not 'merely subjective.' Since the one-sided subjectivity of consciousness has been overcome in phenomenology, the spatial and temporal asunderness of the intuiting intelligence's object are the very forms of the object as 'external to itself' – though the object is so, of course, only 'for' the intelligence. The intuiting intelligence, according to Hegel, is thus the truth of nature's own *Äussersich-sein* as a realm of the *partes extra partes*.

Here we may say a word about attention as a moment of intuition. Of course attention is also a determination of consciousness. In the phenomenological demonstration, however, attention as such is taken for granted in the ego's unself-conscious outer directedness. For the purposes of the phenomenology, the matter of attention is simply not relevant to the logical claims of the subjective certainty. Now that in psychology consciousness itself is the object of intelligence, attention becomes relevant to the constitution of the subject-object identity.

Attention, according to Hegel, is the moment of spirit's making a found content 'its own,' the moment of spirit's 'active inwardization' (*Erinnerung*). As against this inwardness, the necessary opposite moment is spirit's positing the content found as a 'something that is' (*Seiendes*) – but whose very 'being' is at the same time a 'negative' or 'abstract other of itself.' This marks the distinction between the intuiting intelligence and the sense consciousness, whose object is merely an 'in itself.' In attention the intelligence determines its content of feeling as 'self-external' (*Äusser-sich-seiendes*), projecting it into time and space, the forms in which intelligence is intuitive. Thus, unlike the phenomenology, the present standpoint shows intelligence conferring on the material the 'rational determination' of being 'the other of its own self' – space and time being the most primary and abstract modes of this self-externality. Attention is thus at once intelligence's determining the 'found' as separate from itself (spatial and temporal) and positing it precisely in that separateness as intelligence's 'own.'

Hegel's dialectical treatment of attention offers a beautiful example of the simultaneous going-outwards and going-inwards that is the mark of spirituality in all its development (see above, p. 169). Only through attention, says Hegel, does spirit come to be 'present' in the subject matter, and hence this spiritual activity is the necessary beginning of all mental cultivation and knowledge. Attending is a simultaneous separating and uniting of subject and object, a self-within-self reflecting of free mind and "an identical turning of mind toward the object" (*PM*, p. 196). Though attention is in some measure a matter of my free choice, it is not for that reason something easy. It may mean the effort of selective concentration whereby I suppress all sorts of inclinations and distractions, even 'abstract myself from my self' by keeping back those particular interests and prejudices that prevent the *Sache* from coming into its own. In attending, I must hold the object away from myself in order to be closer to it. I must forget myself in order to immerse myself in the other – and in this very self-forgetting I must enter all the more deeply into myself to bring into play my fullest resources. Attending is thus at once a going outward and a going inward. But it is precisely the nature of spirit to unite these opposites in itself. If I am distraught, I cannot fix the object in my attention; but if I am held fascinated by the object in its brute presence, I cannot bring into play all my experience of its ties, references, and possibilities, which first make it the concrete object that it is. How spiritual an activity is 'mere' attending in a philosophical psychology! Intuition, in

Hegel's treatment, is not one 'part' or 'faculty' of the mind, not even 'one stem' of knowledge that has to be united synthetically with another 'stem.' For this reason, Hegel experiences no difficulty in linking his meaning of intuition with the sense we give the word when we speak, for example, of the deep intuition of a Goethe. Intuition, for Hegel, is conceptual thinking in germ; intuition *is* mind, *is* spirit in its living totality, though as yet in undeveloped or deficient form.

Spirit as intelligence, however, cannot remain immersed in this form of its immediacy. In the various stages of representation, intelligence inwardizes the externality of intuition and imagination in memory, and finally overcomes its last vestige of self-externality in conceptual thinking. We can hardly go beyond a bald summary of Hegel's presentation of theoretical spirit in the psychology. A few points of special interest must suffice.

As initially 'recollective' of intuition, says Hegel, intelligence posits the intuited content in its own inwardness, "in its own space and time." Thus taken up into the universality of the ego, the content becomes an image, abstracted from the fullness of the original intuition and isolated from the directly experienced place, time, and context of the intuition. The image thus seems to be but a residue and a loss from the determinate singularity of the intuition. But it is not the case for Hegel, as for Hume, that the image is but a faded copy of the impression. For Hegel, no manifestation of spirit in its activity can be explained merely in terms of loss. In Hegel the intuition is eclipsed as singularity by the actively inwardizing intelligence in transposing the content into its own universal ideal medium. This greatly extends the range of the representing, as against the intuiting intelligence, since I no longer require the immediate presence of the object. Further, in abstracting the intuition from its original setting, says Hegel, representation divests it of what are frequently but the contingent connections of time and space.

Of itself the image is transient, and intelligence as inner attention is its sole 'where and when.' But intelligence is not merely this *Dasein* as but a present awareness of this determinate content; and the image, while inwardized in the intelligence and thus no longer 'existent', has not vanished into nothingness. The image is "unconsciously retained," says Hegel in the "night-like pit" (*nächtlicher Schacht*) of the intelligence. Here are preserved a "world of infinitely many images and representations without their being in consciousness." This preservation of the no longer existent can only be understood, says Hegel, if spirit is grasped in its essence as "ideality" – which means "negation of the real, but a negation in which the real is *aufbewahrt*, retained *virtualiter*, although it does not exist" (*PM*, p. 92). The representing intelligence is the power which, of itself, can bring forth the image to existence from its 'simple night.' In so doing, the intelligence has 'subsumed' the singular intuition under the image as 'the universal in point of form.' Recollection (*Erinnerung*) is thus a merging in the form of a 'recognizing.' In being brought forth from the 'simple night' of the intelligence, the image acquires a *Dasein* and the intuition simultaneously a generality. Through such repeated 'recognitions,' the image obtains the 'liveliness' and 'presentness' of a *Vorstellung*, inasmuch as I can 'set it before' myself without the aid of intuition.

While in imagination I am no longer restricted to the sensuously present, nevertheless as long as my thinking continues to be in images, according to Hegel, I am not yet free. To the extent that my imagining is a kind of roving, I remain subject to contingencies external to my essential being as intelligence. But in imagination itself, says Hegel, intelligence proceeds to sublate this unfreedom and contingency, and the moments of imagination display spirit's advance toward the freedom of conceptual thinking. This advance is marked by the movement from (1) the more or less automatic evocation of images by the 'reproductive' imagination, to (2) the relating of images to one another through general representations by the 'associative' imagination, and finally to (3) the positing of general representations by the 'creative' imagination or *Phantasie* in the sensuous *Dasein* of symbols and signs making up the storehouse of memory.

While Kant poses the problem of knowledge as the "subsumption of an object under a concept," or the "subsumption of intuitions under pure concepts" (*CPR*, B 177-A 138), Hegel takes such terms as 'subsumption' and 'synthesis' as indicating an external conjoining (*Verknüpfung*)[19] that plays a role on the level of representation but is out of place in the notion of pure conceptual thinking. Here, according to Hegel, the determinations are not united in 'syntheses' but "produced by the inherent movement of the moments tending back into this unity" (*SL* [M], p. 75). But such conceptual thinking, Hegel endeavors to show, only first arises in an overcoming of the synthetic connectings of the representing intelligence.

In the reproductive imagination the images are brought forth from the ego's own interiority, thus providing a universal representation for the associative connecting of images. The activity of the associative imagination, says Hegel, is not to be explained in terms of some "force of attraction" (see above, p. 165), but as the intelligence's own power of subsuming and synthesizing its content in its very act of "setting it before" itself in its universality. Association, says Hegel, is a subsumption of singular representations under a universal representation that constitutes their connection (*PM*, p. 209). As the identical power over its 'stock' of images and representations, the associating intelligence informs this stock to its own purport and interest; it permeates the material of representation just as the soul permeated its corporeal determinations with its own subjective purpose in habit. In thus informing its material with its own inner purport, says Hegel, intelligence is symbolizing, allegorizing, poetizing imagination. It is no longer merely reproductive or associative but 'creative' imagination, which Hegel terms *Phantasie*. Inasmuch as the material still comes from intuition, the creations (*Gebilde*) of the *Phantasie* are still syntheses. The essential activity of the *Phantasie* consists in that the *Verbildlichung* of the universal representation is at the same time the *Verallgemeinerung* of the image (*PM*, p. 210). The universal, or inner, permeates the image, or outer, as its very soul, so that the image is rendered but the outer expression of the universal wherein the latter recollects itself and is for itself. This takes place most especially in the imagination as the "signmaking" *Phantasie.*

The *Phantasie*, according to Hegel, is that form of the intelligence's self-externalization in which it brings its inner import and purport (*Gehalt*) to the outer form of image and intuition. In this way the intelligence makes itself *seiend*, as the soul made itself 'being' in permeating its corporeality. The movement proceeds in three stages. The image initially produced by the *Phantasie* is only 'subjectively intuitive.' In the 'sign,' it acquires 'intuitability proper.' Finally in mechanical memory the intelligence completes its bestowal of the form of *Sein* on its hitherto merely inner *Gehalt.*

The 'sign' (*Zeichen*), says Hegel, is some immediate intuition that stands for a "wholly different content" from that which it has on its own account. The sign is "the pyramid into which a foreign soul has been conveyed (*versetzt*)" and preserved (*PM*, p. 213). Hegel here notes the difference between sign and symbol. The latter, in its own proper character as an intuition, has been more or less carried over into the content of the representation, as where the eagle as 'strength' served as a symbol of Jupiter. In symbolization the intelligence is still conditioned by the sensuous, the particular imagery still largely subjective. In the sign, on the other hand, the intuitive content is so removed from its signification that the intelligence is free in its choice and employment of the intuition. But it is only through simile, metaphor, and symbol that the intelligence first raises itself to this freedom in its signmaking. To be sure, in the intuited particular, the symbolizing intelligence has also returned to itself in its universality, if not so completely as in the sign. In identifying the intuited eagle with this invisible god, the *Phantasie* has in effect dissolved the intuition in its sensuous concreteness and taken one essential attribute which is thus raised to an abstract universal: Jupiter *as* strength. But in such pure signs as the *cocarde*, flag, gravestone, says Hegel, the intelligence has so infused the intuition with its own inner purport that the intuition serves as but the incidental occasion for the intelligence's own self-reminder.

In thus taking the intuition out of its immediate spatio-temporal context and positing the image in intelligence's own space and time, the signmaking *Phantasie*, says Hegel, annuls (*tilgen*) the immediacy of the intuition and infuses it with its own soul and meaning, within intelligence's own ideal space and time. In this way, says Hegel, intelligence's sign-creating activity becomes thc 'productive memory.' In the annulling of the immediacy of its being, the original intuition now only 'is' as an *Aufgehobensein.* This sublation of the intuition provides Hegel's deduction of language.

Since the negative power that annuls the intuition in its immediacy is the intelligence itself – which is responsible for intuition's forms of space and time in the first place – the more determinate shape of the intuition-become-sign is that of a "being-there in time" which is just as well the "vanishing of the being-there in that it is." As an external psychical positedness deriving from the intelligence's own natural being, the sign is now a vocal note (*Ton*), "where the inward idea manifests itself in adequate utterance" (*PM*, p. 214). Articulated in speech and language, vocalization endows feelings, intuitions, representations, with a second *Dasein*, spiritually enhanced and higher than before – an 'existence' wholly within the realm of representing.

Language, for Hegel, is the highest product of the signmaking intelligence, the mediator between intuition and pure thinking (*SL* [M], pp. 31–32). In sublating the intuition in the verbal sign, the intelligence is the inner spiritual 'time,' which, like time in general, sublates the asunderness of space (*PN*, p. 33ff). In its creation and employment of language, the intelligence is engaged in that going-outwards that is simultaneously a going-inwards: an investing of its spiritual inner with the form of an outer *Dasein* whose very being is straightway a vanishing. Thus, for Hegel, signmaking is not some specialized function of imagination in 'cognition,' which is mainly the role it obtains in the Kantian schematism. Its role for Hegel is indeed that of a mediator – but one that is the ego itself in its totality as a spiritual life. Language for Hegel is "divine" (*Phen* [M], p. 66). The sublation of the spatial intuition to the vanishing temporal *Dasein* of the vocal sign entails the whole inner spirituality of the intelligence as negative power over its natural being as well as its representations.

In a digression on language, Hegel argues that alphabetical writing is the more appropriate expression for the signmaking intelligence than hieroglyphic. The former has 'dissolved' the concrete signs of vocal language into their phonemes, which are now represented by visual marks. Thus as 'signs of signs,' phonetic writing is a further sublation of the spatially sensuous and hence a more 'inwardized outerness' than picture writing. No matter how rich the content, the 'name' renders it simple for the mind, so that the sign, as a *Dasein* on its own account, offers neither halt nor pause for thinking. In fluent reading, alphabetical writing itself becomes a hieroglyphic but one that retains the pure simpleness of the vocal sign. Nevertheless the word as such, says Hegel, is still an 'outer' that must be wholly taken up into the 'inner' of the intelligence. The *Erinnerung* of the word as sign, says Hegel, is memory (*Gedächtnis*).

In memory the word as such is negated and at the same time preserved, as was the case with the image. As memory, says Hegel, the intelligence inwardizes the sign, thereby raising the single connection of name and representation to a universal one. Since the sign as spoken word is a temporality, its singularity is its transitoriness; through its universalization in memory, it obtains permanence. In thus becoming universal, says Hegel, the name and the meaning are connected for the intelligence; the intuition, which is what the name in the first instance is, is itself made a representation. Meaning and sign are hereby united in one representation. Without need of intuition or image, the reproductive memory now "has and recognizes the *Sache* in the name and the name with the *Sache*." Given the name 'lion,' I need neither an intuition nor an image of the animal; the name alone serves me as the "imageless simple representation."

In thus uniting name and *Sache*, the reproductive memory is even less dependent on the intuition than the *Phantasie*. In the "image-less simple representation," says Hegel, intelligence is self-intuiting. Hence, Hegel maintains, memorization is not a "reading off" of something from a supposed mnemonic tableau of imagination. Learning "by heart" (*auswendig*), as the phrase implies, is rather a bringing forth "from within outwards," from the "deep pit" of the ego. As between memory and imagination, memory is the higher spiritual activity, according to Hegel, since it is no longer dealing with an intuition but with a *Dasein* that is intelligence's own product: with "such a without-book (*Auswendiges*) as remains locked up in the within-book (*Inwendiges*) of intelligence" (*PM*, p. 220).

Although in reproductive memory, name and *Sache* are united, this connecting is of itself no guarantee of the truth or necessity of the *Sache*.

Despite the advance from imagination, in reproductive memory the ego as intelligence is not yet freely with itself in that 'outer' that is such "only *inside* the intelligence itself." This very defect is posited in mechanical memory or rote, where the limitedness of memory discloses itself as the last remaining gap between word and meaning. At the same time the disclosure as such constitutes the possibility for the overcoming of the defect. Hereby we come to Hegel's notion of rote as the consummation of the representing intelligence – a notion whose seeming oddness is in fact no odder than the superiority of phonetic to hieroglyphic writing.

In memory, the linking of the meaning as such with the 'being' (*Sein*) as name is still a synthesis, and the intelligence does not freely return to itself as simple self-identity in this its externality. As the universal and the 'simple truth' of its externalizations, intelligence moves to sublate the synthetic connection of words and meanings. It does so through a final and highest 'inwardization' that is simultaneously a highest 'outwardization.' Intelligence here, says Hegel, "posits itself as the Being, the universal space of the name as such" – that is to say, the "Being of meaningless words."

Thus the completion of memory in its notion, according to Hegel, is rote. In rote, or 'mechanical memory,' the synthetic connection of word and meaning is in the first place dissolved; for that *Verknüpfung* was itself an 'externality' incommensurate with the universality of intelligence in its notion. But the dissolution of the *Verknüpfung* in rote is a making way for something higher; it is the opening up of a 'universal space' in which uniting need no longer be synthetic, as it is in all representation as such. Since the meaning of a synthesis is a uniting of elements that have a subsistence on their own, the overcoming of synthesis can take place only in one way: namely, that the very *Sein* of what is united be divested of its last tinge of externality or self-subsistence, so as to be able to derive internally from nothing but the intelligence itself. Intelligence must therefore make itself to *Sein* – indeed lower itself to the sheer externality and abstractness of a 'universal space.' But at the same time this 'space' is but its own self. In rendering itself an abstract universal space in mechanical memory, intelligence posits itself in its universality as the power over the representation. Rote is despiritualization, but the act of spirit itself: it is the intelligence's way of making the particular outer wholly its own inner by making itself the universal outer in which the particular first obtains subsistence.

Thus mechanical memory, according to Hegel, is spirit's positing itself as objectivity. In intuition, we recall, spirit was external to itself in its 'finding' of itself as determined. In representation it recollected the 'found' within itself and made it its 'own.' In memory it makes itself an externality so that its 'own' once again appears as a 'found': now, however, the finding is a finding of what is intelligence's express creation. In mechanical memory, says Hegel, one of the moments of thinking, "objectivity, is here put in intelligence" as the quality of the intelligence itself (*PM*, p. 222). The movement here recapitulated may be seen as a movement from space to time and from time back to space; or, since space and time are, respectively, the abstract forms of objectivity and subjectivity, from objectivity to subjectivity and back to objectivity. The spatiality of the original external intuition is sublated in the temporal *Dasein* of the vocal sign; now, going back, the transience and contingent subjectivity of the name is itself sublated in an objectivity that is – indeed not that of sensuous space – out the 'inner outerness' of the spirit's universal spatiality: the mechanical memory. As mechanical memory, intelligence puts the stamp of 'being' on all its productivity of representation.

The objectivity which the name acquires in mechanical memory, however, is in the first instance but abstract. If the *Seiende* as name is to be the *Sache* as genuine objectivity, says Hegel, it is in need of the 'meaning' coming from the representing intelligence as the inner. But memory as such, as we have seen, has fallen heir to the inner of representation with its treasury of names and meanings; and as mechanical memory the intelligence contains at once the two moments: outer objectivity and inner meaning. The intelligence, says Hegel, is now posited as 'existence' of this identity. Spirit is now explicitly 'for itself' the identity which it is 'in itself' as reason. The inwardness of intelligence is now 'to itself being' (*an ihr selbst seiend*), subjectivity explicitly one with objectivity. Hereby, says Hegel, intelligence as memory passes over into the activity of thought.

The kinship of memory and thought (*Gedächtnis* and *Gedanken*) is recognized even in language, Hegel claims. Memory as such is thinking in its 'existence,' i.e., thinking in its merely outer aspect or onesided moment. Reason is implicitly identical with this mode of thought's existence, and this means that reason now exists in the subject, that reason 'is' as the individual subjectivity's own activity of thinking.

With the attainment of thinking, spirit has completed its development as theoretical intelligence. Thinking is the explicit and consummated unity of subjective and objective in the sphere of cognition. The movement in the psychology began with spirit as implicit identity of 'being' and its 'own,' and the course of spirit's movement through intuition, representation, and thought has been to posit this identity as 'for' itself. For this reason, Hegel tells us, intelligence has proved to be 'recognitive' (*wiedererkennend*).

Spirit's knowing now means that what it finds in itself it knows both as its own and as being.

Beginning with the formal determinations of 'category' and 'judgment,' thought in 'syllogism' discloses the necessity of the content and sublates the last distinction of content and form. Whereas in understanding the intelligence still deals with a content other than itself, in reason the content as such derives from the thinking intelligence's own distinctions of form. In conceptual thinking (*begreifendes Denken*) the Notion evidences itself in its own proper shape. The universal as form is no longer external to the content but produces the determinations of the latter from itself and is thus the immanent self-developing notion of the *Sache*. The entire movement of subjective spirit has been the demonstration that "the Notion as such can only be apprehended by spirit, of which it is not only the peculiar property but the pure self" (*SL* [M] , p. 618).

That the intelligence now knows that "what is thought, is; and what is, is only insofar as it is a thought," constitutes the transition from the theoretical to the practical sphere. In knowing itself as the "determinant of the content" which at the same time is determined as 'being,' the intelligence is will (*PM*, p. 227). The development of subjective spirit as will comprises the second part of the psychology: practical spirit. This division, however, is not absolute. The will is nothing separate from reason. In man, says Hegel, "there is only *one* reason, in feeling, volition and thought" (*PM*, p. 231).

As will, spirit enters the realm of actuality. Actuality in subjective spirit, however, has to do with the individuality as fulfilling itself, and will viewed psychologically is that of the finite subjectivity. The finitude of will here, says Hegel, consists in its 'formalism'; its self-fulfilling is in the first instance an abstract insistence on its own satisfaction that is not yet "identified with the developed reason."

The pathway of the will in overcoming its formalism consists in its raising itself to a 'thinking will' that has freed itself of its own narrow particularity and now posits itself in its free subjectivity as at the same time objective and universal. The course of the movement displays the three moments of the Notion. As (1) 'practical feeling' the will 'finds' itself determined by its manifold – often contradictory – inclinations and passions, whose contingent and isolated content is lacking in the form of universality that would render it objective and substantial. The would-be harmony of inner feeling with the objectively valid is 'posited' in (2) impulse and choice as a harmony to be established by the will itself through subordination of the manifold impulses to a universal aim. Initially this is (3) 'happiness,' which, however, as a universal, remains only externally related to the particular through the will's arbitrary choice. The genuine unity of individual and universal becomes possible when 'enjoyment' passes over into 'deed' and 'act,' wherein the 'actual free will' knows its object to be but its own autonomous selfhood.

With this emergence of 'free spirit,' the doctrine of subjective spirit is completed. The necessity has arisen that the individual subjective will in deed and act fulfill itself in an objective and universal content. In attaining to free rational will, spirit has actualized its unity as theoretical and practical intelligence. Spirit has thereby fulfilled its notion as subjective spirit and passes over into objective spirit: the realm of right, ethical community, and state.

Hegel's doctrine of subjective spirit above all presents the development of spirit as a *Befreiungskampf*: to permeate its natural corporeal being as its ideality in the anthropology; to overcome the separation of itself and its object in the phenomenology; to sublate its own modes of self-externality in the psychology. The freedom struggle is thus against no foreign overlord but against itself in its own natural being as its limit. The *Befreiungskampf* of spirit marks subjective spirit as finite – but also infinite. This struggle with its own finitude, says Hegel, "constitutes the stamp of the divine in the human mind and forms a necessary stage of the eternal mind" (*PM*, p. 182).

Hegel has purported to demonstrate this movement of spirit scientifically. Indeed he claims that it can only be shown scientifically and, conversely, that only that presentation will be scientific which shows spirit in its inherent truth as a struggle toward freedom. For such a demonstration to be possible, however, there must be an overcoming of that modern form of external reflection which holds that science can only consist in a knowledge of externality. To overcome this external reflection, Hegel advances his 'speculative method,' which, in every philosophic science, is the form of demonstration immanent to the *Sache selbst.* In the realm of spirit, according to Hegel, the *Sache selbst* is spirit's own knowing of itself – to which no external reflection can obtain 'access.'

Hegel sees the overcoming of external reflection as a liberation from the subject-object dualism of consciousness. In post-Cartesian philosophy through Kant, the dualism remained absolute. Thinkers such as Hume said that they made no pretense to a knowledge of the soul or mind in its essential nature but only to what they could observe of its operations.[20] But in this supposedly detached observation, the observer imported such categories as 'force' and 'attraction' from his view of the external world. Thus at one moment the observer separated the realm of the internal from that of the external, and in the next moment he approached the internal with categories from the external. Kant demanded a critique of the forms of thought in general, including those suitable for a knowledge of mind. But while uncritical categories were hereby barred from a knowledge of mind, the nature of the critique was such as to do away with the very enterprise. For Kant, the ego cannot know itself because it can only be subject, never object of its judgments. For consciousness, apodictic knowledge can only be of spatial externality as a realm of phenomena.

Hegel, as we have seen, overcomes these approaches of external reflection by demonstrating that the standpoint of consciousness and its subject-object dualism is a derivative one. Consciousness first comes to be as an *Ur-teil.* The development of the soul-body relationship demonstrated in the anthropology culminates in the subjectivity's act of oppositing to itself its corporeal determinations as an objectively structured external world. Through this movement, subjectivity has first come to be ego. In its relation to its object, the ego of consciousness has the subjective certainty of the object as an independent other. Proceeding immanently in terms of consciousness's own claims, the phenomenology takes up those very categories – thing, property, force – in which uncritical thinking had moved, and demonstrates their sublation as modes of thought in which subject and object are separated.

Only at this point, where the genuine status of object and subject has been established, is a philosophical psychology first possible. Now there is no question of approaching mind with categories imported from elsewhere, or of knowing the mind only in the flux of empirical consciousness. The triad of sciences: anthropology, phenomenology, and psychology, is the overcoming of external reflection and the demonstrated notion of subjective spirit. By virtue of the disclosure of the origin of the subject-object dualism, and by the phenomenological critique of the forms of thought deriving from the standpoint of that dualism, the dualism itself has been overcome. A philosophical psychology that knows finite subjectivity as spirituality in its struggle toward freedom is now possible.

Bernard Baruch College, CUNY
New York

NOTES

[1] For one statement of Hegel's meaning of external reflection, particularly as contrasted with dialectical thinking, see *L* [W, 1892], p. 147.

[2] Hegel intended to write a separate full-scale work on subjective spirit but did not live

to do so. See Friedhelm Nicolin, 'Hegels Arbeiten zur Theorie des subjektiven Geistes' in J. Derbolav and F. Nicolin, eds., *Erkenntnis und Verantwortung. Festschrift für Theodor Litt* (Düsseldorf, Schwann, 1960), pp. 356–374.

[3] *L* [W, 1892], p. 296; *SL* [M], p. 613.

[4] "What merely is (*das Seiende*), without any spiritual activity," says Hegel, is so little the essence of spirit that consciousness is rather the very opposite of it and is only actual for itself by "the negation and abolition of such a being" (*Phen* [M], p. 205).

[5] Referring to Condillac's model of the mind, Hegel says, it shows a successive emergence of mental faculties in a "solely *affirmative* manner" that misses the "negative aspect" whereby the sensible material is "transmuted into mind and destroyed *as* a sensible." (*PM*, p. 183).

[6] *Phen* [M], p. 47; *L* [W, 1892], p. 17.

[7] On the question of the role of the philosopher, see Kenley R. Dove, 'Hegel's Phenomenological Method', *Review of Metaphysics* **23** (1970), 627 ff.

[8] *SL* [M], pp. 51, 60, 63; *Phen* [M], p. 491.

[9] "Psychologists may not expressly speak of parts of the soul or mind, but the mode in which this subject is treated by the analytic understanding is largely founded on the analogy of this finite relation." (*L* [W, 1892], p. 246; see also p. 365).

[10] See R. C. Solomon, 'Hegel's Concept of "Geist" ', *Review of Metaphysics* **23** (1970), 642–661.

[11] See Kant's *Metaphysical Foundations of Natural Science*, tr. J. Ellington (Indianapolis, Bobbs-Merrill, 1970), p. 8; Theodore Mischel, 'Kant and the Possibility of a Science of Psychology' in *Kant Studies Today*, ed. L. W. Beck (La Salle, Ill., Open Court, 1969) pp. 432–455.

[12] *PN*, p. 442 ff; *L* [W, 1892], p. 362; *PM*, p. 9ff.

[13] *PM*, p. 146; see also *SL* [M], p. 766.

[14] *PN*, p. 382 ff; *PM*, p. 76 ff.

[15] See Ernst Bloch, *Subjekt-Objekt: Erläuterungen zu Hegel*, 2d ed. (Frankfurt am Main, Suhrkamp, 1962), pp. 188–189.

[16] *Sämtliche Werke*, ed. Glockner (Stuttgart, Frommann, 1958), Vol. 6, p. 48.

[17] In fact the term 'experience' only occurs thrice, each time in a polemical context incidental to the main exposition of the conception of the phenomenology. (See *PM*, pp. 161, 162).

[18] William James, *The Principles of Psychology* [1918], 2 vols. (New York, Dover, 1950), vol. 1, p. 218 (cited by Aron Gurwitsch, 'Towards a Theory of Intentionality,' *Philosophy and Phenomenological Research* **30** (1970), 359.)

[19] *SL* [M], p. 96; *L* [W, 1892], p. 344.

[20] See Hume's Introduction to his A *Treatise of Human Nature*, ed. L. A. Selby-Bigge (Oxford, Clarendon Press, 1964), p. xxi.

PART TWO

PHILOSOPHY AND METHODOLOGY OF SCIENCE

ERROL HARRIS

THE DIALECTICAL STRUCTURE OF SCIENTIFIC THINKING

I

One of Kant's major contributions to modern philosophy was the recognition that genuine knowledge is never a mere patchwork of items of information, whether gathered from empirical sources or from intellectual, whether inductively inferred or deductively derived from first principles. "If each and every single representation were completely foreign, isolated and separate from every other", he declared, "nothing would ever arise such as knowledge, which is a whole of related and connected elements".[1] Of this fact, Hegel was unshakably convinced. "The Truth", he maintained, "is the whole"; but it is no undifferentiated, featureless whole, no Schellingian night in which all cows are black. "The true form in which the truth exists can only be the scientific system itself" (*Phän* [G], p. 27).

Further, he did not conceive the system as a static or invariable pattern, but as a dynamic system; a system of activity and development inherent in the very relationship of part to whole, or of the rudimentary to the fully fashioned. The relationship is involved in the most elementary physical entities and develops continuously to its fullest self-manifestation in living, mental and spiritual totalities. In its explicit form it is self-conscious, and of the self-conscious mind he says that it is absolute restlessness – a constant activity of thinking the essential principle of which is dialectic. Moreover, it is only in the fully explicit self-conscious form that the whole is adequately realized as a self-maintaining system. "The Truth", he says, "is the whole, but only that essential reality which has fulfilled itself through its own self-development is the whole" (*Phän* [G], p. 24). Moreover, the whole, when fully realized, transpires as the entire dialectical scale through which it has developed itself, and it is to this and the manner of its progression that we must look for the clue to understanding precisely what Hegel means by system.

The whole of which Hegel speaks is not one that can be broken up into mechanically separable parts, and the system that constitutes it is not a mere spatial or mechanical arrangement of such parts. Neither is it a purely formal set of relations between ideas or formulae. It is a continuous scale of forms

R. S. Cohen and M. W. Wartofsky (eds.), Hegel and the Sciences, 195–213.

and categories which generates itself by the dynamic of its own inherent principle of activity, from the most vague, abstract and indefinite to the most complete, articulate and explicit elaboration of absolute reality. Each form, at every level, is, he asserts, the Absolute in some partial, and in some degree inadequate, manifestation of itself. Each logical category, which has a place in the system, is a provisional definition of the Absolute. It is that definition which is characteristic of a particular phase in the process of dialectical development. Hence the progression is not simply from the fragmentary to the complete (though any such progression is one aspect of the process), but is from the vague and confused to the more precise and discriminate. At every stage the whole is in some sense, or manner, or degree, present, for at every stage we have what Kant called a whole of related and connected elements. But the whole is not at every stage explicitly articulated, and the series of forms is constituted by the successive stages of explication engendered from the less to the more adequate self-manifestations of what is ultimately real and true.

The movement is generated by the very inadequacy of the primitive forms. For every entity strives to become self-sufficient; and in those which are finite, incomplete, or in other respects deficient, their very shortcomings give rise to internal conflicts and oppositions that can be resolved only by further self-development and explication. Similarly at the level of development at which consciousness arises and this process becomes explicit as knowledge, finite objects are found wanting just so far as they do not satisfy the mind's demand for, and *nisus* towards, coherent wholeness; and contradictions arise, oppositions are generated, and resolutions are achieved by synthesis and reconciliation. The progression is consequently one that proceeds by the generation of contradictions and their resolution. It is the continual assertion of the negative aspect involved in finiteness and insufficiency and its supersession by a more adequate form in which the particular deficiency is supplied and the negation is negated.

For this reason Hegel calls the principle of the dialectic one of absolute negativity; and it is because of the generation of internal contradiction through inadequacy of the finite form, the generation of consequent opposition to an other which is constituted by what is lacking from the original, that the triadic appearance of the dialectical stages results. 'Thesis, antithesis and synthesis' was not Hegel's own formula for the dialectical movement, but it does summarize what results from it, even though it is not always apparent and by no means always rigidly exemplified in Hegel's exposition.

System, then, is a series or scale of dialectically related forms, each automatically generating the next, by its inherent *nisus* to overcome its own shortcomings and the contradictions to which they give rise. It is system of this kind that constitutes what Hegel calls 'science' or *Wissenschaft*, and it is to this that he is referring when he writes: " . . . knowledge is only real and can only be expounded as a science or system" (*Phän* [G], p. 27). Each form, because the whole is implicit in it, is itself a whole of some sort. Thus the logic of the dialectic is a logic of development and what develops is always a totality.

The gamut of categories that results from the dialectical process is a scale of forms in which each is a specific exemplification of the totality which is developing, but an exemplification at a particular level of adequacy (or inadequacy) or degree of explicitness. At the same time the successive forms are mutually related as opposites because of the difference in adequacy between them: the first as relatively false to that which it is its purport to present, and the other as relatively true. As Hegel puts it, the later is 'the truth of' the earlier. Every category of the Logic is an exemplification of the Idea at a specific level of explicit thought; every category of Nature-philosophy is a specific natural form at a particular level of realization of integral wholeness; and every form of sentience, feeling and consciousness is a specific form of mind (*Geist*) at a definite level of self-development. Yet each of them, as well as each major division of philosophical reflection, is in some respect, or in some sense, opposed to its neighbour, and in some degree to every other. I believe that this can be demonstrated in detail, but my present purpose is only to bring out the fact that the dialectical system is a scale of forms such as R. G. Collingwood describes in his *Essay on Philosophical Method*, in which the forms are mutually related in three ways at once, as degrees of realization of a universal principle, as specifications of that same universal, and as opposites. Such a logical structure Collingwood finds characteristic of all philosophical thinking, and typical of every philosophical system. My object is to show that it is not confined to philosophy but is also the form of scientific theory and scientific advance.

II

Among contemporary philosophers of science there would be general agreement that a scientific theory is a system of some kind. Commonly it would be designated an explanatory system. But the form and character of a scientific system are matters on which there might be more difference of opinion. At least until recently the accepted view was that of a scientific theory as a deductive system constituted as an interpreted calculus. The calculus consists of initial formulae made up from defined and undefined terms from which other formulae are derivable according to specified transformation rules. The interpretation of the calculus is effected by substituting for its variable elements empirical terms indicative, or descriptive, of observable entities; and such substitution converts the formulae into propositions making assertions about the observable facts. When so interpreted, the transformation rules function as rules of inference; but according to the current theory of deductive inference, no new factual information can be derived by its means because all transformations are between logically equivalent statements, and every deduction will therefore be purely analytic.

The source of all synthetic statements included in a scientific system must accordingly be observation and any inference of a synthetic factual nature can be only inductive. The validity of the traditional form of inductive inference has, however, never been established, and serious doubts have been voiced about its use in the sciences which have not been allayed by its proponents.[2] Some writers argue that inductive inference needs no more justification than does deductive, on the ground that validity requires no more than adherence to the stipulated rules. But this argument fails on two counts. First, the rule of induction is a direct violation of the rule of deduction that universal (or general) conclusions cannot validly follow from particular premisses – not even probability estimates[3] – thus induction is allegedly 'justified' simply by arbitrary legitimation of its unjustifiability. Secondly, the only rule ever stated is to the effect that generalization is warranted whenever a constant conjunction between specific events or objects has been observed. The simple observation of an object (let alone an event), however, involves the belief or acknowledgement that certain groups of sensible qualities are constantly conjoined, and thus inductive inference is already implicit in the mere recognition of an identifiable object. Moreover, the assertion that sensible qualities are constantly conjoined involves the placing of events in a time scale, which is impossible without the identification of objects. It follows that the very enunciation of the rule and the knowledge that it is being followed require its prior application, so that even in this form the 'justification' of the inference presupposes the inference itself, the rule for whose validity has not been, and cannot be independently pronounced.

Inference in the natural sciences is accordingly reduced by current doctrine to analytic deduction, which can produce no new knowledge and inductive generalization, which cannot be validated. In that case, the superiority of scientific procedure, over other methods that claim to achieve knowledge of the world – let us say, clairvoyance, divination or feminine intuition – cannot be established by the currently advocated philosophic theories. Nor can these theories be supported by the superior success of science, which is a patent fact, because that by itself is no evidence that the method adopted has been inference by enumerative induction. If it had been, past successes could not justify it as a method, because the only arguments by which the justification could be sustained would have to be inductive and so would inevitably beg the question.

Examination of the actual practice and procedure of scientists will expose the falsity and ineptitude of the above theories of scientific method and, at the same time will reveal a progression of conceptual systems, which answers closely to the account previously given of a scale of forms. Scientific procedure proves, accordingly, to be dialectical, to the demonstration of which fact I shall devote the rest of this paper.

III

A scientific theory, it was said above, is an explanatory system. The current view of explanation is that it consists simply of deduction, within a deductive system, of the fact or hypothesis to be explained, from a higher level hypothesis couched in theoretical terms. The meaning of these terms is specifiable (according to the doctrine) only by translation into the empirical terms descriptive of the observable facts. This is an odd view of explanation, for we normally mean by that word the setting of the *explicandum* in a context that will make it more intelligible. But if the facts to be explained supply the meaning of the explicatory terms, it is difficult to see how the latter can render the former more intelligible. It is sometimes alleged that explanation is the same as analysis; but analysis is itself a vague term with several applications and meanings, not all of them intuitively obvious. If it is taken literally to mean dissection or separation into elementary parts, it is unlikely to provide what explanation demands, especially when the *explicandum* is something whose nature is determined by its structure, for the pattern of structure would be destroyed by such analysis.

If, however, we take our cue from Kant and see knowledge always as a whole of related and connected elements, we are more likely to hold that explanation involves both analysis and synthesis: analysis to lay bare and expose the elements, synthesis to reveal the relations and connexions. And if these again are determined by some general principle of organization, the synthetic view will be more important than the analytic. The psychologist J. J. Gibson is probably nearer the truth when he writes, "The progress of learning is from indefinite to definite, not from sensation to percept" (where 'sensation' may be correlated with atomic element and 'percept' with structured object). If the progress of learning is the same as, or analogous to, the advancement of knowledge, we should then say that it moves from the vaguer and more confused to the more definite and integrated, and this is, in fact, what we find.

The most satisfactory accounts of perception that we have, whether psychological or epistemological, show it to be an activity of structuring vague and confused sentient elements to form coherent and systematically ordered objects,[4] and these again are recognizable and significant to the extent to which they fit into a wider context of mutually related objects. Our awareness of individual entities in our environment develops *pari passu* with our comprehension of that environment as a world of objects ordered in space and time, as well as in other respects. The ordering principles are concepts without which, as Kant taught, there is no coherent perception of objects. This presupposition of a conceptually ordered world implicit in the perception of ordinary material things is prior to, and is involved in, all scientific observation; so that there is no observation wholly free from theoretical presupposition.[5] Science, moreover, is the fruit of an intellectual effort to render the world of common perception – what we call 'the world of common sense' more intelligible. It arises out of puzzling experiences and develops as the resolution of initial problems gives rise to further questions. Human knowledge, in short, is always a system of some kind, but is never a fixed system established for all time. It is a constantly developing process – a continuing activity of discovery and enlightenment.

From what has been said it will follow that the logical character of this process of discovery will be that characteristic of the development of a whole. Ordinary perception – the common sense awareness of the world – is already a whole of related and connected elements. Incoherences in this structure give rise to problems that stimulate the scientific enterprise; and this consists in successive efforts to discover the coherent structure underlying the apparent discrepancies, and generates a series of theories of increasing explanatory power. Let us look at examples of this process in more detail.

Common sense accepts the familiar fact of motion as natural and unproblematic, and yet does not reject the conception of a static world as a possibility. We perceive things at rest and also things in motion and transfer the resulting ideas of them at will without undue shock. The Biblical poet speaks of the mountains skipping like lambs and hills like lambkins, and his reader, while he may be incredulous, does not find the image impossible. Further our own sense of effort in moving both our own bodies and others gives rise to a conception of a cause of motion. It seems obvious that something should act on a body at rest in order to move it; for rest seems to be the more natural condition which calls for no explanation, while motion carries with it the idea of activity.

Common experience seems to show that to be moved a body must be pushed or pulled, but falling bodies move without apparent impulsion. Here is a discrepancy in the manifest behaviour of things which demands explanation. Such an explanation was devised by the Ancients and set out systematically by Aristotle. It was to the effect that all bodies have a natural place in the universe from which, if they are to be moved, they must be violently dislodged, and to which, if they are dislodged, they will naturally return. The motion of falling bodies is, therefore, their natural motion in returning to the centre, which is the natural place of earth, whereas forcible impulsion is required to move them from rest in their natural setting. We are all aware that, far-reaching as it was, Aristotle's theory led to contradictions when applied to the behaviour of projectiles, and the subsequent efforts to resolve these conflicts gave rise to the theories, first of impetus and later of inertia and gravitation. A projectile like a flying arrow apparently moves independently of push or pull, so Aristotle's theory provided no explanation of its continued motion after its initial propulsion. The theory of antiperistasis, which was proposed to account for this, led to immediate contradiction for it presumed a rush of air behind the projectile to prevent a vacuum, which, while it might keep it moving, would provide no cause for its eventual cessation. This was likewise attributed to air pressure, so that one was committed to an appeal to the same process and the same force acting in diametrically opposed directions. To resolve this difficulty Buridan introduced the theory of impetus – a kind of induced natural motion originated by the original impulsion, which, like the loss of heat in a glowing body, gradually died out. This conception was still in use, though in suitably modified form, by Galileo in his attempts to explain the motion of falling bodies, and it merges, as we shall see, into that of inertia.

Thus a common sense notion leading to contradiction stimulates a theory, in which, as it is applied, new conflicts arise; and these in turn are resolved by successive theories, each applying to the same domain of facts, each ordering the facts by means of laws, but successively doing so more coherently.

As another example, the familiar experiences of permanence and change among things and their qualities prompt the ideas of underlying permanent substance with ephemeral qualities. But these are immediately in conflict, for if a substance is permanent in character, how can (and why should) its properties vary with time? Various hypotheses were advanced by the earliest scientists. Anaximander suggested a vibrant motion that separated opposites from the boundless substrate, opposites later to combine in varying ways and proportions; Anaximenes proposed condensation and rarefaction of the original substance, leading to a quantitative explanation; Anaxagoras conceived the basis of change as mixture in varying proportions of innumerable, infinitesimal seeds; and so we proceed to the atomic theory which explains all appearances in terms of aggregation and separation of minute elementary particles.

In this outline I have skimmed over the intervening contradictions, to which Zeno, in particular, drew attention. If rarefaction and condensation are to be intelligible, equal spaces must be conceived as occupied by different amounts of matter or substance. Matter must then be assessable quantitatively in units, which must either be divisible *in infinitum* or consist of ultimately indivisible parts. Both alternatives, as Zeno showed, lead to intractable paradoxes, a solution to at least some of which the atomic theory seems to offer. But the atomic theory will explain qualitative change only if atoms are constantly, or at least commonly, in motion; and motion is explicable only in terms of impulsion. Accordingly impact between atoms has to be envisaged as the mutual cause of their movements. But unless they are elastic, how could impact cause them to move, except while in mutual contact? And if a moving atom coming into contact with another does impart its own motion to the second, what is the cause of motion in the first? Elasticity, however, implies compressibility, which again involves contraction between parts and so requires a further division of atoms, contrary to their defining character, into lesser parts. So the effort at explanation, beginning from contradictions in common sense conceptions of things, proceeds by hypothesis, via new contradictions, to further hypotheses.

Whether an assumed explanation be termed an hypothesis or a theory usually depends on one or both of two considerations. If the fact, or facts, which it claims to explain, are few and of restricted scope, it may be regarded as an hypothesis rather than a theory; if it concerns a wide range of phenomena, the reverse. Or, alternatively, if it is a mere conjecture, as yet unproven, it will be called an hypothesis, but if it has been confirmed by numerous tests and is considered 'established', it is a theory. Nevertheless, no theory is finally established, and some writers (like Karl Popper) maintain that none can be established at all, though they may be falsified. So there is good reason for holding that all scientific theories are in this sense no more than hypotheses.

The function performed by hypothesis, by virtue of which it explains, is to organize the experienced phenomena into a system, and the degree of its success depends on the range of phenomena to which it can consistently be applied and the absence of conflict in the results of its application. The long hegemony of the Aristotelian system resulted from the eminent success with which it coherently ordered the common experience of movement, change, growth, sensation, thinking and moral conduct, in terms of a single principle of organization: the matter-form relationship. The collapse of Aristotelian physics resulted, nevertheless, from the contradiction which arose in the attempt to apply its theory of motion to the flight of projectiles.

Likewise, most of the visible phenomena of movement in the heavens were very successfully organized by the conception of spheres revolving around a central earth, except that the aberrations of planetary motion could not be consistently included. First Eudoxus worked out a model of spheres each revolving on a different axis, which carried the different heavenly bodies, each of which communicated its motion to the next. Callipus elaborated this system, increasing the number of spheres and introducing some revolving in opposite directions. This system was adopted by Aristotle, and later Ptolemy modified it still further by introducing epicycles between the spheres to account for aberrations which still defied explanation. Each successive correction of the system to overcome the discrepancies was a new effort at organization, each succeeded in some measure, but none completely, until the hypothesis of Aristarchus of Samos that the sun and not the earth be taken as central was adopted. The heliocentric theory was delayed by fourteen centuries because the Aristotelian system, with which it conflicted, was so comprehensive and explained so many fields and departments of common experience, that theorists could not afford to abandon it without a better alternative. In the Middle Ages it even became the philosophical support for Christianity so that a challenge to its authority threatened the very dogmata of the prevailing religion. It was eventually despair of accurate calculation of the annual occurrence of religious feasts and festivals which prompted Copernicus to revert to the hypothesis of Aristarchus in the hope of a more coherent theory.

This change was not prompted by the empirical discovery of new phenomena, either in ancient times or in the fifteenth century, but was the fruit of an intellectual endeavour to unify and integrate the system. As a consequence of its adoption, in Copernicus' words, "not only do their phenomena [of planetary motion] presently ensue, but the orders and magnitudes of all stars and spheres, nay the heavens themselves, become so bound together that nothing in any part thereof could be moved from its place without producing confusion of all the other parts of the universe as a whole'.[6] Each theory is a whole of related and connected elements; it relates and connects the factual elements by principles of order so that they constitute such a whole.

The common-sense view of the world is already a more or less systematic totality, but is rather less so than more. The conflicts which become apparent in it on reflection lead to the scientific endeavour to explain what is puzzling. This produces more consistent and more tightly knit systems ordering the observed facts in an intelligible way. But these systems again are partial, being restricted in scope, and (frequently for that very reason), as their implications are developed they evince new contradictions. The efforts to resolve these lead not only to more coherent, but also to more comprehensive and more widely interconnected systems, uniting the lesser ones and removing their inner conflicts. The effect of this progressive integration is the development of an hierarchical structure in scientific theory. Thus Newton's system united the areas of celestial and terrestrial mechanics, Maxwell's system united optics and electro-dynamics, Einstein's combines all these in a single system of physical geometry. The sciences, as they structure and elaborate their own fields, tend to unite and to constitute a single system. In each case we have a whole of related and connected elements, and knowledge advances through a series of such wholes, each at a higher level of integration, articulation and clarity – each successive system being more widely and more penetratingly efficient as a means of explanation.

The next point we must notice is that the establishment of a new hypothesis is not effected by a search for (or discovery of) a large number of similar confirmatory observations, but is achieved by the construction of a body of varying evidence, the diverse factors in which are interdependent and mutually supporting. The term 'corroboration' is more applicable to such mutual support of diverse phenomena than to the mere repetition of similar cases. The true empirical method actually practised by scientists is that of marshalling diverse pieces of evidence which dovetail into one another to form a system, and it is better described as construction than as induction. It may be carried out by a series of mathematical calculations based on empirical records, as Kepler constructed the orbit of Mars; or by a series of observations prompted by pertinent questions, like those Harvey made to work out the function of the heart and the circulation of the blood; or by a series of experiments directed towards proving a specific relationship of processes, as were conducted by Lavoisier in his discovery of oxidation in combustion and his disproof of the phlogiston theory.

In none of these examples do we find the scientist accumulating like instances of a general rule. Kepler worked with calculated positions of the planet none of which could strictly be observed, for nobody can directly observe the position of a planet in space relative to the sun, and Kepler could only work this out from angular measurements made of appearances in the heavens. When he had done so he had to construct the orbit more or less as one constructs a jig-saw puzzle out of displaced pieces. The various calculations did not give a collection of similar instances of a frequent conjunction of objects or events, but a range of differences among which a pattern had to be found. Similarly Harvey did not simply observe a number of like instances of vascular activity. He examined the anatomy of the heart and its valves, the position of the vessels and the valves in the veins, the action of the heart in living creatures, the behaviour of the blood in a ligatured limb, and numerous other diverse phenomena. He calculated the rate of flow and the quantities of blood passing through the heart in a given time. And all the varied evidence so gathered he welded into a structure of interlocking parts 'so bound together than nothing . . . could be moved from its place without producing confusion in all the other parts.' Uranus had been seen twenty times before Sir William Herschel 'discovered' the planet, yet the repetition of observation led to no recognition of a new planet. It was only after its position had been charted over a period and its orbit calculated that the 'discovery' became an established fact – only, in short, when the phenomena had been fitted into the astronomical system.

The method of science, then, is one of constructing systems. A scientific theory is a systemization of empirical facts according to one, or a few, principles of organization (cf. Newton's laws of motion), and it is established by the marshalling of a system of evidence, diverse in its details, which are interlocking and mutually corroborative. The history of scientific advance is a progression from one such system to another, each structured by a conceptual scheme that orders the facts, and which, in large measure, by ordering, constitutes them – makes them what they are. The supersession of one system by its successor is occasioned by the discovery of conflicts within the former, as its implications are developed and its details worked out, and by the consequent attempts of thinkers to devise a modified conceptual scheme that will organize the phenomena more coherently.

The special conceptual schemes may be of narrower or wider scope, but at any one time they fit (more or less loosely) together into one system covering the entire world and involving definite conceptions of its elementary constituents and substances as well as of the laws that govern their movements, combinations and changes. The process of scientific advance is thus a succession of wholes or systems, like the Hegelian dialectical system, and the series of conceptual schemes that emerge in it are related to one another as the Hegelian forms and categories are related. Like them each is a conception of the universe – a provisional definition of the Absolute.

The progress of scientific ideas further conforms to the dialectical progression in ways we have already listed as typical. In the first place, the development, despite appearances in some cases, is continuous. Secondly, thc consecutive systems are specifications of a universal – each is an example of a general type of scientific theory. And thirdly, each system comes to be in opposition to the one it supersedes. They are thus at once degrees in a scale of progressively more satisfactory explanations, species of a special kind or area of explanatory theory, and opposite or rival theories of a certain range of phenomena. The progress of science thus constitutes a scale of forms in a manner comparable to that demonstrated by Collingwood in the case of the history of philosophy. I shall attempt to give some illustrations of this thesis.

(i) Continuity

It is common to represent the transition from Aristotelianism to Copernicanism as a *volte face* or revolution, and likewise the transition from Newtonian classical dynamics to Einsteinian relativity; and, as I shall presently show, the changes are in a significant sense revolutionary. But the transition process is nevertheless continuous and each step leads naturally to the next. The Copernican revolution was no sudden *renversement*. Not only was the heliocentric hypothesis a revival from Aristarchus of Samos, which had forerunners among the Pythagoreans, but the Copernican system itself was full of Aristotelian and Ptolemaic survivals, of which it took more than a century of successive adjustments to purge it. Copernicus followed Aristotle in making the planetary orbits circular and Ptolemy in retaining some epicycles to keep the resulting motions close to observation. The notion of natural motion persisted and Galileo maintained that the Earth being spherical would naturally rotate on its axis.[7] The conception of inertia, which became a cornerstone of Newtonian mechanics was anticipated by, and developed from the Aristotelian notion of natural motion by way of the conception of impetus put forward in the thirteenth century by Buridan and his Parisian colleagues. The view that the earth rotated on its axis was anticipated by Nicole Oresme. Even after the heliocentric hypothesis had been advocated by Copernicus, it took much time and labour to develop it by stages into a system that could account for all the facts with sufficient comprehensiveness and coherence to persuade scientists to abandon the Aristotelian and Ptolemaic system in its favour. Kepler had to work upon the laws of planetary motion and Galileo on those of falling bodies. Descartes had to develop the notions of vortical rotation and conceive the idea of inertia, and Gilbert had to develop the conception of magnetic attraction, before the genius of Newton could combine all of these discoveries into a single all-inclusive explanatory system conceptually whole and self-consistent.

Despite superficial appearances, the Einsteinian revolution grew in like manner from seeds planted long in advance. The introduction into physics, by the mid-nineteenth century, of the concept of the field, was already the beginning of revolutionary change. It shifted the focus of attention from mass-points and forces to the space between charged particles and the time of propagation of electro-magnetic waves. Thus space, time and velocity became prior in importance to particles and impressed forces. Philosophers like Berkeley and Kant had even earlier called in question the absolute character of space and time, and later difficulties which increasingly beset the notion of an aether stimulated new theories. Mach, whose influence Einstein acknowledged, had already propounded relativistic concepts, and asserted in particular the relativity of gravitational, coriolis and centrifugal forces, and the Principle of Relativity itself was enunciated by Poincaré prior to the exposition of the theory by Einstein.

Similar continuity can be illustrated in the cases of other scientific revolutions – Harvey's in physiology was continuous with Galen's through a series of stages from Vesalius, who doubted the permeability of the septum, Colombo, who discovered the lesser circulation from the heart through the lungs, and Fabricius who first revealed the presence of the valves in the veins; Lavoisier's in chemistry followed the work of the British pneumatic chemists and their discoveries of 'fixed air' (CO_2), 'dephlogistigated air' (oxygen), and hydrogen, identified by Cavendish with phlogiston. Darwin's revolution in biology, likewise, was a continuous growth from the leads of his grandfather, Erasmus Darwin, of Cuvier, Buffon, Geoffroi St. Hilaire and Lamarck, as well as from those of von Baer and of Charles Lyell.[8] And Heisenberg's quantum mechanics has grown from the progressive development of theories put forward by Max Planck, Niels Bohr, the de Broglies and Schroedinger.[9] Space does not permit a proliferation of detail here, nor does my present purpose warrant it. What is important, however, is not just the demonstrable fact of continuity but its nature and direction.

(ii) Graded Specification

Each successive hypothesis is an attempt to explain the same range and class of facts, so that each is an example of the same general type of theory:

(a) one of celestial mechanics: Ptolemy's geocentric theory with circular orbits, deferents and epicycles; Copernicus' heliocentric theory with circular orbits and epicycles; Kepler's heliocentric theory with elliptical orbits, and finally Newton's gravitational celestial mechanics. (b) one of vascular physiology: Galen's implied doctrine of oscillating blood flow with the heart as a cooling mechanism, its main action in diastole; Vesalius' implied modification of this in the denial of the permeability of the septum; Colombo's theory of the lesser circulation; Fabricius' demonstration of the valvular obstruction in the veins to the flow of blood away from the heart; and then Harvey's coordination of all these in *De motu cordis* into a theory of general blood circulation, with the heart functioning as a pump, its principal action in systole. (c) one of combustion: Stahl's theory of phlogiston; the theory of 'fixed air' put forward by Hales and Black, Priestley's theory of 'dephlogisticated air', and finally Lavoisier's demonstration, following upon Guyton de Morveau's of augmentation, of the process of oxidation.

In each case we have progressively more adequate examples of the same universal type of principle of explanation. Later theories explain (i.e. fit into the coherent system) facts which in the earlier theories gave rise to conflicts and remained intractable, and they do so with fewer discrepancies of their own. They may, therefore, be regarded at the same time as degrees in a scale and also as specifications of a universal concept.

(iii) Opposition and Reconciliation

But because it is from initial contradictions that the new efforts arise, the successive theories are in mutual opposition, the later being corrections of the earlier, which by contrast appear as errors. The newer theory embodies a new principle of order and establishes a new conceptual structure. It is, therefore, genuinely revolutionary. Moreover, not merely is the new synthesis in opposition to the old, it reconciles oppositions within the old. Aristotle had argued that the heavenly bodies must move in a circle because it was the perfect figure, but to accommodate the motions of the planets Ptolemy had to introduce epicycles. The path of a planet turning on an epicycle which simultaneously moves along a deferent is not, however, circular and may well turn out to be elliptical. Here is a conflict within the conceptual scheme. Kepler indeed found the orbits to be elliptical and Newton's laws explained this fact coherently: but the circle itself, as a special case of the ellipse, was not wholly rejected. Moreover, the heliocentric system still entailed, now without contradiction, the apparent circular motion of the heavens, for this is the result of the earth's rotation on its axis – in short, a genuinely circular motion.

This is but one of innumerable possible examples of the way in which supervening theories, though in opposition to their predecessors, nevertheless preserve within themselves the elements of truth that their predecessors contained. In Hegelian terms, the earlier theory is *aufgehoben*, both annulled and at the same time preserved and sublated. In this way, the Aristotelian conceptions of natural and violent motion reappear in the Newtonian system in the guise of inertial and accelerated motion despite the obvious opposition between the two sets of concepts. Aristotle's natural motion was purposive, whereas inertial motion is purely mechanical, but the intervening concept of impetus is a kind of natural motion imposed by impulse. Though it was originally conceived as short-lived, it was also easily conceivable as persistent so long as not opposed by countervailing forces. Buridan suggested that God might have set the heavens in motion with a primordial shove, and that they could then have continued to revolve indefinitely by their own impetus. From such an hypothesis it is but a short step to the concept of uniform inertial motion in a straight line as Galileo's reasoning reveals. Aristotle's accompanying idea of violent motion is then seen as accelerated, and the opposition is resolved in the notion of a body moving under the influence of impressed forces.

The Ptolemaic system was geocentric, the Copernican in contrast was heliocentric, but for Einstein the choice between the two becomes no more than one between alternative co-ordinate systems and the opposition is reconciled. The classical (Newtonian) conception of the physical world was particulate; Maxwellian electrodynamics, in contrast, concentrates on fields and wave-motion. In contemporary physics the wave and the particle are complementary concepts and are combined in the notion of the wave-packet. Looking to another branch of science, Lamarck believed in the inheritance of acquired characteristics; Darwin, Mendel and their modern successors, in chance mutation and natural selection. Contemporary geneticists like Waddington give an organismic account of the genome and of genetic activity which preserves the Mendelian theory modified in the direction of Lamarck.[10] Also, Jacques Monod, who insists on the Darwinian account of evolution as natural selection of chance variations, nevertheless maintains that a free choice of the animal in pursuing a certain line of habitual behaviour may determine the direction of selective pressures and so orient the course of evolution towards the development of characteristics increasing the efficiency of the chosen behaviour.[11] This is Lamarckism in modern form – admission of direction of evolution by acquired habit.

The history of science thus presents us with a progress which is genuinely continuous, but in which successive main theories relate to each other as opposed and revolutionary, yet at the same time the later sublates (*aufhebt*) the earlier by preserving and transforming significant elements in it. The next succeeding revolution tends to unite these sublated elements so as to resolve the opposition that was most strident in the earlier theories. Consequently, we have an Hegelian triad, but this structure need not be, and should not be regarded as, rigid; for the actual candidates for the places of thesis, antithesis and synthesis may well be varied. Newton synthesizes Aristotle and Buridan, Einstein synthesizes Ptolemaic Aristotelianism and Newtonianism. It is not the triadic relation that is fundamental but the successive negation of negation involved in the generation of oppositions (through inadequacy) and their resolution in more coherent systems. The triad is the natural result, but it is incidental.[12]

The progress of science, therefore, generates a scale of forms, each being a theory of some field of experience whose function is to organize the relevant phenomena into a self-consistent system. In the first instance the principle (or principles) or organization are enunciated more or less in the abstract, or in a general form which requires articulation and development. This it receives in the course of application, in the attempt to iron out minor difficulties, during the period that Thomas Kuhn calls 'normal science', or the period of 'puzzle-solving'. Some of these difficulties can be removed with relative ease and without serious modification of the system. But others are more stubborn and if they are overcome it is usually at the cost of considerable revision. Frequently this sort of change reveals a crack in the fabric of the theory, which, in the course of time, brings to light intolerable contradictions. It is at this point that scientists cast around for new hypotheses. They are usually variations of the old ones suggested by the discrepant facts[13] (as Kepler's ovoid orbit was a variation on the circle produced by means of an epicycle turning in the opposite direction and prompted by the discrepancy of 8 minutes of arc), and they succeed when they bring to light a new pattern in which the formerly recalcitrant phenomena find harmonious place. This new pattern is the next in the scale of forms, related to its predecessor, first as an alternative theory, secondly as a more adequate explanation of the facts and so a higher degree of truth, and thirdly as a correction of errors (hence a theory opposed to the faulty one), yet a correction that does not totally reject what was earlier accepted, but preserves, modified and improved, the elements of truth contained in the prior view.

A scale of this kind is dialectical and the logic that impels it from one stage to the next is a dialectical logic – a logic of construction and of development appropriate to a continuous movement of evolution. It is that principle of absolute restlessness of which Hegel speaks, inherent in rational self-reflection and critical thought. In short, it is the principle of that systematic self-development which is science, "the true form in which the truth exists" (*Phän* [G], preface, p. 27) – though we must remember that, for Hegel, empirical science is but a part of the total system, an essential part, no doubt, but one that leads on through the same perpetual drive of self-reflection to the philosophical sciences of logic, philosophy of nature and philosophy of mind. The last-named presents natural science as a stage, which it undoubtedly is, of the development of consciousness – a phase of our awareness of the world, in which we view it as an external object to be observed and described in terms of coherent categories of rational understanding.

IV

Hegel's own account of the scientific process in the *Phänomenologie* shows observant reason as reflecting upon the common-sense view of the world (*gesunde Menschenverstand*), progressing from indiscriminate observation and minute description to classification, thence to the notions of cause and law, and so to ideas of unobservable imponderables that are the precise opposite of the material objects with which the investigation began. The opposition is resolved in the conception of organism specified in the processes of living functions, which are teleological. Mechanism and teleology are then further sublimated in mental functioning, and empirical science, which began with physics and mechanics, ends with formal logic,[14] empirical psychology, physiognomy and phrenology.

Much of this analysis is still applicable to, and true of, science in general, even as we view it today. But Hegel's account is naturally limited to the evidence that the science of his own day provided, much of which is outdated and obsolete. Hegel, however, was treating scientific ideas as typical of specific forms in the development of consciousness and not the actual history of scientific discovery. He comes nearer to doing this in the *Naturphilosophie*, but there he is more concerned with the conception of nature as such; and, of course, the only reliable source of that is empirical science, viewed now from the philosophical viewpoint, after reflection upon both the forms of consciousness and the categories of experience (in the *Logic*).

What I have tried to do in this paper is to show that the methods of discovery, confirmation and advance in the empirical sciences, both of his day and of our own, are dialectical in the Hegelian manner, rather than cumulative of empirically derived factual information coupled with merely formal deduction. Philosophy of science, if it is to give a faithful account of its subject matter as the scientific disciplines are actually pursued and conducted by practising scientists, must, it seems to me, return to Hegelian notions of logic and abandon the narrow empiricism to which it has hitherto been addicted.

NOTES

1 *Kritik der reinen Vernunft*, A 97.

2 Cf. my discussion in *Hypothesis and Perception* (London, Allen & Unwin, 1970), Chs. II and IV.

3 Cf. *ibid.* pp. 32–42.

4 Cf. Sir Frederic Bartlett, *Remembering* (Cambridge, Cambridge University Press, 1961), Ch. II; Floyd Allport, *Theories of Perception and the Concept of Structure* (London, Routledge and Kegan Paul, 1955) *passim*; Merleau-Ponty, *The Phenomenology of Perception* (London, Routledge and Kegan Paul, 1962) and *The Primacy of Perception* (Evanston, Northwestern University Press, 1964); Brand Blanshard, *The Nature of Thought* (London, Allen and Unwin, 1939), Bk. 1.

5 Cf. Harris, *Hypothesis and Perception*, Ch. VII.

6 *De revolutionibus orbium coelestium*, Prefatory letter to Pope Paul III.

7 *Vide*: *Dialogues Concerning the Two Chief World Systems* (Berkeley, University of California Press, 1953), p. 134. Cf. also pp. 31 ff.

8 Cf. Harris, *Hypothesis and Perception*, pp. 220–223.

9 Cf. Louis de Broglie, *The Revolution in Physics* (London, Routledge and Kegan Paul, 1954).

10 Cf. C. H. Waddington, *The Strategy of the Genes* (London, Allen and Unwin, 1957), Ch. V, and *The Nature of Life* (London, Allen and Unwin, 1961), pp. 96–97.

11 Cf. *Le hasard et la nécessité* (Paris, Seuil, 1972), pp. 141–142.

12 Hegel does not himself make this point, but it is no less true of his triads. For instance, the opposition of Being and Essence is reconciled in the Notion, yet that, as concrete, is opposed to either of the other two as abstract, and abstract and concrete are united and reconciled in the Absolute Idea. Again, Organism synthesizes Mechanism with

Chemism, but the organic is opposed to the inorganic and finds synthesis with it in the psycho-physical.

[13] It must, however, be borne in mind that there are no 'facts' independent of, or neutral with respect to theory.

[14] For Hegel 'the natural history' of thinking.

ERNAN McMULLIN

IS THE PROGRESS OF SCIENCE DIALECTICAL?

The claim that science advances according to a dialectical logic is not an unfamiliar one. It has always been a staple of Marxism-Leninism. And now that the empiricist account of scientific change in terms of the gradual accretion of law-like generalizations of ever wider scope has been discredited, Marxist philosophers are wont to say, somewhat smugly, that their dialectical principles prevented them from falling into the logical empiricist trap in the first place. Long before Kuhn and Toulmin came along with their talk of revolution and evolution, Marxists knew that change in science is *dialectical.* They did not need the critiques of empiricist theories of meaning, the appeals to the history of science, the analyses of theory-replacement, that now propelled others to new accounts of scientific change. Dialectical materialism had assured them, on antecedent grounds, that all development, whether of matter or of knowledge, must be dialectical in character. The dialectical model, in consequence, could resolve the deep differences that separate Kuhn, Lakatos, Popper, *et al.* So the story goes, at least.

The trouble here is with the notoriously vague term 'dialectical', and with the methodological structure of the argument for the dialectical thesis. We are fortunate, then, to have a detailed essay from Errol Harris in which this thesis is presented and defended.[1] Fortunate too that Harris takes Hegel rather than Marx for his inspiration, since Hegel is the source of the thesis and Marx has very little to say about it.[2]

Harris concludes his paper in these words:

> . . . the methods of discovery, confirmation and advance in the empirical sciences, both of [Hegel's] day and of our own, are dialectical in the Hegelian manner, rather than cumulative of empirically derived factual information coupled with merely formal deduction. Philosophy of science . . . must . . . return to Hegelian notions of logic and anbandon the narrow empiricism to which it has hitherto been addicted (p. 212).

In point of fact, Harris has little to say about the questions of discovery and of confirmation. And indeed there is very little in Hegel's work that would be of help in addressing the quite specific issues that philosophers of science have been debating in recent years in these two areas. It is not clear what a 'dialectical' mode of confirmation would be like, much less a 'dialectical' method of discovery.[3] Harris concentrates his attention, instead, on the more

R. S. Cohen and M. W. Wartofsky (eds.), Hegeland the Sciences, 215–239.

tractable question of scientific change. Progress in science, he says, constitutes a "scale of forms" (p. 206), a "succession of wholes or systems like the Hegelian dialectical system". The successive theories in a given scientific domain are "related to one another as the Hegelian forms and categories are related" (*ibid.*). This is the claim I want to investigate in this paper.

But first, one preliminary. Harris writes as though logical empiricism were still in possession of the field. He several times describes it as the 'current' view. He has little difficulty in locating the shortcomings of some of its characteristic doctrines, such as its dichotomy between theoretical and observational statements, for example.[4] But it is *not* the current view; its inadequacies were pointed out long ago. And this was done, it should be said, not by defenders of dialectical logic, whether Hegelian or Marxist-Leninist, but by philosophers of language like Quine and Wittgenstein, by philosophers of science from within the empiricist tradition like Popper and Feyerabend, and by historically-minded critics like Kuhn and Toulmin.

A Hegelian looking at the demise of logical empiricism in the 1960's would be likely to say that it generated its own contradictions. But if there has been a dialectic at work here, its outcome can scarcely be described as itself a 'dialectical' philosophy of science, as Harris takes it to be. The 'sublating' of logical empiricism has not left a single orthodoxy, whether dialectical or other, in possession. It has left us with choices to make between Kuhn and Popper, between Lakatos and Toulmin, between Feyerabend and everybody. And these choices have quite far-reaching consequences, for epistemology, for example. It seems doubtful that all the above-named would accept the generic title, 'dialectical', for their philosophies of science. And if they would, the term would have to be extended to near-vacuity to accommodate such radical differences about the manner and significance of scientific advance. But perhaps Harris would say that dialectical logic constitutes an *alternative* to all these other views, and one which is preferable to them as an overall philosophy of sciences? If this is his position, we shall require a fairly detailed specification of what constitutes a change as a 'dialectical' one, distinguishing it, for example, from the sorts of evolutionary and revolutionary changes proposed by Toulmin and Kuhn respectively. And we shall need some convincing arguments for preferring the 'dialectical' account to those given by other contemporary non-Hegelian, non-positivist philosophers of science.

Let us begin with those aspects of the Hegelian dialectic that would seem most relevant to recent developments in philosophy of science. Hegel emphasized the *historical* and the *systemic* character of knowledge and reality. True, his claims in these regards cannot be separated from the uncompromisingly idealist metaphysics in which they were embedded, a metaphysics which would find little support today. But the directions he took in reacting to a dominant empiricism may prove to have some analogies with the directions taken by the critics of logical empiricism in recent decades.

1. SCIENCE AS HISTORICAL

There is a trivial sense in which science, like any other human product, is historical. It appears in history; it is constructed by historical individuals in ways that are contingent. Furthermore, it does not remain the same. It advances, accumulates, progresses, changes, whatever one's favorite verb here may be. No one has ever denied these claims. Eighteenth-century science contained elements both of earlier deductivist views (which would make scientific change something like the derivation of new theorems or the discovery of new intuitively necessary axioms) and of inductivism (making scientific change primarily an addition of new empirical generalizations or an extension of older ones). In either case, science would proceed in a cumulative manner. Earlier states would be conserved in the later ones. The addition of new elements (laws or theories) would leave the 'established' earlier ones untouched.

Hegel's views on how thought and reality develop suggests a much more radical account of scientific change, one which may be called 'historicist' or 'transformationist'. At the level of what he calls 'understanding', precise and determinate concepts are developed in order to grasp the structures of experience. But these are inevitably inadequate; in the overcoming of the ensuing tensions, revisions of the conceptual system are made.[5] The system is never definitive then; no part of it is immune from challenge. Its warrant lies in its overall coherence, not in the priority of some foundational statements, whether axiomatic or observational. Its 'scientific' character lies neither in the intuitive necessity of its principles nor in its empirical basis but in the degree to which it has attained the status of reason. "The goal is this, that Spirit came to consciousness of itself or made the world congruent to itself – for these come to the same thing".[6]

In the empiricist climate of eighteenth-century science, it had been assumed that the concepts corresponding to the primary properties of matter (length, mass, and so on) were anchored safely in pre-scientific experience and were thus unproblematic. Since these were the concepts on which Newtonian mechanics was built, this system seemed to be immune from basic conceptual revision. It could be added to, of course, and could find new applications in areas like chemistry. But its grasp of the fundamentals of motion appeared to be completely adequate. A dialectical account of knowledge, on the other hand, would presumably imply that an irrevisable science of mechanics is not attainable, that contradictions are bound to be generated at all stages. Thus, Hegel can readily be made to appear the forerunner of the sort of historicism that has become the accepted view in recent philosophy of science.

Unfortunately the matter is not so simple. Nowhere in his writings (to the best of my knowledge) does Hegel explicitly treat the sciences of nature as a continuously evolving conceptual system. Nowhere does he attack the empiricist assumption that the basic observational concepts of length and time are already definitely grasped. Though the implications of his dialectical logic for the system of the natural sciences are clear to *us*, he did not (so far as I can see) think of them himself in the transformationist terms we might be tempted to attribute to him. Society, yes; organisms, yes; the life of mind, yes. But the science of nature: no.

It is important to see why this was so. For Hegel was not at all a historicist in the modern style. The function of dialectic was to reveal the *necessary* connections between elements. And once these came to light, no further contradiction could occur. It is instructive to recall what he has to say about Newton's laws of motion. They:

> are immortal discoveries which redound to the greatest honour of the analysis of the Understanding. The next step concerns their *proof* independently of empirical methods; and this proof has also been furnished by mathematical mechanics, so that even a science based on empirically ascertained facts is not satisfied with the merely empirical *pointing out* (demonstration). The *a priori* proof in question rests on the presupposition that the velocity of a falling body is *uniformly* accelerated; the proof, however, consists in the conversion of the moments of the *mathematical* formula into *physical* forces, into an *accelerating* force . . . and a force of *inertia* . . . determinations utterly devoid of empirical sanction and equally inconsistent with the Notion (*PN*, p. 57).

So Hegel rejects what he takes to be the *a priori* proof of Galileo's law given by 'mathematical mechanics', and goes on to propose his own proof:

> The law of descent of a falling body is a *free* law of Nature, i.e. it involves an element which is determined by the *Notion* of body. Since it follows that the law must be deducible from this Notion, what has to be done is to show the way in which Galileo's law . . . coheres with the determination of this Notion. This connection lies simply in this, that because it is the Notion that now determines motion, so time and space (as determinations of the Notion) become *free* in regard to each other; that is to say, their quantitative relationships conform to their *notional* determinations. Now seeing that

time is the moment of *negation*, of being-for-self, the principle of the One, its magnitude . . . in relation to space is to be taken as the unit or denominator. Space on the contrary is asunderness, and its magnitude is no other than that of time . . . (*PN*, pp. 58–59).

The proof continues in this vein and concludes: "This is the proof of the law of descent of a falling body as derived from the Notion of the Thing." A few pages later he goes on to give equally conceptualist proofs of Kepler's three laws, proofs which display his dialectical ingenuity at its furthest extension. Newton's formulation of mechanics he disregards, since despite its admitted mathematical convenience, it is without physical meaning, being "steeped in an unspeakable metaphysics" that runs "contrary to the Notion" (*PN*, p. 67).

Is Hegel's mechanics an *a priori* one? Not in the Kantian sense, certainly. It requires experience; it does not dispense with the work of the experimentalist. Hegel is not saying that Kepler's laws might have been obtained in advance by a philosopher reflecting on the Notion of the Thing. Rather he is saying that once they *have* been obtained empirically, the philosopher must set out to show their logical necessity (taking 'logical' here in the dialectical sense):

The philosophy of nature takes up the material which physics has prepared for it empirically, at the point to which physics has brought it, and reconstitutes it so that experience is not its final warrant and base. Physics must therefore work into the hands of philosophy, in order that the latter may translate into the Notion the abstract universal transmitted to it, by showing how this universal, as an intrinsically necessary whole, proceeds from the Notion (*PN*, p. 10).

Hegel's "absolute mechanics" (as he calls it) is thus *a posteriori* in the sense that empirical discovery of its laws must come first. But it is not *a posteriori* in the sense that its laws appear as the 'working out' of the Notion in a necessary way. But it is not analytic in the stricter sense, since the 'working out' is not just a matter of deductive exploration of the implications of a definition. The connectives are dialectical, not deductive; the philosopher reconstructs the sequence needed to show a series of dialectical relations leading from the Notion to the desired terminus. To the non-Hegelian, it might seem that the constructive imagination plays an important role in demonstrations of this kind, and that the adjective 'dialectical' severely qualifies the term 'necessary' in describing the status of the outcome.

Be that as it may, the important thing for us here is to note that Hegel, despite his emphasis on the historical character of reality and knowledge, maintains that philosophy of nature has already arrived at scientific truths possessing the character of 'intrinsic necessity.' Since nature is posited by the Notion, it displays an inner necessity which can, in principle, be fully grasped by a progressive and careful application of reason to the Notion. When it is thus grasped, is the resultant piece of necessary knowledge unchanging? Contemporary discussions of modality have revealed senses of 'necessary' which would make it possible for 'necessary' truths to cease being necessary (or truths). And Hegel's notion of sublation might suggest something like a constant transformation at the conceptual level.[7] But it seems more correct to interpret what he has to say about the 'intrinsic necessity' of the propositions of philosophy of nature as implying necessity in the traditional strong sense. Even if at a later time the philosopher constructs a more comprehensive dialectical sequence with which to show the truth of such a proposition, the necessity attributed to the earlier sequence would be such as to exclude the possibility that the original proposition might have to be modified.

Far from being historicist in the sense in which Kuhn or Toulmin are, Hegel is still working with something like the classical Aristotelian understanding of science (i.e. philosophy of nature) as demonstration leading to necessary and unchanging truth. It is true that for him history, dialectical development, plays a positive role in science that Aristotle did not envisage. But his historicism is balanced by his assertion of the inner necessity of the Notion, of which Nature is the externalization. On this necessity the entire functioning of his dialectical logic depends. And it leads him to hold that the philosophy of nature is moving towards the status of necessary truth, and that at some quite basic levels this status has already been attained. Insofar as there are barriers to its further attainment, they derive not so much from the contradictions that history still has in store, as from what he calls the "impotence of Nature", the fact that it preserves the determinations of the Notion only abstractly. This "sets limits to philosophy", as he puts it. One cannot expect the Notion to comprehend the infinite variety of forms which are the "contingent products of Nature" (*PN*, p. 23). A proper science of such products is thus not just difficult or slow in the making; it is impossible (*PN*, p. 24). This is closer to the Kantian distinction between a 'pure' (*a priori*) science and such empirical quasi-sciences as chemistry, than it is to the historical continuum characteristic of recent discussion.

2. SCIENCE AS SYSTEMIC

One theme that appears again and again in Hegel's writings (Harris reminds us) is that of the interconnectedness which is constitutive of system. An element of the system is specified by its relations with the other elements. Thus there cannot be simple concepts since to conceptualize is to interrelate.

A concept is necessarily bound with other concepts.[8] In the early empiricist tradition, it had been supposed that the basic terms of natural science could be given their significance by linking them directly to sense-contents. These observational terms were thus unproblematic and independent of one another, even though they could of course be interrelated in a meaningful way in scientific laws.

Newton's *Principia* had already implicitly called this sort of empiricism into question, although the challenge had not, in fact, been clearly perceived. The opening Definitions of that work (mass as "quantity of matter" and momentum as "quantity of motion") strongly suggested that the new mechanics was still dealing with familiar empirical notions, rooted in our everyday experience. But this was, in fact, not the case. Not only was there a sharpening of these notions which carried them quite far from their original meanings, but this sharpening was done by interlocking them in an intricate way with the notion of force by means of the three 'Laws'. The warrant for the new meanings of force or mass was not a direct appeal to some simple experiences of effort or of stuff. Rather, it was the explanatory success of the system taken as a whole. To the many who found his notion of force unsatisfactory, Newton could only point to its crucial role in a system whose predictive success was admittedly impressive. This answer was unsatisfactory to a phenomenalist like Berkeley, who insisted on the empiricist requirements for meaning. But what Berkeley and Hume did not see was that even the more innocuous Newtonian terms 'mass' and 'motion', and even 'length' and 'time', had also been shifted; they could now be understood only as elements in a tightly-connected postulational system, the warrant for which had to be found in the success of the system taken as a whole.[9]

Hegel never explicitly makes this point in regard to mechanics. He assails Newton's concept of force as illegitimate, as we have seen. But he does not underline the dependence in meaning of each term on the other, and the consequent shift in meaning that will occur across the system if one element be changed. Nonetheless, it can be said that this was a consequence of his treatment of concepts as integral elements in a system, rather than as linked one-to-one to sense-contents. Further, his emphasis on Reason tends to orient science to explanation rather than to mere empirical generalization. Thus there is a shift (in our terms) away from the empiricist concern with *law* to a more far-reaching and holistic concern with *theory*. Since the systemic character of concepts is much more evident in theory than in law, this is a further important modification in what we would now take to be the 'right' direction.[10]

But once again, one has to be careful not to make Hegel *too* modern in this regard. Balancing his stress on coherence and integration of disparate empirical elements is his hostility to explicitly hypothetical reasoning utilizing atoms, waves, ether, light-rays, or the like. He argues, for example, that water must not be thought to *consist* of oxygen and hydrogen; rather these latter are only "different forms assumed by water". The "greatest contradiction thus appears", he goes on, "when through the abstract thought of identity, the thing is still held to exist". He rejects the trust in the sciences of his day to explain the behaviour of a complex entity in terms of the presence within it of specified real constituents. Further:

It is a similar conception which looks on heat, water of crystallization etc., as reduced to latency. Heat, for example, is no longer seen, felt, and so forth; yet it is said to be still there, though not perceptible. But what is not subject to observation does not exist in this sphere; for to exist is precisely to be for another, to make oneself perceptible; and this sphere is precisely the sphere of existence (*PN*, § 286, *Z*, pp. 117–118).[11]

Hegel goes further than did later positivists in that he opposed the use of theories employing such elements even as heuristic devices. They are fictions of the Understanding and cannot but mislead. His preference for phenomenological or descriptive theories[12] thus leads him to reject Dalton's atomic hypothesis, notions of latent heat, Newton's theory of color, and so on. He can see no possibility of Reason's discovering *necessity* in such postulations, hence they cannot be regarded as the roots of exteriorization the Notion could create in its constitution of Nature. Nor do they have a direct empirical warrant.

Despite the acuteness of his perception of the systemic character of the sciences, then, Hegel was blocked from an appreciation of the ampliative power this character bestowed on theoretical reasoning. The warrant for the existence of such entities as atoms, as we know, lies not in the necessity of the conceptual connections nor in observation directly, but mainly in the fertility over an extended history of the conceptual system employing them.[13] The systemic character of science enables it to extend ontology indefinitely. And this is done, not in terms of constructions that can be made to take on some sort of necessity after the fact, nor in terms of static determinations inevitably doomed to be rejected, but by means of tentative models whose metaphoric power enables them to grow. The organic image here is one that Hegel could appreciate. But the science of his day still gave little evidence of the ontological power of retroduction that is so plain in the structural sciences of the past century. And his idealist standpoint had too much in it both of phenomenalism and of rationalism to allow him to see just how great the resources of hypothetical systemic thought could be.

In these two sections, I have argued that Hegel's emphasis on *history* and on *system* suggested some needed correctives for the philosophy of science of his day, but that his insights cannot be neatly transposed into the context of contemporary discussions because of the shape given them by the 'necessitarian' character of the Dialectic. It may be worth adding that these insights of his had remarkably little impact on later nineteenth-century philosophy of science, in part perhaps because of the difficulties noted above, but for a whole host of other reasons as well.[14] When tracing the origins of the historicist and systemic theses in contemporary philosophy of science, it would be easy to overestimate the actual historical influence of Hegel in their regard.

3. THE DIALECTICAL CHARACTER OF SCIENTIFIC ADVANCE

To say that science is a historically-developed conceptual system is not yet to say that its advance is specifically *dialectical*. Now we come to the main point of this paper, which is to investigate Harris's claim that science advances dialectically, according to the prescriptions of Hegelian logic. The weakest version of the 'dialectic' thesis would be that science advances by conceptual transformation and systemic revision, rather than by straightforward accumulation, as the original empiricist account suggested. If science consists of empirical generalizations of ever wider scope, the earlier generalizations are directly incorporated in the later ones and science is descriptive rather than explanatory in any stronger sense. This view still prevailed in Hegel's time (in part because the paradigm of science was assumed to be mechanics); one sees it in Comte, Ampère, Herschel, and Mill.

As we have seen, Hegel's logic directly opposed this sort of inductivism, although it was Kant's critique of it (especially as furthered by Whewell) that proved the more effective. Insofar, then, as the 'dialectic' thesis amounts to a rejection of the view of scientific change as accumulation, it would be accepted by Kantians and by most theorists of science of our own century, including the logical positivists, who were well aware of the importance of theoretical terms and of the fact that they are often modified, even drastically modified, in the course of time. This is clearly too broad a notion of 'dialectic' to be of much service, so we must see whether a more specifically Hegelian theme can be isolated.

Harris suggests three criteria of a properly Hegelian notion of dialectic in this context. Scientific change is continuous; the consecutive systems are specifications of a universal; each system comes to be in opposition to the one it supersedes (p. 206). It is the third of these on which I intend to focus, but a word about the other two first.

Continuity: On the face of it, it seems simply false to claim that the development of science is continuous. Instances of discontinuity are everywhere: the Wegener continental drift hypothesis, which, in its plate-tectonic form, has revolutionized geology in the past decade; the Planck-Einstein quantum hypothesis; Lavoisier's oxygen hypothesis – the list could be extended indefinitely. What, then, can Harris mean? He maintains that "despite the appearances in some cases", scientific change is continuous. But then he goes on to say that many of the major changes in science "are in a significant sense revolutionary" (p. 206). So that his criterion (which at first might seem to lean to Toulmin, say, and away from Kuhn) begins to seem rather elastic. 'Continuous' for him appears to mean (1) that there are hints in advance of even the most novel-seeming hypotheses, and (2) that "each step leads naturally to the next" (*ibid.*). I would want to argue that neither of these alleged requirements can, in fact, be maintained, in anything other than the vaguest senses of the key terms.

It would be agreed by all that scientific discoveries do not occur in a vacuum, that even the most creative thinker makes use of ideas of which he was not the discoverer. He may modify them or juxtapose them in new ways, but no system is *entirely* new, if by that is meant that it makes no use of elements previously available.[15] But surely this is trivial, and is not enough to constitute continuity in any meaningful sense. There was no hint in advance of the Planck quantum hypothesis; it did not develop out of some 'more-or-less' quantum views held before-hand. To describe Aristarchus' heliocentric hypothesis as an antecedent of Copernicus' is perfectly correct. But to generalize from this that 'hints' of this explicit sort must always be available is to misunderstand the creativity of the scientist. Of course, there are anticipations of Galilean mechanics in earlier thought. But it is risky to say that Galileo used a 'modified' conception of Buridan's notion of impetus in his account of falling motion, one that 'merged' into the concept of inertia (p. 201).[16] This is to diminish unduly the basic shifts that occurred in early mechanics, though there were undoubtedly analogies at work of the sort Harris suggests. Likewise, to say that "the Einsteinian revolution grew ... from seeds planted long in advance" (p. 207) is in one sense trivially true, but it is nonetheless a dangerous formulation if this is taken to imply that there was nothing in the Einsteinian formulation that did not have a specific antecedent in earlier thinkers. Even more dangerous, if this be proposed as a generalization covering *all* discovery.

The notion that each step in science leads 'naturally' to the next is equally open to question. It becomes either trivial or false, depending on the sense given the term, 'naturally'. There is a sense in which each stage sets the problems for the next and suggests parameters for the kinds of solutions that are likely to be effective. Once again the scientist requires a very considerable continuity of language and method, even for the most creative leap ahead. There is no disagreement about this (although the extent and nature of the continuity has been vigorously debated of late, Feyerabend in particular limiting it quite sharply). But it would be wrong to suppose that there is some sort of 'natural' path from stage to stage, a 'logic of discovery', as it has been called, that would allow one to say, after the fact, that, for example, the situation in spectroscopy in the early 1900's led 'naturally' to the Bohr hypothesis.

What Hegel himself had in mind might rather better be described as a progression in which each stage contains the earlier ones implicitly; they are 'sublated' and their truth is manifested in the manner of their retention. This is reminiscent of the logical empiricist thesis that each successive theory must be able to explain all that the earlier one it replaced did, plus something more. But for Hegel, the 'sublation' supposes a transformation that is incompatible with the simple retention the empiricists had in mind. And Kuhn and Feyerabend have challenged the empiricist continuity thesis, pointing out that not infrequently there is a loss of explanatory power, something which the earlier theory *did* handle but the later one cannot.[17]

The trouble with the 'sublation' notion lies in the vagueness of the phrase, 'contains implicitly'. It can be trivialized as: "there must have been some hints in earlier thought", so indefinite as to be incapable of test. Or it can be put in terms of empirical predictive power, in which case it runs into the 'Kuhn-loss' objection. Or it can be made to correspond with a substantial retention of theoretical entities, of the sort that most construals of scientific realism seem to demand. But this encounters the familiar objection of the phlogiston/caloric sort. And more seriously, it is incompatible with the Hegelian refusal to allow ontological significance to such entities. In short, until the 'sublation' idea is spelled out in quite specific terms, it is difficult to know what it would amount to in the context of scientific change, as we know it.

Specifying a universal: The effect of the dialectic is to make a universal become progressively more specific. Harris puts this as follows: "Each successive hypothesis is an attempt to explain the same range and class of facts, so that each is an example of the same general type of theory" (p. 208). The familiar charge: 'vacuous or false' can once again be made. To say that the progression of theories belongs to the same general 'concept', e.g. mechanics, is analytically true, since if it counts as a 'successor' of T_1, T_2 would automatically be described by the same very general category; it presumably explains most (at least) of the facts that T_1 explained. But in an important sense, the supposed requirement cannot be admitted. Very often T_2 will unite classes of 'facts' hitherto seen as disparate; 'interfield' theories (as these have been called) are of enormous importance in the history of science, and the manner of their 'linking' to form a new type of theory is significant in any general account of scientific advance.[18] If 'specifying a universal' be taken to be a gradual process of focussing on a domain whose boundaries are more or less set from the beginning, then it is an inadmissible requirement.

Yet Harris can scarcely have intended the 'specification' of a universal to be taken as restrictively as this. He elsewhere counts it a merit of the Hegelian scheme that it suggests a "progressive integration" in which previously unrelated areas (like optics and electrodynamics, for example) would be brought together in a single scheme, thus organizing wider and wider sets of observations (p. 204). It is hard to reconcile this with his statement of the 'specification' principle as requiring successive theories each to apply to the "same domain of facts"; the progression would then (as he says) be one of 'coherence' in the laws by which the domain is organized (p. 201).

Hegel himself has relatively little to say on this topic. But he certainly does envisage some kind of progressive integration. He criticizes Berzelius's unification of the disparate domains of electricity and chemistry, not because such unifications are misguided in principle, but only because Berzelius's 'electro-chemistry' has (in his view) outrun the facts. He shows himself here, as elsewhere, fairly knowledgeable about the empirical details of galvanic electricity and chemical action as they were known in his day. And he admonishes the chemists not to overlook the distinctions between the electrical and chemical aspects of the galvanic process:

> Perhaps this higher demand, addressed to the instinct of Reason, to grasp the course of the galvanic and the electrical process as such, conceived as a total of natural activity, is responsible in part for the fact that the lesser demand, viz., simply to take notice of empirically demonstrated *facts*, has up till now met with scant compliance (*PN*, § 330, p. 246, see also § 313).

But could the unification of the chemical and electrical aspects of nature be brought about under a *single* universal, a universal that had been inchoately present from the beginning of the scientific inquiry? Hegel's system oversimplifies the resources for conceptual *innovation* that science possesses. The unification in this case occurs through the postulation of underlying structures which are characterized by means of newly-developed complex concepts. These are validated retroductively not primarily by the internal properties of coherence and simplicity of the resultant system but rather by the success of the system in prediction and more especially in the guidance of research into new areas. The process here simply cannot be characterized as the specification of a single universal, dimly grasped from the beginning.

Progress through contradictions: The most characteristically Hegelian feature of a dialectical movement is that it progresses by generating contradictions whose resolution carries the movement onward stage by stage. In the *Logic*, Hegel tries to show that contradiction belongs to the very nature of our most basic categories. The positing even of such a category as *being* already involves one in an incoherence that forces one to a further level (*determinate being* in this case), and that in turn to a further (*infinity*). The new concept resolves an internal contradiction in the original category; one is forced to it by an attentive consideration of that category. It is not a question of empirical conflict with the facts. Rather, it flows from the internal logic of the concepts themselves, and thus determines the ontology, which for Hegel is the *embodiment* of Spirit rather than something over against it.

A second sort of dialectic is found in the historical development of society, which also is powered by internal contradictions. The historical civilizations are successive realizations of Spirit, each stage being a fuller expression of reason than the previous one. At each level, a people labors to develop a particular social and political form. In doing so, they expose its internal contradictions, and thus prepare the way for its demise and the origination of a new form which will incorporate that which is worth retaining of the earlier one. It is the "cunning of reason" that the Idea should thus make men, all unknowingly, work for its progressively fuller embodiment. But the philosopher, looking back at the sweep of history, can see the internal logic that carried it in the pattern of concept, contradiction, and resolution from one stage to the next.

Hegel does not, however, make explicit use of this pattern to illuminate the history of natural science. Was this simply for want of time or opportunity? Or because the historical development of the sciences had not progressed very far by his day? The reasons, to my mind, go deeper than this. They derive, I would suggest, from two fundamental optimisms in the Hegelian program, both of them based on pervasive ambiguities in his thought. One is the assumption (rooted in his powerful analysis of political forms) that a historical dialectic must follow the dialectic of the reconstructive reason. The other is the assumption that the empirical is the occasion for, and not in any ultimate way the warrant of, the assertion of the theoretical scientific reason. He writes:

> To the observing consciousness, the truth of the law is found in experience, in the same way that sensuous being is [an object] for consciousness; is not in and for itself. But if the law does not have its truth in the Notion, it is a contingency, not a necessity, not in fact a law. The fact that it is essentially in the form of Notion, not only does not conflict with its being accessible to observation, but rather for that very reason gives it necessary existence, and makes it [an object for] observation (*Phen* [M], § 249).

This view of the role of the empirical in the development and warranting of theory would make it very difficult indeed for him to make sense in anything other than a very selective and schematic way of the history of the natural sciences between Galileo's day and his own. This is where Harris boldly takes the plunge and asserts that the Hegelian dialectic *can*, in fact, be successfully applied to the illumination of the history of science, just as it can to the history of civilizations. "My object is to show that [the dialectic] is not confined to philosophy but is also the form of scientific theory and scientific advance" (p. 197). Let us see whether he can bring this off.

How, in his view, does scientific change occur? Each theoretical system "automatically [generates] the next, by its inherent *nisus* to overcome its own shortcomings and the contradictions to which they give rise" (p. 197). These "internal conflicts and oppositions . . . can be resolved only by further self-development and explication" (*ibid.*). When scientists continue to explore a theory, "intolerable contradictions" invariably appear (p. 210) which force the abandonment of the theory and the formulation of a new and more all-embracing hypothesis, the successive theories being "related as opposites" (p. 197), "related as opposed and revolutionary" (p. 210). This certainly sounds like Hegel. But how well does it respect the history of science?

Change in science can be of many sorts. First it can be nomothetic: a new empirical law is proposed, an older one extended or sharpened. Such advance is crucial to science and constitutes by far the largest part of day-to-day scientific work. Harris himself mentions (p. 204) that the "true empirical method" involves this kind of patient accumulation of skillfully varied observational evidence; he alludes to the work of Harvey and Lavoisier as examples. What he does not seem to see is that the motive force behind advances such as these is not *contradiction* in previously known empirical law, i.e. something a theorist would recognise as a discrepancy. When the experimenter extends the gas laws to even lower temperatures or higher pressures, for example, he is not responding to any sort of perception of incoherence in these laws as they stand.

Harris adopts two different strategies in the face of this obvious objection to his thesis. The first is to weaken it by using terms like 'inadequacy' instead of 'contradiction', and to propose that a change in science can occur only when some inadequacy is perceived. This is true, but is of course vacuous. The scientist will presumably not move from stage A to stage B unless he has some reason to suppose stage A incomplete, inadequate, in *need* of change or supplementation. Once again, we are back to the objection of vacuity. But even apart from the fact that this would deprive the thesis of any interest, it should be obvious from what we have seen of Hegel's views above that this is not at all what *he* was saying. His notion of contradiction had much more 'bite' to it.

Harris' second strategy is to suggest a Kuhnian distinction between 'revolutionary' changes in science and those that occur in the process of 'normal science', only the former qualifying as properly 'dialectic' (pp. 210–211). Apart from the fact that this could restrict the scope of the dialectic thesis to a rather small number of developments in the history of science, the objection that Kuhn himself would be likely to put is the important role played by the accumulation of empirical anomaly in such 'revolutions'. But might these not be called 'contradictions' in only a slightly extended use of the Hegelian term?

Before answering this, let us look at some other sorts of changes in science. There can, as we have just seen, be 'revolutionary' ones where a theory that articulates an entire domain is replaced by a substantially different theory. Or there can be less shattering changes where a set of empirical laws for the first time receives some sort of theoretical unification, or where conflict between a number of competing theories in a relatively restricted domain (e.g. origin of the moon, nature of capillary attraction) is finally resolved. This latter case seems to involve 'contradiction' of a sort; there is surely a sense in which alternative theories put forward to account for the same set of empirical regularities 'contradict' each other.

But here again, vacuity lurks nearby. It *is* perhaps worth saying (since the early empiricists so often missed this point) that the acceptance of a theory entails *in some sense* the rejection of the earlier one, if there was an earlier one in regard to roughly the same domain. Harris puts this by saying that "the successive theories are in mutual opposition, the later being corrections of the earlier, which by contrast appear as errors" (p. 208). But this notion of a 'rejection' has to be taken carefully here, as critics of Feyerabend's use of 'refutation' in this same context have pointed out. Newtonian physics is still used for the vast majority of physical problems today; to say that it has been 'rejected' is to say that as an overall explanatory system it has been proved to be inadequate. But a relation of inadequacy of a theory in regard to its successor is, as we have just seen, hardly an interesting sense of 'contradiction', and is certainly much weaker than the relation claimed by Hegel.

There is one important type of change in science to which the notion of dialectic contradiction is peculiarly inapplicable. Sometimes a theory proves unable to predict some domain of phenomena which the scientist believes it *should* be able to predict. Thus, for example, the original Bohr theory of the H-atom proved unable to explain the fine-structures of the spectroscopic lines produced when the emitting hydrogen was subjected to a magnetic field (Zeeman effects). It was not so much that it predicted them wrongly as that it did not predict this cluster of effects at all. It would be entirely misleading to describe this as a 'conflict' or 'contradiction' within the theory; it would also be wrong to describe the relationship between successive theories (or successive versions of the theory) here as one of conflict or opposition.

Quite the opposite, in fact. A good theoretical model is expected to have the resources of a good metaphor. That is, it is expected to *guide* research, not just by the predictions it makes, but by further conceptual innovations it may suggest in the face of anomaly or a range of phenomena (like the Zeeman effects) which *ought* to prove tractable. The original gene theory of heredity was quite vague about where and what the gene was. But it provided a research program within which the gene came to be more and more closely linked with a location on the chromosome, on the basis of very sophisticated experimental testing of alternative sub-hypotheses. The primary impetus here was provided not by contradiction nor even by anomaly but by the imaginative resources of the original model which at each stage pointed the way to the alternatives that had to be explored in order that the structures of the model itself might be progressively more sharply defined.

One might want to say with Hegel that in cases like this one a universal is being progressively specified. But what the Hegelian system does not account for is just why this happens. It is not contradiction or incoherence in the original concept that leads to the formulation of an opposed alternative.

Rather, it is an unresolved set of choices within the original metaphor (a 'neutral analogy', in Hesse's terms) that provides the way ahead. Harris's flat statement that "the supersession of one system by its successor is occasioned by the discovery of conflicts within the former" (p. 205) cannot finally be admitted.

Internal contradictions do, of course, occasionally lead to the abandonment of a theory. Harris mentions several such cases: the contradictions that kept cropping up within Aristotelian physics throughout the Middle Ages, which gave Galileo part (but *only* part) of the ammunition he used so effectively in his *Dialogue on Two Chief World Systems*;[19] the contradictions that from the beginning were noted within the Newtonian formulation of the 'absolutes' of space and time and which might be said to have pointed the way for Einstein. But one cannot properly generalize from cases such as these, which are characteristic of the most general formulations of mechanics, to a claim about scientific change generally.

In practice, however, Harris seems quite often, in his references to history, to be settling for a lesser claim. It is empirical anomaly that forces the scientist to reconsider his theory, some observational datum that just does not fit. The history of early astronomy, for example, is one of successively more complex systems of epicycles, equants, and the like, the motive for each stage being some "observations that still defied explanation", 'discrepancies' between the theoretical scheme and the observations (p. 208).[20] It is of no importance (as Harris seems to think in this context) whether the aberrant phenomena were 'new' ones or not. What *is* important is that the 'contradiction' in such cases was not in any sense an *internal* one. It was not generated by the internal logic of the system, but by the '*phenomena*', new or old.

This is really the crucial point. The anomalies that so often decide the direction of theoretical change cannot be anticipated. And when that change *has* occurred, it cannot, after the fact, be made to seem like the inexorable march of conceptual dialectic. It is plausible enough to suppose that the pure position of the philosopher may generate its opposite, and that the ensuing dialectic will carry inquiry forwards. It is plausible that, in some cases at least, the embodiment of a particular political form in a society will provoke its opposite, and that in the resultant clash, through the efforts of individual men, a synthesis embodying something of both may emerge. But it is *not* plausible to represent in this manner the challenge of empirical anomaly to theory. The theory does not generate the anomaly, as anomaly. It is observation, theory-impregnated of course, but nonetheless observation that does this. Internal criteria of coherence and simplicity are important, and often decisive. But by themselves, they are not sufficient to account for the kinds of change that occur in natural science.

Harris himself appears to recognize this when he says that "the establishment of a new hypothesis . . . is achieved by the construction of a body of varying evidence", leading to "corroboration" from the "diverse phenomena" (p. 204). Unfortunately, he then goes on to construe the notion of construction as a means of ordering the facts, "which in large measure, by ordering, constitutes them, makes them what they are" (p. 205). This enables him to return to the Hegelian internalist thesis: theories change as their "implications are developed and their details worked out". What must be done (he continues) is to "devise a modified conceptual scheme that will organize the phenomena more coherently" (*ibid.*). It is as though the set of phenomena is somehow already there, and the criteria of the successive theories are those of coherence only. Theory does not simply *constitute* fact; if it did, empirical anomalies would never arise. Theory shapes the way in which fact is expressed. But that is not nearly enough to allow the primacy of reason that the Hegelian dialectic demands.

In this connection, it is important to notice that Hegel did not recognize the importance of retroductive patterns of inference. There were, on the one hand, the empirical inductions of the inorganic, and on the other the most general laws of mechanics and the general interrelations of the organic world. It was only in the latter that he could trace the lineaments of the Idea. The structural sciences of chemistry and geology were only beginning to display their power and range. The ability of structural models to stretch out beyond the necessities of the rationalist or the empirical generalizations of the inductivist was still not plain. So that Hegel, to the extent that he saw the inadequacy of the inductivist model, could be forgiven for assuming that his own rationalist alternative was the only other one which could account for the onward march of science.

Harris can less easily be pardoned. He rejects the inductivists' account of scientific change in language sufficiently ambiguous, for the most part, to count as an endorsement either of rationalist dialectic or of retroductive inference. But he cannot have it both ways. If empirical anomaly and corroboration by the variety of empirical evidence play the role he allows them to do, he has opted for retroduction. To retain the language of Hegel is at the very least misleading once the substance of the rationalist metaphysic underlying it has been compromised.

4. THE WARRANT FOR THE DIALECTIC THESIS

This raises one last issue. On what did Hegel's claim for the dialectic rest? Taylor has pointed out a fundamental ambiguity at this point in Hegel's argument between the 'contrastive' sense, in which we cannot have the shape-concept *square* without contrasting it with other concepts like *round* which 'negate' it, and the 'interactive' sense in which beings struggle to maintain themselves in the face of others and hence 'negate' each other in an active way.[21] It is essential to Hegel's argument for dialectic as a universal pattern that he elide the differences between these two senses. A determinate being can only be defined by reference to another with which it is contrasted. This other is its negation. But Hegel then goes on in a way which (as Taylor says) "arouses our suspicion", to suppose that this negation is to be understood not merely in its contrastive sense but also in that of interaction. Thus the determinate being has to *struggle* against its negation; the outcome can only be the modification of both.

The confusion here is manifest. It is true that everything does have a contrastive frontier with others. Its frontier is constitutive of it:

> Hence in containing it, each contains what negates it as well as what essentially constitutes it. If we now shift to the other sense of frontier, that of interaction, we can give this "negation" a concrete as well as just a contrastive logical sense, and it looks as though each entity essentially contains the seeds of its own destruction. But, of course, however much we may be tempted to speak of something as "containing its frontier" in the contrastive sense, when we move to the frontier at which things "negate" each other by interaction, it is just false to say that each contains its own negation. Quite the contrary, to the extent that they maintain themselves, they hold their "negations" off. If they fail to do so, of course, they go under, but they are not essentially determined to do so by the very way in which they are defined.[22]

There is another difficulty, already alluded to, facing anyone who attempts to apply the dialectic thesis to the history of science. Hegel assumes that the *historical* dialectic follows the same path as the *logical* (or ontological) one. This may have some plausibility in the politico-social realm. But why should one expect it to be the case that *scientists* in their work retrace the lines of the necessary dialectic of Reason? Hegel makes a sharp distinction between empirical science (which he identifies with the systematic observation of nature) and philosophy of nature (which takes empirical science as its presupposition, reflects on it and shows how nature is an externalization of the Idea). This distinction is inadequate on several scores, but leaving this aside, how is Hegel to show that the gradual development of empirical science itself follows a dialectical pattern? This in no way follows from the logical dialectic which the philosopher of nature may be able to discern in his reflection on the results of the scientist.

A couple of examples of this latter dialectic may help to make this point:

Darkness, as immediately the negative of light, is the opposition to light's abstractly identical ideality; it is this opposition in itself. It has material reality and falls apart into a duality: (1) corporeal difference, i.e. material being-for-self, rigidity; (2) opposition as such which . . . is merely sunk within itself and is thus a dissolution and neutrality: the former is the *lunar*, the latter the *cometary* body (*PN*, § 279, p. 99).

Thus, moon and comet are both dialectically derived from darkness. Their "peculiarity as relative central bodies" in the solar system "is based on the same Notion as their physical peculiarity: they do not rotate on their axes". They arc two logical sides of an opposition, so that their presence in the solar system is not a chance matter once we have grasped the nature of the Notion. "They constitute the self-subsistent moments of the dissolving earth: the moon is the earth's hard interior, the comet is the earth's atmosphere which has acquired independent existence, an enduring meteor".

Later, he rejects the chemical notion of an element as an 'abstraction' and proposes instead a four-element theory based on an intricate set of oppositions:

Air corresponds to light; it is passive light which has sunk to the level of a moment. The elements of opposition are Fire and Water. Rigidity, the lunar principle, is no longer indifferent, existing by itself alone; but as an element entering into relation with another, namely the individuality, it is an active unstable being-for-self which is in ceaseless process, and thus is liberated negativity: Fire. The third element corresponds to the cometary principle and is Water. The fourth is the earth again . . . (*PN*, § 281, p. 106).

From oppositions of this imaginative sort, light and darkness, active and passive, rigid and unstable, magnetisim and electricity of opposite polarities, and so forth, he constructs an immensely complex logical dialectic. My purpose here is not to comment on the validity of this dialectic as an explanation of nature. Rather, it is to make the simpler point that it has no bearing whatever on the manner in which the empirical scientist develops his laws and theories. The oppositions on which Hegel bases his dialectic of moon, comets, physical elements, are not reflected in the history of astronomy nor of chemistry. Nor, indeed, does Hegel claim that they should be. But if the history of empirical science cannot be represented by the logical dialectic revealed in the philosophy of nature, what grounds are there for supposing that this history *is* in fact dialectical?

We have earlier questioned the validity of Hegel's argument for the universality of the dialectical process. The suggestive analogies which carry his analysis of history and of ontology are not available in the history of science. Is there any other source of evidence for the dialectical character of this history? Harris spends much of his time pointing to episodes in the history of science which he takes to exemplify the dialectical principle. We have argued at some length above that most of them do not do so unless the notion of 'dialectical' is extended to the point of near-vacuity. But suppose they *did* exemplify it. Are they then being presented as the *warrant* for our accepting it? Harris nowhere develops Hegel's own argument for the dialectic, and the impression the reader might easily receive is that the reason for accepting his thesis is that it does in fact describe the realities of the history of science so exactly.[23]

If this is the case, it is of course an *a posteriori* inference of a rather un-Hegelian sort. One would then have to ask (which he does not) whether there might not be other alternative modes of description that might not do the job better. But perhaps the references to history are intended only as illustration, not as evidence? It is hard to be sure. But either way, it would seem that the dialectic thesis is in trouble.

We began this paper with a reference to the Marxist version of the dialectic thesis, so perhaps it is appropriate to turn once more, very briefly, to this before ending. What kind of warrant can Marxism-Leninism produce for its dialectical claim for the history of science? Because it rejects the idealism of Hegel, it is deprived of the central argument he could call on: that nature is an externalization of mind and thus the same structures must govern the processes of both. For Hegel, as Taylor puts it, "the motor force of movement is contradiction, that between the external reality and that which it is meant to realize".[24] But in a materialist metaphysics, this tension does not exist, since nature is not meant to realize anything. Only a teleology that verges on the (forbidden) idealist one would seem to justify the reimposition here of a dialectical scheme on nature.

Marxist-Leninist writers make much of the presence in nature of 'contradictions'. Leaving aside the question of the appositeness of such a term to describe such polarities as that between electrical attraction and repulsion,[25] the more pressing question for us in this context is: how is one to get from such 'contradictions' to the claim that thought about nature, because of the inevitable occurrence of internal 'contradictions', must follow a progressive dialectical pattern? It would be just as much a fallacy to infer from the fact (if it *is* a fact) that material nature is dialectical to the dialectical character of the history of science, as it was for Platonism to argue from the unchanging character of true science to the (necessarily) unchanging character of the objects of science.

Marxism-Leninism cannot point to the kind of coherent presupposition for the dialectical quality of mind and nature alike that Hegelianism can. It cannot, therefore, undertake the sort of ambitious logical dialectic of categories in terms of which Hegel tried, in the *Philosophy of Nature*, to construe the science of his day as the embodiment of dialectic reason. Marxist-Leninists must then depend either on the results of an analysis of cognitive development generally, or else on direct evidence for the dialectic thesis drawn from the history of science. Both ways have been tried,[26] but it would take us too far afield to analyze these efforts in the detail they deserve. Suffice it to say that the same tendency to broaden the notion of 'dialectical' to cover all sorts of scientific change, evolutionary and revolutionary, cumulative and transformational, which we have already seen in the case of Harris, can be found among Marxist-Leninist writers also.

To describe the history of science as a 'dialectical' process can be a legitimate shorthand way of repudiating the inadequacies of the classical empiricist account of this history. But as an analytic instrument which would be of service in the complex debates now going on around the exact nature of scientific change, this term seems, finally, to be of very little use.

University of Notre Dame

NOTES

[1] 'The Dialectical Structure of Scientific Thinking', in the present volume.

[2] In his introduction to *A Contribution to the Critique of Political Economy* (1857) and in his Preface to the German edition of *Capital* (1873), Marx discusses the dialectical character of science. He refers his views on this to Hegel, and argues that the specifically idealist cast Hegel gives the topic can be disregarded as unnecessary and unacceptable 'mystification'. But he declares himself in full agreement with Hegel's dialectical model of how theoretical science proceeds.

[3] He speaks of a dialectical logic which 'impels' science from one stage to the next, a "logic of construction and of development appropriate to a continuous movement of evolution" (p. 211). But this is not at all a 'logic of discovery' in the sense in which this has recently been debated. It would give no hint, for instance, of the sorts of strategies that would be used by scientists when faced by anomaly or when seeking to generalize a result. To call it a "logic of construction" or a "logic of development" is misleading, since it might suggest that it specifies how constructions are made or how development occurs. But 'dialectical logic' does not, in fact, do this, as will become clearer below.

[4] But he is much too rapid and schematic in his discussion of Carnap's proposal to treat "theory as a deductive system constituted as an interpreted calculus" (pp. 197–198). In

particular, it is not correct to render thus the 'correspondence rules' which played such a vital role in this proposal: "The interpretation of the calculus is effected by substituting for its variable elements empirical terms, indicative, or descriptive, of observable entities". Were this to be the case all reference to unobservable entities would have to be eliminated, which was very far from Carnap's intention. Harris' conclusion that "inference in the natural sciences is accordingly reduced by current doctrine to analytic deduction, which can produce no new knowledge and inductive generalization, which cannot be validated" (p. 198), is too sweeping. The main fault with this Humean characterization is that it lumps under 'induction' a number of modes of confirmation that have to be taken separately. But to explore this would lead us too far afield. See E. McMullin, 'Structural Explanation', *American Philosophical Quarterly* **15** (1978), 139–147.

[5] " 'The understanding' is the name of the attitude of mind in which notions are regarded as being fully determined, with a precise meaning and with a precise and restricted range of extension so that they can be applied rigidly and clearly to the differentiation of experience . . . It is part of the abstract and formal function of the *Logik* to show that to stick to such fixed notions involves contradictions which can be overcome by the development of more adequate, less rigid notions. This generation of contradictions which shows up the inadequacy of the understanding is the process of dialectic" (R. Plant, *Hegel* (London, 1973), pp. 140–1).

[6] *Reason in History*, transl. R. S. Hartman (New York, Liberal Arts Press, 1953), p. 74.

[7] M. J. Petry in the introduction to his three-volume translation of the *Philosophy of Nature* (London, Allen and Unwin, 1970) asserts both that Hegel seems to suggest that a science of nature is open to constant revision, and that it also is ultimately rooted in absolute and unchanging principles (**I**, pp. 33, 90). He reads perhaps more of the 'infinite revisability' theme into Hegel than the text will actually bear. See G. Buchdahl, 'Hegel's Philosophy of Nature', *British Journal for the Philosophy of Science* **23** (1972), 257–266; see p. 262.

[8] C. Taylor notes that to put it in this way is to give it a more contemporary expression than is actually found in Hegel's text, but he argues that it is a fair rendering of what Hegel had in mind (*Hegel* (Cambridge, Cambridge University Press, 1975), p. 305).

[9] For a fuller discussion of this issue, see the Introduction to *The Concept of Matter in Modern Philosophy*, ed. E. McMullin (Notre Dame, Ind., Notre Dame University Press, 1978), especially pp. 47–52.

[10] See E. McMullin, 'The Ambiguity of 'Historicism' ', *New Directions in the Philosophy of Science*, ed. P. Asquith and H. Kyburg (E. Lansing: PSA, 1979).

[11] See also Petry, *op. cit.* **1**, p. 108.

[12] The philosopher of nature must 'submerge his freedom' in the content and let it be moved by its own nature. Though the object of study follows a dialectic movement, the philosopher (scientist) must stay as close as he can to a purely phenomenological method, and intrude no purely hypothetical entities. See K. Dove, 'Hegel's Phenomenological Method', *Review of Metaphysics* **23** 1970, 615–641.

[13] See E. McMullin, 'The Criterion of Fertility and the Unit for Appraisal in Science', *Boston Studies in the Philosophy of Science* [Lakatos Memorial Volume] Vol. **39** (Dordrecht, Reidel, 1976), pp. 395–432.

[14] Much historical work remains to be done on the development of philosophy of science in the first half of the nineteenth century. In Larry Laudan's detailed review of the sources for this period, it is interesting that Hegel is not even listed among the several

score who are counted there as having contributed to the philosophy of science during this time. See 'Theories of Scientific Method from Plato to Mach', *History of Science* 7 (1967), 1–63.

[15] See E. McMullin, 'Creativity and Scientific Discovery', *Freedom and Man*, ed. J. C. Murray (New York, Kennedy, 1965), 105–130.

[16] This issue is treated by several authors in *New Perspectives on Galileo*, ed. R. Butts and J. Pitt (Dordrecht and Boston, Reidel, 1978).

[17] P. Feyerabend, 'Explanation, Reduction and Empiricism', *Minnesota Studies in the Philosophy of Science* 3 (1962), 28–97.

[18] See, for example, L. Darden and N. Maull, 'Interfield Theories', *Philosophy of Science* **44** (1977), 43–64.

[19] It is incorrect nonetheless to say that "the collapse of Aristotelian physics resulted . . . from the contradiction which arose in the attempt to apply its theory of motion to the flight of projectiles" (p. 203). The matter was *far* more complex than this. It involved not merely the inability of Aristotelian physics to handle the phenomena of (relatively) free fall but also the direct anomalies that were revealed by the telescope.

[20] Harris, oddly, goes *too* far in this direction when he says that "it was eventually despair of accurate calculation of the annual occurrence of religious feasts and festivals which prompted Copernicus to revert to the hypothesis of Aristarchus in the hope of a more coherent theory" (p. 203). All the most recent scholarship in the tangled Copernican story would indicate that this was one of the cases where the perception of empirical anomaly did *not*, in fact, play the crucial role.

[21] Taylor, *op. cit.*, p. 234.

[22] *Ibid.*, p. 236

[23] Elsewhere Harris utilizes an argument of an interestingly different, and more complex, kind. In 'Hegel and the Natural Sciences' (*Beyond Epistemology*, ed. F. G. Weiss (The Hague, Nijhoff, 1974), 129–153), he claims that it was Hegel's "conviction of an immanent dialectic in nature" which enabled him "to see, at times, in the forms and phenomena of nature, what the science of his day had not yet discovered, but what has since become sound scientific doctrine" (p. 151). Thus, for example, Harris continues, he anticipated both the theory of evolution and the theory of relativity in his conceptual dialectic of nature. This utility to *anticipate* later developments furnishes (in Harris's view) an impressive additional warrant for the validity of his metaphysical insights.

It could, however, be objected that a great many of the dialectical constructions in the *Philosophy of Nature* were *not* borne out by later developments in science, so that this is a rather vulnerable strategy of argument. And Hegel's opposition to the theories of evolution of his own day cannot be explained away by claiming (as Harris does) that it was because these theories were unduly speculative and that "his faithful adherence to what in his time was scientifically respectable led him to reject the notion of biological evolution, convenient though it would have been for him to adopt it", thus leading to a "discrepancy" between his "metaphysical insights and what he accepts as scientifically supported fact". Hegel is not slow elsewhere to enter into a critique of the sciences of his day, on the grounds of their inability to structure the facts along lines that he considers acceptable.

His objection to evolution came from within his own system, as he makes clear in the *Philosophy of Nature*:

> A thinking consideration must reject such nebulous, at bottom, sensuous ideas as in particular the so-called origination, for example, of plants and animals from water, and then the origination of the more highly developed animal organisms from the lower, and so on. . . . The Notion timelessly and in a universal manner posits all particularity in existence. It is a complete empty thought to represent species as developing successively, one after the other, in time. Chronological difference has no interest whatever for thought (*PN*, § 249 and *Zusatz*, p. 20).

It would seem to require some degree of partisanship to claim as a warrant for the validity of Hegel's dialectical analysis that he anticipated the modern theories of inorganic and organic evolution. One might at least as readily claim the reverse.

[24] Taylor, *op. cit.*, p. 391.

[25] See N. Lobkowicz, 'Materialism and Matter in Marxism-Leninism', *The Concept of Matter in Modern Philosophy*, ed. E. McMullin (Notre Dame, Ind., University of Notre Dame Press, 1978), pp. 154–188.

[26] See, for example, the essay by V. A. Lektorsky, 'Dialectic of the Subject and Object and Some Problems of the Methodology of Science', *Philosophy in the USSR*, transl. R. Daglish (Moscow, 1977).

HERMAN LEY

SOME 'MOMENTS' OF HEGEL'S RELATION TO THE SCIENCES

Hegel's dialectic seems to possess the property of provoking again and again either adherence or vehement rejection. In this the question whether the philosopher had a troubled relation to the exact sciences, or, on the contrary, exercised a stimulating influence, remains just as much at issue as the possibility of reaching an adequate understanding of his intentions. At the 15th World Congress of Philosophy in 1973 at Varna, at least four positions emerged in the sessions of the Institut Internationale de Philosophie alone. Mario Bunge assured the meeting, in essence, that science was present only where the last vestige of dialectics had been eradicated in order that the principles of exact philosophy might be complied with. Paul Ricoeur related dialectics solely to human practice, differing in this evaluation from Mihailo Marković. In an interpretation drawn from Marx and Engels, I. S. Narski accepted the validity of objective and subjective dialectics, to be interpreted materialistically. Under these circumstances, it seems advisable to consult Hegel's texts on the various points in order to recall to memory how his ideas, which refer mainly polemically to Kant, are to be understood. That it is worthwhile to seek to complete Hegel's methodological and epistemological structure at this late date results from a fact that simply cannot be neglected: in recent times a turning back to Hegel's ideas has begun within predominantly empirical disciplines, which has confirmed an unmistakable relationship to the most modern problems.

In a controversy by correspondence with William James, Charles S. Peirce professes to a kind of thinking in which three modes of consciousness are contained, which he classifies as feeling, experience and awareness of the future.[1] The theoreticians, alluded to but not named by Pierce in this correspondence, are characterized as Thomists, Hegelians and intellectualists. In this connection only the absence of an aversion to Hegelian intentions is of interest. One of the more recent editors of Pierce, Karl-Otto Apel, places the 'objective logic' of evolution, already postulated in 1868 and worked out explicitly after 1890, within a framework of influences that extends from Hegel and Schelling to Darwin and Lamarck.[2]

That an influence of Hegel on Charles S. Peirce definitely exists can be concluded from, for instance, Peirce's *Lectures on Pragmatism*. In the

R. S. Cohen and M. W. Wartofsky (eds.), Hegel and the Sciences, 241–270.

tradition of Aristotle, Kant and Hegel, a category is understood as an element of the phenomena having the highest rank of universality. Peirce cites the list of Hegel's particular categories drawn from the structure of the *Encyklopädie*. Hegel's three stages of thought are designated by Peirce as universal categories, and he considers them quite generally as the correct table of universal determinations of thought.[3] As will be seen, while thinking through the categories as Hegel understands them, Peirce adopts an epistemological position that at least in part finds itself in agreement with Hegel: "We should hardly find today a man of Kirchhoff's rank in science saying", Peirce attempted to prognosticate, "that we know exactly what energy *does* but what energy *is* we do not know in the least."[4] Peirce does not use the expression, which has become current since, that we have to test the hypothesis in experiment; he confines himself to pointing to the application of, say, the dynamical equations, where the application is conceived as an instance of testing. In Peirce's considerations the acceptance of a real universality of the laws of nature is combined with the real possibility of experimental knowledge, and this is one of the transmissions of Hegel's critique of Kant, concerning the so-called unknowable character of things-in-themselves.

How far-ranging is the evidence for Hegel's confidence in rational theoretical thought can be shown in a field to which it is usually assumed that Hegel had hardly any concrete relation. This is the field of biology and, in a narrower sense, genetics. A few years *before* winning the Nobel Prize for Medicine together with François Jacob and Jacques Monod, André Lwoff began his catalogue of the discovery of hereditary laws and genetic information with a remark from Hegel's *Lectures on Aesthetics*, which he presented in various forms during the third decade of the nineteenth century. As Lwoff mentions, determinants are spoken of in the French test, published in 1835, which exist in the seed as extremely simple forces and determine living reality. There is nothing in the tree which has not been previously in the seed [*Keim*].[5] Hegel already clearly uses the same image in the Preface to the *Phenomenology of Spirit*. Particularities "slumbering as if in a still enclosed seed" first appear in the *Aesthetics*, which completely derives the aesthetic value of nature's beauty from the natural substructure, which will by no means seem paradoxical today. For Lwoff the birth of the modern concept [of genetics] begins with Hegel, and in this he recognizes that Hegel discusses the problem of life in a penetrating way. Thus, the differing points of departure, which eventuated in the ideas of Peirce and Lwoff, are philosophically relevant and rooted in Hegelian philosophy. We find side by side the relation both to the logic of evolution, and to the categories, which for Lwoff retains the constancy contained in the seed throughout the various stages of existence of ontogenesis. Without actually discussing it, Lwoff seems to sense the paradox, that the dialectician Hegel finds it necessary to emphasize the constancy of living organisms more than, say, Kant, who points to a possible hypothesis of the evolution of animals and plants. In contrast to Kant, the evolution of nature is absent in Hegel's philosophy, while for society an evolutionary trend is worked out in consciousness and in the economic-technical infrastructure, one that does not relate solely to the phenomena of mind. While nature is considered to be the realization of absolute spirit without its own immanent evolution, i.e., for nature as a whole, yet Hegel was able to seize upon the constancy of the determinants and of the simple forces which condition the individual development of the objects of the species whose change of form is visible. Even though a not inconsiderable indefiniteness attaches to the dialectic, this does not hinder Hegel from emphasizing the constancy of the relation between the seed and the fully formed object. In the references to Hegel which have become relevant in more recent times, these opposing sides contained in the dialectic gain in significance. The relatively constant is related to a process of well-defined change of form, and belongs in the domain of the phenomena of the empirically accessible reality within an objectively understood multiplicity; Hegel conceives of this domain as dialectical.

Perhaps the most enthusiastic advocacy of Hegel's evolutionary thought comes in more recent times from the biologist Adolf Meyer-Abich, who designated phylogeny as an essentially historical science indebted to Romanticism, and especially to the historical thought of the Hegelian school.[6] As a dialectical synthesis he mentions, among others, the vitalism-mechanism problem. This kind of problem is characterized by Niels Bohr as ideals of knowledge, as deep truths, of which Bohr says that "their opposite also contains deep truths."[7] If Meyer-Abich deals too boldly in some respects with Hegel's dialectical triad of thesis-antithesis-synthesis, which is hardly rigidified into woodenness in Hegel, still, in any case the holistic theoretician is able to link his views to Bohr. The compensations of phylogenetic epochs are conceived as complementaries *in statu nascendi*. The emergency of antinomical states of affairs is seen to be the result; these have their domain in the history of society, but are extrapolated by Meyer-Abich to the natural sciences, whose structures as developing sciences he sees as approaching human history.[8] Whereas Helmholtz saw the Hegelian identity philosophy founding on natural science,[9] Meyer-Abich assures us that the natural scientists of that time thought in a completely unhistorical manner, so that this decisive misunderstanding is comprehensible. Historical reality (with respect to nature) first begins with complementarity.[10] If Meyer-Abich considers Hegel's logic to be the logic of phylogenetic thought, still the paradox remains that Hegel recognized no evolutionary development in nature, and sought to diagnose the dialectic supposed to exist there, solely synchronistically as it were. In Hegel's dialectic, the diachronic structures found their place, strictly speaking, only within the successive configurations of consciousness, from which deductions back to nature were made, while in the meantime, evolutionary development asserted itself in nature's various domains. As Kant, in contrast to Hegel, had already sought to present evolution in society and nature, and had assumed the presence of antagonisms as the driving force in the latter, going back to Hegel for the understanding of evolution can hardly be understood in terms of anything other than Hegel's relatively well elaborated state of dialectical categories. But by no means least important, the ancient outlook, construed as pantheistic and monistic, seems to have conferred renewed credit upon the Identity Philosophy. To begin with, Hegel derived the laws of his dialectic exclusively from society, and made them independent in a doctrine of categories understood as logic of development, the possible generalization of which was not to be interpreted as within things, but had to find its possible confirmation in the empirical material. Since Hegel considered that evolution had *not* been proven for nature, he confined himself to the synchronic aspect of patterns, whose epistemological value he expressly reassessed, in opposition to Kant. Thus far, however, the systematization of the dialectical categories retained their significance as heuristic determinations. They are also understood as such by C. F. von Weizsäcker, who refers expressly to Hegel.[11]

C. West Churchman directs attention to the necessary tension between formal logic and dialectics. Due to the needs of operations research, ways of posing questions became current that had already ripened in the philosophical debate revolving around problems emerging in natural science and technology. In the first place they concerned the reduction of multiplicity to models, the interpretation of existing models with respect to previous, present and future movement of their contents, but, also, together with this the prediction of fields of probability, as well as decisions about goals and measures to be taken. The criterion for the quality of decisions never lay solely in the logical stringency of an inference, but in the material fulfillment of purposes and assigned goals. As the representation of the processes to be treated in terms of mathematical structures, whether of a material or an ideal nature, is carried out in anything but an automatic fashion, but instead has its precondition in subjective activity, Churchman, to begin with, goes back to Kant's thesis: data are not movements of nature, but partially our own movements, created by our judgment.[12] In any case, that is how Churchman formulates the state of affairs, and he emphasizes the objective and subjective participation in the production of data, which corresponds to Kant's confrontation of Platonism and Epicureansim in his *Critique of Pure Reason.* Churchman regards the production of data as a creative process. Since he cannot receive sufficient enlightenment about the inner structure of this process from either Berkeley or Kant, and also knows that he cannot receive sufficient information from contemporary philosophy, he falls back to the point of departure of recent discussion, which he recognizes as the content of the confrontation between Kant and Hegel. It is astonishing that this resumption of a discussion that is generally assumed to be concluded comes from a theoretician who, together with others, organizes 'teamwork' through which applied mathematics and the most modern techniques are placed at the disposal of processes of management and production. Also to be noted, aside from this, is the state of controversy in which at least two positions confront each other: the assumption that future events can be precisely predetermined and carried through on the basis of a stringent calculus; the other assumption being the prediction of transitional probabilities, the concrete fulfillment of which unconditionally involves human activity, and must keep open *any* system of human beings or of human beings and machines.[13] Churchman related his theoretical generalizations to the domain of empirical research, and confirms that in their effective execution these go in principle beyond the boundary set by the strict rules of formal logic.[14] The procedure of those philosophers like Hegel, who subsume under 'logic' the whole process of developing an understanding of world, as well as of those natural scientists who use the term 'logic' to cover all the methods by which they reach their conclusions in empirical research, may be understood as parallel modes of thought. Accordingly, it becomes understandable that Churchman includes Karl R. Popper's logic as an argument for the validity and justification of Hegel's logic, although Popper can also be considered as the antipode of the Hegelian dialectic. In another passage Churchman calls Hegel's result a dialectical theology. However, it is possible to use the dialectical method more significantly in order to describe a sociological, materialistic process without God.[15] What deserves to be especially noted is that the dialectical method can be applied to nature and to society, and, as we may see from Churchman's own account of the state of affairs, actually is being applied within the sciences. Whereas Hegel constructed a teleology of the entire world-process, the dialectical method can be confined to an area within this world. Insofar as in society, by contrast to nature, goals are set for which the precondition is human action, these reach ahead into the future, but do not transcend the world. The transcendence of formal logic and its present status has another significance for society than it does for natural objects. However, in both cases synchronic and diachronic schemata and corresponding patterns have to be taken into consideration, as far as the state of affairs requires.

Once it has been shown that there is renewed interest in Hegel, we may investigate those of Hegel's thoughts which furnish the precondition for the renaissance of his ideas. First of all, some material from the Preface to the second edition of his *Science of Logic*, which was formulated, so to speak, as the last word on November 7, 1831, may serve this purpose (*SL* [M], pp. 31–42). Hegel emphasizes the function and the objective content of the categories,[16] which he treats as 'moments' of the development of the sciences and makes the object of analysis. This Preface and the earlier Introduction to the *Logic* (*SL* [M], p. 43ff) must be considered an authoritative statement of his most essential thoughts. Abstraction from opposition and transition to universals mean something similar for Hegel. The categories or determinations of thought are not considered to be a merely subjective product. There is no difficulty for the idealist Hegel to insist on the unity of the objective and the subjective, since mind externalizes itself in the objects, and a correspondence must necessarily be considered possible. When Hegel speaks of categories, then, it is not merely instruments of thinking which are meant, but also specific properties of objects to which the processes that take place within or between them, or which become possible, are considered to belong. In this *Science of Logic* Hegel wants to represent "the realm of thought philosophically, that is, in its own immanent activity or what is the same, in its necessary development" (*SL* [M], p. 31). It is thinking that distinguishes man from the animals for Hegel (*SL* [M], p. 31). But to give logical determinations a merely subjective significance, Hegel calls fear of the object.[17]

If Hegel occupies himself with thought, then the world of objects is always also included. If Hegel forces object and subject together into the mystifying externalization process of the absolute world-spirit, then it is also granted to thinking to be able to reverse the process. The development of object *and* subject is mirrored in thinking and its categories. If the categories are still grasped merely subjectively, then the result is, in Hegel's view, that the object continues to be attached to them in a merely external way, because it appears to them as something beyond thought. If, to understand consciousness, the subjective categories include the objective in their determinations, then they have incorporated the substantial content within themselves. That means that consciousness has fused the concrete into itself in the process of appropriation, and with that lacks the stuff and substance of sensuousness and the concrete, *because*, in a dialectical sense, it has abolished – and transcended [*aufgehoben*] – both in itself. With that, the self-evidently existing difference between object and subject, as the world of objects and the world of consciousness, becomes identical in its very differentiation. Naturally, this does not mean the concept of identity of formal logic, but rather the fusion of what belongs to content with the determinants of consciousness, which thereby become appropriate for making statements about contents and make possible the capacity for applying the categories, For Hegel fear of the object means an unfinished standpoint in the epistemological process. Only if we assume that it is by means of concepts that the thing-in-itself is overcome as something beyond, and appropriated, can the categories of the purely-rational, as Hegel calls it, be known. Since the categories are comprehended in their relation to content through a dialectical identity with the object, they belong to science. Hegel assumes that science overcomes opposition to the object. Accordingly, Hegelian logic *and* the special content of the sciences belong together, a product of development. Hegel's dialectic wants to be assured of the object and of the epistemological competence of the categories.[18]

The existence of universal laws in the objective world means for Hegel the existence of understanding and reason, with which the bridge to the categories has been erected. Laws and categories are recognized to have objective value and existence through which, at the same time, the difference between objective and subjective is to some extent attenuated. Since Hegel uses the existence of evolutionary development at least within consciousness as the starting point of his reflections, the laws also refer to the change of categories, and thereby they quite definitely are *not* considered to be limitations upon knowledge, as Kant believed must be assumed.

Hegel does not concern himself with the question, how knowledge can be gained. He assumes that knowledge is to be gained by the procedures under formation in the various disciplines in question, and instead he investigates what it is that is thereby made available to human consciousness. Hegel objects to any underestimation of thinking and theoretical work; he sees such underestimation as arising from bare empiricism *and* from the doctrine of the unknowability of real objects, on whatever basis this may be founded. For Hegel, the objectivity of concepts which emerge from the development of thought consists in the limitation imposed on human beings at that time, but in the sense that thereby there can be actual knowledge corresponding to the current stage of development. If, for Hegel, the categories and the concepts of objects are delimited with respect to each other, on the other hand they are congruent with respect to their reference to contents, the range of which goes beyond merely formal correctness. Hegel arrives at a formulation which formally French Structuralism could lay claim to for its use, but which receives a different evaluation in Hegel's context.[19]

So also Francis Bacon's "*natura non nisi parendo vincitur*" receives a corresponding modification in Hegel when applied to the process of knowledge. By setting object and subject as approximately equal, it would seem obvious that the thesis of the domination of nature by man or the purely instrumental function of concepts should be accepted. But quite the contrary: Hegel emphasizes the objectivity of object and subject as confronting man, whose product is explicitly conceived of as the achievement of concepts. In the categories and concepts of objects, what Hegel calls the universal presents itself, with which man, for his part, has to comply. It is far removed from Hegel to conceive this dialectic between man, the products of thought and the confronting objects in an indefinite manner; the objectivity of the products of thought achieved by man, even with regard to man himself, is maintained with all definiteness.[20] Not to go beyond the nature of things means for Hegel not to be able to elevate oneself above the understanding of things. If Hegel says the objective concept of a thing constitutes the thing itself, then philosophically a limit is given which is practically effective, but which also must be understood historically. With such a term as 'the objective concept of things', taken by itself, it appears that philosophy is proclaiming a closure of knowledge in its own time. But just as Hegel sets about the determination of the categories historically, so the concept of things is no less historical. Forms of thought and knowledge are conceived as determining and dominating man, because in them, as it were, the horizon is revealed which contemporary man of each period has created, and which only he will be capable of transcending when the forms of thought and the rest of his cognitive activities have absorbed other contents.

While Kant insists on absolute original 'root' concepts of pure understanding, Hegel insists on their changing character, which appears as a 'moment' of becoming and therefore signifies a limitation that continually strives to go beyond itself. The dialectical method seeks to grasp the historical process but surely not as history, not only because it comes after the fact like the owl of Minerva and begins its flight in darkness, but because it universalizes becoming itself. Hegel saw himself placed in a position to thematize evolutionary development, because it had come to be visibly underway. The creation of higher relations of thought, as Hegel expresses himself, is traced back directly to the "advance of culture generally, and of the sciences in particular."[21] As the empirical disciplines were only at the beginning of their ascendancy, while mathematics was already at hand as a developed science, Hegel explicitly raises the question of the relationship with them. Dialectical philosophizing should not replace the other disciplines, but generalize the visible process which spreads over a broad field, and, in his view, is ready for conscious comprehension.

There will be polemics against any scientific viewpoint that would select one or several among the special sciences and make the others appear dependent upon it. Hegel's 'logic' tries to point to common features that are delineated within the various phenomena and requires an interpretation that goes beyond the opening up of a partial field of objects and domains that have become accessible to consciousness. With that, every existing special science is recognized, without consideration of its possible ramifications; while at the same time their common influence on the so-called higher relations of thought is also recognized. But with that, Hegel postulates, without reservation, the dependence of the phenomena of the mind that are becoming manifest, and thus also the dependence of philosophy, on the complete understanding of nature, of the relations of human beings with each other, of the externalization of these in the world of objects, as well as of their reincorporation into consciousness.

In his reflections, Hegel in fact mentions physics, within which the category of polarity had recently come to play a role, and incidentally was being, *à tort et à travers*, forced into everything, even into [the phenomena] of light (*SL* [M], p. 32). Hegel demands that opposition be reflected if it is articulated in the sciences; it should not be down-graded. What remains worthy of note is that Hegel warns against seeking to insert this polarity into everything at all cost; and in keeping with this, he opposes the forced construing of categories into things, for in his understanding this comes with the thing itself.

Consequently, objectivity is taken seriously, as is the relation between the sciences and the dialectical method, the determinations of which are themselves subordinated to further development. Not to tailor the determinations of thought according to our arbitrary decision or freedom is one of Hegel's foremost requirements and insights. If some modern theorists find that Hegel's categories have only become relevant in the twentieth century, that only means that Hegel's analysis was able to discover aspects in the undeveloped material of his time, which only achieved broader validity with delay, and which suggest modification of the philosophical generalizations widely held and considered valid until then.

As an identity philosopher and philosophical idealist, Hegel hesitates to simply accept the 'nature of things' which he considered. He does not subjectivistically deny the existence of things, but sees in the "object . . . nothing else but our notions of it" (*SL* [M], p. 36). Even if we reject a relation of our thoughts to the thing, still Hegel seeks to avoid the suspicion that the concept present in the subject is in any way made independent and does not permit any distance between the subject and its objects, which always act only corresponding to the level of consciousness.

With that Hegel seeks to clarify his own position and that of critical philosophy. Thoughts should *not* stand between us and things, nor should the medium [of thought] shut us off from things.[22] Insofar as Hegel thinks of a mediation, it does not shut consciousness off from things but instead "joins us with them". In Hegel's understanding the identity is mediated by the concept of the thing and by the categories, and therefore directed toward something definite. If thinking, in Hegel, belongs to the abstract form of labor, yet he also occasionally comes to the more concrete means, to which externality is explicitly ascribed. As against "our very own, innermost, act" (*SL* [M], p. 36) the external means is markedly contrasted; precisely through this characteristic, the external means reaches a meaningful significance which at first is negligible to Hegel, as it is his main concern to bring subject and object more closely together, in order to re-evaluate the constructive form of thought, as well as to bring theoretical thinking into closer contact with objects epistemologically. With that, Kant's concern to draw closer methodologically to the active operation of the understanding is given adequate scope philosophically; but precisely in this way the positive and dialectical form of mediation is emphasized. In Hegel's understanding, the categories, insofar as they intervene mediatingly into the process of thought, prove to be the vehicle which permits us to get at the thing itself. But as, in addition, the development of knowledge also remains to be treated in terms of its own mechanism, the external means appear as a complement to the categories, which also contain, as it were, a measuring function, and aside from that also intervene in the overall process.[23] Hegel sees them entering into the process in combination with the goal-setting activity of man.

Without moralizing or going so far as to postulate goal-free activity, as the Preface asserts at least partially with respect to scientific activity, nevertheless goals belong within the domain of mediation. Of finite content, it inserts another object between itself and the result, as Hegel expresses it, whereby the 'cunning of reason' asserts itself. To characterize the total difference between it and formal logic, and the merely formal function of its content-determined categories, Hegel's content-determined logic takes this other object to be the means of the external middle-term of the inference; and the middle-term is the carrying out of the goal. Since we are concerned with finite goals, as Hegel presents it, the objects of the external middle-term of the inference are also not to be understood in any absolute sense. They change and turn out to be concrete means, without being something rational in-and-for-themselves [*an und für sich*]. Themselves 'moments' of mediation, they mediate something higher. If the categories and concepts of the thing itself were conceived as limits which would not have to serve as means of domination by man, then, following Hegel's stylized way of speaking, the case would be different with respect to objects, which are an 'external middle' term and should properly be conceived as subordinated, if one were to understand them within the framework of Hegel's understanding.

As contrived as this train of thought may seem, and as mannered as the composition of Hegel's text may appear, so too, with closer explication or later completion of the stages of thinking, are we far removed from the exactitude demanded, say, by Mario Bunge. Since Hegel subsumes structures of various kinds under what he wants to consider included under 'essence', these structures can, so to speak, be of internal and external configuration. In the case before us, where Hegel speaks of that mediation which emerges from finite purposes, in his view rationality asserts itself in externality.[24] Given that Hegel has reduced the thing-in-itself to the concept and sees the categories arising from this, which fundamentally are not restricted to the number found either in Aristotle or in Kant, and which in their turn again dominate consciousness, nevertheless, on the other hand, toward the end of the *Science of Logic*, human activity is by no means confined to merely inner activity. Completely in keeping with the understanding of the seventeenth and eighteenth-century Enlightenment, from the two Bacons through Diderot, and including Descartes, it is tools that give men power over external nature. But in contrast to these aforementioned periods of the Enlightenment, Hegel analyzes a reciprocal relation that unites the position of the world-outlook with the methodological position, and utilizes analysis for that purpose. The relationship of the 'thing itself', the categories and consciousness are construed in terms of the world-outlook, while with respect to a dialectical understanding, the dialectical identity, which is not that of formal logic, is brought forth only methodologically. The goals are, as it were, products of the 'anthropological' constitution of the individuals and of what Hegel understands as civil [bourgeois] society. With the 'plow', however, Hegel is not concerned with technical problems nor with the unity of mental and physical activity. Hegel assumes it to be self-evident that if we speak of human beings, instruments are required to achieve goals, where we deal with the foundation of life-supporting activity. And so he means concrete labor with the corresponding tools, and not merely recourse to abstract labor. But for this it is not simply the objects which we represented by the 'plow' that are addressed, but the relations posited together with this. As means, the plow relates to nature and to man. If the plow signifies domination over nature, then through the consideration of the category as a limiting value, an overarching 'moment' is reached, which reacts back on man, but as barrier becomes fluid due to the continually newly set goals. The state of affairs under discussion cannot be described as a Hegelian triad, or only in a most extremely forced manner, but it does contain characteristics of structures, conceived as a set of elements, and aside from this offers patterns of scientific explanation.[25] If more recent structuralism, in its several variants, exploits its methodological procedures – used in linguistics and ethnology – to deal with social processes, it must not be forgotten that first of all Hegel found a sporadically applied approach, which finally expressed itself in the mathematical procedures of modern economics right down to the balance of complex interacting interests. Lenin calls historical materialism "one of the applications and developments of the ideas of genius – seeds existing in embryo in Hegel."[26] Hegel offers a catalogue of variables which stand related to each other and which, for the purposes of mathematical analysis, are in each case to be investigated for transitivity, reflexivity and symmetry, in the logical sense. As Hegel first discovered the presence of certain parameters and their interconnectedness, he had to also secure information about the nature of their interconnection. The purpose set by man remains, preserved as the domain of departure. The domain of relations determined by each parameter, however, remains both variable and constant. The arching functions of the plow remain preserved, although the instruments of production change through human activity, and, aside from this, the limits of conceptual adequacy for the possibilities of understanding expand up to the extent that development takes place. Since theoretical philosophical investigation treats empirical material and secular experience, the evaluation of various parameters related to each other and their reciprocal influence signifies considerable progress beyond William Petty's *Political Arithmetic* and Quesnay's *Tableau économique*. We must also note that, *before* Hegel, quantification in accordance with probabilities had already been undertaken by Petty, and the influence of the superstructure had been admitted, but the conceptual analysis still remained in its beginnings. Compared with later conceptions that mathematical representation of economic expectations should be contemplative and exclude the concrete subject,[27] Hegel offers an analogous but superior position which does not cut off the course of scientific development. As Hegel postulates the dialectical unity of quantity and quality, he does escape from the dilemma of selecting *either* quantitative *or* qualitative procedures and making the corresponding decisions. Hegel's theory of mediation in principle reserves for itself the choice with which procedures the process of reciprocal involvement of circumstances is to be treated. As in the understanding of content-determined 'logic', any arbitrary phenomenon can be comprehended as the expression of quality and quantity, the verbal representations, which are formulated qualitatively, can also be quantified. Contrary to the conception that Hegel rejected 'formal logic' or 'mathematics', his theory of knowledge instead contains the basic structure, from which quantification would be considered possible if an appropriate technique is available. However, it was only after the middle of the twentieth century that quantitative approximation of material and ideal domains of objects was achieved, and in this, evaluation scales exercised a not inconsiderable influence. In Hegel's understanding these possess possibly an objective character. In this, Hegel draws on Leibniz. The second proposition from Leibniz's *Dissertatio* states that "it is not altogether improbable that matter and quantity are in reality the same."[28] The antinomian 'moment' of quality and quantity is repeated in the relation of continuity and discreteness, a falling apart into opposing 'moments', which Kant presented in his cosmological antinomies, but which Hegel considered demonstrable for every concept.[29] The practical significance of such results, which had been obtained by hard intellectual work only becomes fully visible, when practical access has actually been created. Hegel's theory of mediation can be conceived as a model in which consciousness, activity and the over-arching product of goal-satisfaction manifest themselves; it becomes possible to represent them together, but the function of the means is brought out. The state of affairs is understood as conforming to laws by Hegel, but only under the supposition that an essential relationship is involved,[30] which Hegel assumes to be present in the thing itself and which he seeks to reflect.

What is to be conceived as law for Hegel is not a matter of a specific model from natural science, to be derived from the disciplines existing in his time. It remains undecided whether something of the type of Newton's mechanics is to be taken. Since law is characterized as an relationship which should be subject to the same dialectical categories as the other conceptual determinations, it also possesses, according to Hegel, the property of quantifiability. In relation to the theory of mediation, Hegel already discussed the problem – which later is discussed again and again – of the *other* relationship between beings, consciousness and extra-human structures. The objection concerning contemplation relates to those categories, recognized in each instance, which are conceived as bonds. One of the phenomena described as alienated in the modern sense are the parameters put forward by Hegel in the theory of mediation in which the overarching 'moment' in each case receives a recognized overarching meaning. But in Hegel's understanding this is not a bond, but rather a 'moment' of freedom. On the other hand, given the objections that have been raised, the viewpoint is brought forth, that it is just in becoming conscious of the determinations of relationships "that the subject" is transformed "into a mere organ of conception of the known chances of lawfulness." However, in confrontation with later polemics, Hegel's other thought content is made plain and begins to draw near to future rationality in a mystifying language which is considered to belong to Romanticism. Under the presupposition of a dialectical identity philosophy, Hegel transcends the 'moment' of agnosticism present in Kant's philosophy, and does so in every phase of the development of the entire system of consciousness, categories and means, without allowing the human being to be worn out by the structures. Differing here from more recent philosophy, there is in the means an alienation that *can always be overcome*. What is later called pervasive rationalization of reality, in Hegel belongs to the human products that are at the same time the transforming and overcoming [*Aufhebung*] of alienation. For that reason they are a necessary stage, they can neither be eliminated nor neglected. As Hegel formulates it: "the movement of essence is in general the *becoming of the Notion*" (*SL* [M], p. 526). By way of the means and mediation, concrete confrontation with reality is included under this. The formula, which sounds idealistic and scientistic, subsumes under the [so-called] "becoming of the Notion" not only the action of thought but also its realization in reality. In this context, 'notion' does not mean only the end of a process, but at the same time refers to the beginning from which reality is to be traversed. Identity philosopher that Hegel is, he appears hostile to subjective idealism, and sees in the materiality of things the realm through which human beings must pass. Accordingly, Hegel expressly praises the scepticism of antiquity, which demonstrated the antagonisms within concepts and with that discovered an objective phenomenon. Therefore the concept by itself means only "immediate, passive existence."[31] As the means is included in the concept, Hegel also points to the contradiction between the means inserted between man and nature, which alone let becoming come to totality. Thereby the merely passive concept appears as one side of the contradiction, which, when without realization and without retraction, remains an empty and unfulfilled demand; it thus lacks the other side of the contradiction. Hegel's dialectic is already confronted with the later critics of technology in that they, just like the theorists of 'post-industrial society', fixate on only one side of reality. But as Hegel speaks of *becoming*, then knowledge does not mean the exclusion of history, nor does the domain of means signify the destruction of humanity. Becoming consists in the unity of the diverse moments, which are formulated as the task of human beings and of society, and only evading these tasks can represent withdrawal from this historical process.

How broadly Hegel interprets the term 'concept' is shown when he points to natural, scientific and mental development, in which the natural, in contrast to the others, has been moved to first place. When Hegel speaks of 'concept', the material side, and also that side which is in the process of becoming accessible to consciousness is included in his meaning. However, since what is involved is always a becoming that includes the negation and overcoming of the negation, the structures of what is real develop no less than their images, and refer to each other. Accordingly, merely to be a concept also means for Hegel a germinating and embryonic existence, which must yet make its way to the full concept and complex manifestation. But it is from this too that Hegel's remark comes about the equality of the whole and its parts. The parts are accepted as representation of the whole, and the whole as that of the parts, without the other dialectical category thereby being revoked or overcome [*aufgehoben*], namely, that as against its parts the whole possesses a special quality. Hegel came to the assumption that a tautology was expressed by this, which only gains its truth through mediation (*SL* [M], p. 515f), and thereby receives an additional meaning. It appears simultaneously as an inner and an external movement, which comes to expression in things [*Sachen*] *and* in instinctual activity.

In the 'Preface' Hegel does not let freedom of the mind be dependent solely on consciousness. The so-called unmediated unity with the subject serves as the beginning. Freedom of the mind begins when what is conscious is brought to objective status before the subject, and no longer remains merely split up among the categories. In the dual signification of Hegelian diction, the logical business, as he calls it, is meant, *and* the transition into reality, as well as further knowledge made fruitful by this. What is involved here, side by side, is the fundamental determination of the reality of the mind and of reality itself, which man appropriates. According to this, it is doubly efficacious categories that concern Hegel. They are only considered to be appropriated by the understanding, if they have to the same extent passed through confirmation in reality.[32] The fixed points and nodes are considered to be the life and consciousness of mind; and in the same way they are considered to be an essential relation, i.e., law. Since becoming and objectivity have a specific function in Hegel's philosophy, a further dialectical antinomy appears in this, which belongs to the state of affairs to deny which means a so-called prettifying of things. If law is defined as essential relation, then it is not to be demanded of it, that it be valid always and everywhere in all possible worlds; rather proof of the existence of certain essential relations suffices, and the domain of their validity will then follow as a possible further question from this.

What is brought to consciousness can be the state of affairs as such, as well as the universality of the claim to validity. In the categories of the dialectic Hegel finds fundamental determinations of the reality of mind; these have had to emerge and thus are also a product of development. *Becoming* itself, however, appears as the inevitable state of affairs, which can never be dispensed with. If the categories are expressly affected by this, then their development applies to the domain of overarching mediation, in the first place, for example, society, and then, natural, scientific and intellectual development, all of them states of affairs in which development is linked to the emergence of new conditions.

No matter that one or another discipline may have had a more rigorous conception of modern science than was available for Hegel, for the discovery of a logic of development is the achievement of this philosopher. Although both the natural sciences and social theory could already show the beginnings of a concept of evolution, the generalization of these attempts has not appeared till then. As the revolutions of the bourgeoisie and the Third Estate – now brought more or less to a conclusion – had shown there is development in society, it was time for a hypothesis about the course of human history accomplishing itself in this development, whether contingent or necessary. Hegel sought for a historical understanding of society, and in this he encountered the historical character of the determinations of thought, the reciprocal connections of which he treated, in modern technology, synchronically and diachronically; and by this he sought to represent the truth in process, but not in the manner of formal logic. If the formal logicians are shocked by Hegel's application of a sequential algebra of truth, it must be remembered that he distinguishes between correctness and truth (*SL* [M], p. 38). However one might otherwise determine truth, "it is unjust to fail to recognize that they (the rules of inference and indeed the main use of the understanding) are neutral [*gleichgültig*] means, used at least as much by error and sophistry" (*SL* [M], p. 38); and so Hegel finds it necessary to develop the hitherto neglected complexity of content. Truth appears to him to lie in the transition, in the mediation, in the becoming. As these processes had till then not been elevated to the rank of determinants of thought, out of them emerges a heuristic methodology of dialectics, which probably also had its effects in those disciplines where no obvious indication of the mediating function of Hegelian philosophy is to be found. Hegel's influence fans out far beyond Feuerbach to Moleschott and perhaps the [materialist] Monists, by way of Marx to Lenin, by way of Kierkegaard to the broad field of the Existentialists, and in addition to those tendencies which were singled out in their relationship to Hegel at the beginning of this paper. To elevate transition, mediation and development to the rank of categories, and to label them as 'contentful' truths, had yet to attain its status as a philosophical innovation. If it initially met with approval, only then to flow out into diverse directions, which transformed it, turned it upside down and thereby subjected it to partial testing in the special sciences, it still remained present. It can be maintained that dialectics then also enters into the discovered structures of various disciplines, whenever only 'development' or 'evolution' is the subject of discourse. Hegel grasped this possibility in his thesis of the sliding scale of becoming conscious of nodes [of experience] that initially are tied up in acquired instincts, but which finally become conscious categories. Understood heuristically, the Hegelian dialectic has the possibility of being applied to past and future, and can be comprehended as a method which also contains definiteness and indefiniteness immanently as a dialectical pair of categories. The definite reproach brought forward against the dialectic, that it is not exact enough, raises the question, whether it is not precisely its content-determined relation to development which first gains for it its scientific character precisely through the mediating partial category of indefiniteness. As the dialectic distinguishes itself from mechanistic thinking and the exactitude of a philosophically generalized Mechanics, it is not subject to the temptation to try to prognosticate a point by point prediction for every future event.

What Hegel conceives to be incompleteness, when considering thought, is to leave truth "on one side" (*SL* [M], p. 39). Hegel sees truth not only in the external form, but also in the content.[33] If the Aristotelian categories of content and form arouse the suspicion of being stuck in medieval scholasticism – although a hardly appropriate judgment is pronounced upon the schools of the latter in that way – then the assumption of fluid categories proves to be suitable to pry the categories of content and form loose from their mutually unrelated confrontation. If content and form were related to each other as matter and entelechy, or material substance and configuration [*Stoff and Gestalt*], then Hegel would seem to want to include various categories in both traditional determinations of thought. Since content is not to be debased to empty form, and as 'moments' of content are granted to form, while at the same time a change of content and form should be grasped *in a formal sense* as well as developmentally, Hegel can claim an enrichment of the empirical material that enters into thinking about objective things. Content and form no longer confront each other dualistically and cease to incorporate spiritualistic dualistic elements within themselves. The original root concepts of the understanding do not remain in the transcendent space of the *a priori*, nor settle in the same realm as a collection of performed possible formations. The Aristotelian separation of content and form meant a dualism, which, Hegel felt, was contradictory to his identity philosophy in its approach; for the same reason it had already lost its cogency in the tradition of the young Schelling.

Hegel seeks to approach content and form in accordance with their structure, by proceeding from the object of investigation, in which content and form are conceived as fused, and thus no actual separation occurs. From this Hegel concludes that the separation of content and form is an activity of the understanding, something worth reflecting upon, and indeed simple self-reflection can give rise to notable results. The separation of content and form by the understanding does *not* create, so Hegel assumes, any content without form neither in the absolute unformed Aristotelian matter nor in Spinozistic substance. Since Hegel, without exception, imputes to things a heuristic dialectic and process of becoming, and thinking thereby is induced to find such states of affairs, so form must be found in so-called content, a form whose inspiration and shape, as Hegel figuratively expresses himself, can always be found in what can be called content. If the categories, as determinations of thought, are to be distinguished from the 'thing' and to be advanced before consideration in thought, then the result is the definite concept in which, in fact, dialectically linked content and form are to be found, and constitute a substantial unity.[34]

If the manner of expression seems more or less awkward, still Hegel explains why it is that scientific concepts are capable of conferring knowledge and do not come to a standstill *beyond* the thing. A so-called 'substantial' shape and form is acknowledged in the concepts. And with the term 'substantial' we return to the traditional conceptual framework, which, however, had been pressed into Spinoza's intellectual framework, and there had assumed an antidualistic position. When Hegel designates concepts as substantial, he reflects precisely this antidualistic point of view, and notes that no content and no component of form, which in principle resisted being known and thereby brought the property of being unknowable from thinking into the products of thought, could enter into concepts. Then the concept would be debased and falsified. Hegel reproaches Kant in this respect with having retained dualism.

Since Hegel maintains substantial unity of scientific concepts, then thinking and its results always remain within the sphere of its own business, which is to get at the thing and to act upon the means within its schema, that attributes to mediation and concept their binding connectedness with the objective objects. In Hegel's conception, the intellectual objects never lose the possibility of applying thought to things and of becoming effective in the process in which the means, which act in an overarching manner, find their application and have their function.

But with that Hegel draws action and thought closer together. Hegel explicitly turns to a digression by way of social labor, and the result is his assumption of a functional correction, by which the concept remains adequate to the extent that the 'means' permit. To this extent, however, an impulse comes from Hegel's philosophical idealism which gives a stimulus to materialism, and can influence the modern development of the methodological instrumentarium. From that time on, concepts were linked with the things referred to, even when philosophic reflection temporarily lost its awareness that scientifically secured concepts, or expressed differently, terms laden with meaning, have reference to objective processes. The growing function of conceptual thinking in the progress of the sciences is, otherwise than Hegel believed, accompanied by a new flowering of work in formal logic. At the same time, openness to the hypothetical imagination of the creative scientist, and to the transition from projected structures in thought to nature isolated in experiments or to structures represented in models which seek to grasp the relations between man and man or man and medicine, all these allow Hegel's intellectual approach once more to become relevant. The [scientific and practical formulas and] abbreviations that work against the particularity of things are realized as the determinateness of concepts which makes the indeterminateness and variance of the individual processes accessible to theory and praxis. Along with the emphasis on the determinateness of concepts, Hegel expressly holds that the indeterminations of the individual object will not escape thought. Against naked empiricism, Hegel emphasizes the grasp possible with theoretical thinking. Whether the data and facts secured by observation or the basic propositions necessary for theoretical derivation possess absolute validity, or are to be understood as relative, remains a secondary problem at Hegel's stage of identity philosophy. Determinateness and at least the coming-to-be of the 'means' and of consciousness, including the categories, are regarded as properties of *one* thing. As Hegel subsumes contradiction and [evolutionary] development under dialectic, he does not come into conflict with his simultaneous respect for freedom of thinking from contradictions defined by formal logic. In Hegel's understanding, the analogical procedure that differentiates various applications of models already unambiguously signifies the dialectical contradiction within the same mathematical structure. Hegel sees another dialectical contradiction in the difference between the individual and the universal, which characterizes *every* statement formulated in consciousness, even when it is supposed to refer expressly to an individual object, but especially when generalized theoretical discussions are involved, whether these be formalized or not. Insofar as awareness seems to be intensified, it is seen that methodologically secured rigor at the same time requires creative imagination, fantasy, within the domain of fundamentally scientific expertise.

This active experimental approach to concrete objects and domains of objects, in Hegel's understanding, aims to obtain the corresponding structures by hard work. How Hegel gets from the multiplicity of empirical objects to structure is treated with a certain naiveté which belongs to a problematic since then taken up again and again, from time to time. Obviously, for Hegel the concept stands above the concrete object. If the concepts are what is real, then Hegel seems to stay completely within classical realism of universals.[35] On the basis of the conception of this identity philosophy, however, the relation to objects shows itself to have the same import as old conceptual realism and naive copy-theory [of perception].[36]

In contrast to the classical conceptual realism, as Hegel sees it, sensuous, spatial and temporal, palpable existence are just as far from complete as the concept, because both contain 'moments' that can only be elevated to the 'idea' together. Among these, Hegel understands the unity of the 'concept and reality'. In connection with this, 'reality' points to that palpable existence which meanwhile has been conceived as externality, and as far as Hegel's text is to be understood, contains laws and structures as a field of internal relations. If, to begin with, Hegel speaks of reality as the external, he then goes on to relate reality primarily to inner systematic coherence, to be characterized more or less as the universal, or approximately, say, as a 'white box', as an approach which permits adequate characterization. How this is to be achieved occupies Hegel because of the obvious denigration in his time of scientific concept or, as we would say now, of theoretical thought as against [immediately] palpable existence.

In the opposition of the concrete and the abstract, Hegel offers a quite simple, but not unimportant delimitation of the two categories. If the palpable counts as more admirable, as superior to the concept, then the concept meets with the reproach the "it lacks so much of this kind of material."[37] But Hegel assumes that scientific procedure operates in just that way. Since concepts are credited with a unified connection with reality, the various terms utilized at the same time mean something like representation, or image, but in which the constructive 'moment' of the understanding and its activity find adequate consideration. Palpable existence is to be found in the attributes to which reference is made. They do not function under the criterion of any completeness required by the thing, but only that of being sufficient. It is assumed that the object or the domain of objects can be reduced in the multiplicity of its properties and characteristics to a smaller number, without impairing its worth. Without any particular added speculation, Hegel assumes that scientific thinking omits a greater number of known properties and characteristics, at least in the formation of concepts, and does so, indeed, *not* for subjective reasons, although these can be decisive too, but for objective reasons which can be derived from the thing itself. It is expressly attested for the understanding, that it is not a question of its inability to appropriate the overwhelming multiplicity that confronts it. Rather, it is the reverse, a property of the understanding to proceed in an appropriate manner, and thereby just allow that adequacy of thing and concept, which Hegel does not understand as a literal total mirror-image. Nor is the Aristotelian equality of thing and concept accepted, for it too is subordinated to the process of becoming, namely to the development of the scientific concept itself. Accordingly, which features are of interest is subject to a catalogue of influences in the Hegelian dialectic, through which the value and dignity of the object only begin to assert themselves; thereby something is assumed which is similar to distinguishing the essential from the unessential parameters in working out structures in systems analysis. As Hegel considers the existence of things to be given, he does not require any additional speculation about the range of the capacity of the understanding. He considers it sufficient to secure the process of knowledge enriching and developing itself. As criterion of validity, cogency of theoretical thinking, attention to the thing itself, and the means as a mediating 'moment' are presupposed once and for all, so that no stage in the progress of conceptual consciousness, which might have to be distinguished, would be justified in claiming the world to be something unknowable. With its observation of properties, empirical work is granted sufficient significance, but theoretical thinking is considered as of equal value, if not superior. When Hegel speaks of philosophy he means self-understanding about the total state of affairs; when oriented toward all intellectually applied means, we then see that none of the special sciences by itself will be granted the ability to grasp the thing itself. In any case, Hegel does not equate knowledge of objects with the transfers of all the features and properties into the understanding. Intellectual construction receives no direct copy-function, but rather the representation of structures which are not equated with the "fleeting and superficial phenomena of the world of sensuous particulars";[38] and indeed these do not correspond completely either, but give a more accurate impression of the thing than is offered by superficial semblance.

For the beginning of the nineteenth century in which Hegel wrote, the argument was convincing, that religion held to external appearance just as little as did the concepts of science. Some objections might be raised about this reciprocal relationship, but these will not be discussed here. Adhering to external semblance, to the phenomena, may be useful to religion or not. It is similar in the sciences. The Copernican revolution contradicted the reigning conceptions of religion and science. Both are subject to changes in their principles. The speculative world design of a Genesis, a phantasmagoric projection of social and natural reality onto a dualistically conceived horizon, may on occasion be preparatory stages for an understanding of the world, but may also block access to the world, if petrified into tradition. Hegel is a Pantheist, and therefore, in his character as identity philosopher, possesses, effectively and potentially, the entire heretical tradition that was linked throughout history with mass movements *and* with attempts at an objective understanding of nature. As has been noted, this relation is preserved in Hegel's philosophy but what one must not forget, it is displaced into the understanding of the becoming of various stages of social phenomena. Since at that time becoming as process of development of nature *and* of society was *not* accepted, the epistemological reflections provided analyses of these *processes*. Incidentally, they contain a historical critique of religion, which was understood as atheistic by Bruno Bauer.

Historically appropriate, the becoming and developing of scientific concepts is actually preceded by 'feeling and intuition' ['*Gefühl and Anschauung*'] which have accompanied further development throughout the long epochs down to the present, and do not necessarily become empty now, as they continue to be capable of stimulating access to new fields. However, it was decisive for scientific concepts to have a non-sensuous mode of expression which finds its structures behind the phenomena even when they derive, say, from the surface phenomena of statistical data. In the hypothesis of Democritus's atomism, the model of scientific procedure was presented which conceives the speculatively formed objects of thought, i.e. particles, as the determining objects of sensually accessible reality. With the same procedure, Hegel arrives at the 'seeds' which induce the manifestation of form in biological objects and which struck André Lwoff as a modern idea. However one may look at Jacques Monod's discussions of social philosophy, his analysis of modern genetics shows that the categories of chance and necessity have actually changed just as much in the meantime as the forms of intuition, space and time have changed since Newton and Kant, since Einstein, and lastly but by no means in conclusion, through Ne'eman, Chew and Gell-Mann, further developments already appear to be assured.

Churchman's relative sympathy for Hegel appears to be determined by the 'moments' of the objective significance of subjective evaluation and scales of evaluation. Leaving aside the dialectical character even of micro-development, which itself, with the growth in the magnitude of productive opportunities, demands a qualitative change of technology (in principle precisely processes described by Hegel as dialectical), the dialectics of development probably has the greatest significance. However, what should never be lost in Hegel's dialectic is that it holds to the objectivity of mental constructions, which on the one hand do *not* have to be palpably observable in nature or society, and thus are not to be brought to immediate perception, and on the other hand may be a directly material substrate like DNA and RNA. Hegel recognizes objectivity in both cases which, expressed differently, can be material substrate and, say, relations expressing themselves in terms of fields of probability. Access to these states of affairs, Hegel calls overcoming sublation [*Aufheben*] and reduction of reality-as-mere-appearance to the essential, and this manifests itself only in the concept.

Thus Hegel's epistemological ideas put the terms feature [*Merkmal*] and sign [*Zeichen*] not only in concrete phenomena.[39] As they are also results of theoretical thinking, they lead to the category of synthesis *a priori*, which Hegel calls a most important idea of Kant.[40] Once again intuitive thought is decisively rejected. But for Hegel synthesis *a priori* does not signify apriorism as a limit to knowledge, but as a form of deductive approach which can have its basis in empirical reality, but in any case can refer to empirical reality. Hegel seeks to establish a relationship between nature and thought, inclusive of social objects, without sacrificing the independence of thought nor of concrete things [*Dinge*], nor playing these off against each other. The synthesis *a priori* incorporates multiplicity into the concepts without losing itself in the former. Any regression to ultimate ideal or material objects continues to be avoided and is even expressly excluded, because bringing the thing to its concept only requires recourse to a space of structures, through which the field that is being treated is made pragmatically accessible. For Hegel it is not the concrete results which the special sciences have furnished that are of primary concern, but rather 'the nature of the concept', thereby securing an assurance for human thought, which at the same time is set forth as a basis for a theory of action. To analyze thinking and acting as two sides of a thing remains the concern of Hegel's philosophy, in order to surmount fragmentation into externally separated domains and yet without obscuring the difference between these. Hegel was able to display in his own work those multi-stratified structures, which in his own time offered points of departure to the most diverse and opposite positions, and the possibility for such points of departure also seems to be offered in our day. It may be surprising, but it is not incomprehensible that we have newly uncovered access to constellations of problems in fields concerning which it was generally agreed that Hegel certainly had nothing to say.

Humboldt University
Berlin, G.D.R.

NOTES

[1] Charles Sanders Peirce to William James, Oct. 3, 1904: "More or less explicitly, some writers, namely the Thomists, the Hegelians, and other Intellectualists, together with some scientific thinkers not too much sophisticated by reading philosophy, recognize with me (until I shall have studied your views, which I don't believe will carry me entirely away from this anchorage) three modes of consciousness, that of feeling, that of EXPERIENCE (experience meaning precisely that which the history of my life has FORCED me to think; so that the idea of a struggle, of not mere twoness but active opugnancy is in it), and thirdly, the consciousness of the future (whether veridical or not is aside from the question) . . . " *Collected Papers of Charles Sanders Peirce*. 8 vols. (Cambridge, Mass., Harvard University Press, 1931–1958), Vol. VIII. *Reviews, Correspondence and Bibliography*, ed. by Arthur W. Burks, 1958, p. 202.

[2] K.-O. Apel, 'Peirces Denkweg vom Pragmatismus zum Pragmatizismus', in Charles S. Peirce, *Schriften II* (Frankfurt a. M., Suhrkamp, 1970), p. 18.

[3] Charles Sanders Peirce, *Collected Papers* . . . Vol. V: *Pragmatism and Pragmaticism*, ed. by C. Hartshorne and P. Weiss, 1934. p. 31f.

[4] *Ibid.*, p. 128. Peirce refers to Kirchhoff's *Vorlesungen über mathematische Physik*, vol. I, Vorrede, Leipzig, Teubner, 1974–76. Apel notes that Peirce goes beyond the reality theory of the first and second period; he is no longer content with the postulate, critical of the senses held by the eschatological consensus of all scientists, but demands beyond that – in the interest to a certain degree of a methodological orientation of the research process – that the real universality of the laws of nature, which as real possibility is the condition for experimental knowledge, can be 'perceived' already now. K.-O. Apel, *op. cit.*, p. 388, notes 62 and 64.

[5] "The tree is a living reality, says the German philosopher. In its germ the determinants exist as potentialities. Nothing is in the tree which was not already in the germ, and yet in the germ one does not see anything, even with the microscope. We can visualize the determinants as existing in the germ as extremely simple forces" (A. Lwoff, *Biological Order* (Cambridge, Mass., MIT Press, 1962), p. 14). Lwoff mentions the English philosopher, Herbert Spencer, as of the year 1864, Ernst Häckel, as of 1866 and Weismann as of 1892. The quantitative studies began with Gregor Mendel in 1865. A passage similar to the one mentioned by Lwoff, but not identical, can be found in Hegel's *The Philosophy of Fine Art*, where Hegel says with respect to the beautiful in art: "There is no end to the caprice of the human features. Speaking generally, we would associate with this ground the fact that the beauty of children most arrests us. In their faces we find all pronounced idiosyncracies slumber as it were beneath a quiet veil [*Keim*] ; no dominating passion as yet ravages their soul; not one of the thousand interests of the grown man has engraved for ever the expression of its necessity on these mobile features" (*The Philosophy of Fine Art*, trans. F. P. B. Osmaston. 4 vols. (London, G. Bell, 1920). Vol. I, p. 207. cf. *Vorlesungen über die Ästhetik*. Vol. I, pp. 210–211 (*Jubiläumsausgabe* **12**, 1927)).

[6] A. Meyer-Abich, *Geistesgeschichtliche Grundlagen der Biologie* (Stuttgart, Fischer, 1963), Vorwort (no pagination).

[7] *Ibid.*

[8] *Ibid.*, p. 237: "Corresponding to this completely, the complementarities too of natural science in physics and biology stand side by side as antinomical theses, they distinguish themselves from truly historical complementarities, like those of the history of philosophy, already mentioned, only by the fact that in them the clearly defined synthesis is lacking. In this sense we regard the complementarity in natural science as weak historical complementarities that are only beginning. In reality this synthesis can also be found in the complementarities of natural science. In physics, in the case of waves and corpuscles, physics as such in its totality. In biology, in the case of the form-function complementarity the same is true of physiology, in the case of the heredity-adaptation (environment) complementarity, of developmental physiology, and in the case of environment-inner world, of ecology."

[9] H. v. Helmholtz, *Ueber das Ziel und die Fortschritte der Naturwissenschaft* (Innsbruck 1869). In *Populäre wissenschaftliche Vorträge* **2**, 1971.

[10] Meyer-Abich, p. 233: "The philosophers of the time, however, already treated the natural sciences as if they were dealing with historical structures. From that nothing but incoherence could result. All that has become fundamentally different today . . . and indeed we have phylogeny as the history of organisms. With that Hegel's logic becomes the logic of phylogenetic thinking."

[11] C. F. v. Weizsäcker, *Die Tragweite der Wissenschaft* (Stuttgart, Hirzel, 1964), p. 145.
[12] C. West Churchman, *Prediction and Optimal Decisions, Philosophical Issues of a Science of Values* (Engelwood Cliffs, N.J., Prentice-Hall, 1961), p. 168: "This metaphysical theory is borrowed without benefit of the scrutinies of critics of the intervening centuries from the empiricism of John Locke. Anyone adopting such a theory must at least try to answer the devastating theses of Berkeley and Kant: data are not moves of nature; they are in part our own moves, created by our judgement. What neither Berkeley nor Kant provided to our full satisfaction is the theory of this creative process; but the study of such a theory seems far more fruitful than the study of strategies against an indefinable Great Mother."
[13] See H. Ley and Th. Müller, *Kritische Vernunft und Revolution. Zur Kontroverse zwischen Hans Albert und Jürgen Habermas* (Cologne, Pahl-Rugenstein, 1971), pp. 210–232.
[14] C. West Churchman, *On the Ethics of Large Scale Systems* (Berkeley, September 1966, working paper), Part III, Ch. 1, p. 118: "On the other hand philosophers have often extended the meaning of logic to cover many things, that the formal logicians do not consider. Hegel, for example, in his writings in the last century described 'logic' as the whole process of developing an understanding of the world. This idea of logic was captured in his 'dialectical method.' Similarly scientists often use the term 'logic' to refer to the methods by which they reach their conclusions in their empirical investigations, and they frankly confess that these methods go beyond the rigorous principles of formal logic."
[15] "Of course, there were a few diehards in this long process of killing the Age of Reason and its Gods. There was Hegel who suggested a dialectical theology; the end point of the process of reality is an Absolute Mind. But Marx showed it was far more significant to use the dialectical method to describe a sociological, materialist process without a God" (*ibid.*, p. 148).
[16] "Because of the fixed reality of natural objects the study of nature compels us to fix the categories which can no longer be ignored in her, although with complete inconsistency towards other categories which are also allowed to remain valid; and such study does not permit the further step of abstracting from the opposition and indulging in generalities as so easily happens in the intellectual sphere" (*SL* [M], p. 33).
[17] "The critical philosophy had, it is true, already turned metaphysics into logic but it, like the later idealism, as previously remarked, was overawed by the object, and so the logical determinations were given an essentially subjective significance with the result that these philosophies remained burdened with the object they had avoided and were left with the residue of a thing-in-itself, an infinite obstacle, as a beyond. But the liberation from the opposition of consciousness which the science of logic must be able to presuppose lifts the determinations of thought above this timid, incomplete standpoint and demands that they be considered not with any such limitation and reference but as they are in their own proper character, as logic, as pure reason" (*ibid.*, Introduction, p. 51).
[18] "But inasmuch as it is said that understanding, reason, is in the objective world, that mind and nature have universal laws to which their life and their changes conform, then it is conceded that the determinations of thought equally have objective value and existence" (*ibid.*).
[19] "Consequently it is much more difficult to believe that the forms of thought which

permeate all our ideas – whether these are purely theoretical or contain matter belonging to feeling, impulse, will – are means for us, rather than that we serve them, that in fact they have us in their possession; what is there more in *us* as against them, how shall *we*, how shall *I*, set myself up as *more* universal than they, which are the universal as such" (*ibid*., Preface to the Second Edition, p. 35).

20 "Or again, to speak of *things*, we call the *nature* or the *essence* of things their notion, and this is only for thought; but still less shall we say of the notions of things that we dominate them, or that the determinations of thought of which they are the complex are at our service; on the contrary, it is our thinking that must accommodate itself to them, and our caprice or freedom ought not to want to mould them to suit itself. Since, therefore, subjective thought is our very own, innermost, act, and the objective notion of things constitutes their essential import, we cannot go outside this our act, we cannot stand above it, and just as little can we go beyond the nature of things" (*ibid*., p. 35f.).

21 *Ibid*., p. 32: "The advance of culture generally, and of the sciences in particular, gradually brings into use higher relationships of thought, or at least raises them to greater universality, and they have thus attracted increased attention. This applies even to the empirical and natural sciences which in general employ the commonest categories, for example, whole and parts, a thing and its properties, and the like)."

22 "The way critical philosophy understands the relationship of these three terms is that we place our thoughts as a medium between ourselves and the objects, and that this medium instead of connecting us with the objects rather cuts us off from them. But this view can be countered by the simple observation, that these very things which are supposed to stand beyond us and, at the other extreme, beyond the thoughts referring to them, are themselves figments of subjective thought, and as wholly indeterminate they are only one single thought-thing – the so-called thing-in-itself of empty abstraction" (*ibid*., p. 36).

23 "Further, since the end is finite it has a finite content; accordingly, it is not an absolute, nor simply something that in its own nature is *rational*" (*ibid*., p. 747).

24 "But the *means* is the external middle term of the syllogism which is the realization of the end; in the means, therefore, the rationality in it manifests itself as such by maintaining itself in *this external other*, and precisely *through* this externality. To this extent the *means* is *superior* to the *finite* ends of *external* purposiveness: the plough is more honourable than are immediately the enjoyments procured by it and which are the ends. The *tool* lasts, while the immediate enjoyments pass away and are forgotten. In his tools man possesses power over external nature, even though in respect of his ends he is, on the contrary, subject to it" (*ibid*.).

25 E. Nagel, *The Structure of Science, Problems in the Logic of Scientific Explanation* (New York, Harcourt, Brace and World, 1961), p. 15: "As we shall see, such explanations may be offered for individual occurrences, for recurring processes, or for invariable as well as statistical regularities."

26 V. I. Lenin, 'Conspectus of Hegel's Book, *The Science of Logic*' in *Philosophical Notebooks*, vol. 38 of *Collected Works* (London, Lawrence and Wishart; Moscow Foreign Languages Publishing House, 1961), p. 190.

27 Georg Lukacs, *History and Class Consciousness. Studies in Marxist Dialectics*. Trans. Rodney Livingstone (Cambridge, Mass., MIT Press, 1971), p. 129f.: "It is important to realise that if we take action in the sense indicated above to mean changing reality, an orientation toward the qualitatively essential and the material substratum of action, then

the attitude under discussion will appear much more contemplative than, for instance, the idea of knowledge held by Greek philosophers. For this 'action' consists in predicting, in calculating as far as possible the probable effects of those laws, and the subject of the 'action' takes up a position in which these effects can be exploited to the best avantage of his own purposes. It is therefore evident that, on the one hand, the more the whole of reality is rationalised and the more its manifestations can be integrated into the system of laws, the more such prediction becomes feasible. On the other hand, it is no less evident that the more reality and the attitude of the subject 'in action' approximates to this type, the more the subject will be transformed into a receptive organ ready to pounce on opportunities created by the system of laws and his 'activity' will narrow itself down to the adoption of a vantage point from which these laws function in his best interests (and this without any intervention on his part)."

28 "Non omnino improbabile est, materiam et quantitatem esse realiter idem." See also *SL* [M], p. 189f. where Hegel continues: "In fact, the distinction between these two concepts is simply this, that quantity is a determination of pure thought, whereas matter is the same determination in outer existence."

29 "But profounder insight into the antinomial, or more truly into the dialectical nature of reason demonstrates *any* Notion whatever to be a unity of opposed moments to which, therefore, the form of antinomial assertions could be given. Becoming, determinate being, etc. and every other Notion, could thus provide its particular antinomy, and thus as many antinomies could be constructed as there are Notions. Ancient scepticism did not spare itself the pains in demonstrating this contradiction or antinomy in every notion which presented itself in the sciences" (*ibid*., p. 191).

30 "Thus law is *essential relation*" (*ibid*., p. 511).

31 "This makes itself apparent in all natural, scientific and spiritual development generally and it is essential to recognize that because something is at first only *inner* or also in its Notion, the first stage is for that very reason only its immediate, passive existence" (*ibid*., p. 526).

32 ". . . when the content of the interest in which one is absorbed is drawn out of its immediate unity with oneself and becomes an independent object of one's thinking, then it is that spirit begins to be free, whereas when thinking is an instinctive activity, spirit is enmeshed in the bonds of its categories and is broken up into an infinitely varied material. Here and there in this mesh there are firm knots which give stability and direction to the life and consciousness of spirit; these knots or nodes owe their fixity and power to the simple fact that having been brought before consciousness, they are independent, self-existent Notions of its essential nature. The most important point for the nature of spirit is not only the relation of what it is *in itself* to what it is *actually*, but the relation of what it *knows itself* to be to what it actually is; because spirit is essentially consciousness, this self-knowing is a fundamental determination of its *actuality*."

33 "The inadequacy of this way of regarding thought which leaves truth on one side can only be made good by including in our conception of thought not only that which is usually reckoned as belonging to external form but the content as well. It is soon evident that what at first to ordinary reflection is, as content, divorced from form, cannot in fact be formless, cannot be devoid of inner determination; if it were, then it would be only vacuity, the abstraction of the thing-in-itself; that, on the contrary, the content in its own self possesses form, in fact it is through form alone that it has soul and meaning . . ." (*ibid*., p. 38f.).

34 "But in this connection we can be reminded that *there is* a multitude of objects [*Sachen*]. We have, however, already said how it is that restrictions are imposed on this multitude, that the Notion, simply as thought, as a universal, is the immeasurable abbreviation of the multitudes of particular things which are vaguely present to intuition and pictorial thought; but also *a* Notion is, first, in its own self *the* Notion, and this is only one and is the substantial foundation; secondly, *a* Notion is *determinate* and it is this determinateness in it which appears as content: but the determinateness of the Notion is a specific form of this substantial oneness, a moment of the form is totality, *of that same Notion* which is the foundation of the specific Notions" (*ibid.*, p. 39).

35 "But the truth is that it is not the material given by intuition and representation that ought to be vindicated as the *real* in contrast to the Notion" (*ibid.*, p. 587).

36 "In both these actions the Notion is not the independent factor, not the essential and true element of the prior given material; on the contrary, it is the material that is regarded as the absolute reality, which cannot be extracted from the Notion.

"Now it must certainly be admitted that the Notion *as such* is not yet complete, but must rise to the *Idea* which alone is the unity of the Notion and reality; and this must be shown in the sequel to be the *spontaneous outcome of the nature of the Notion itself.* For the reality which the Notion gives itself must not be received by it as something external but must, in accordance with the requirement of the science, be derived from the Notion itself" (*ibid.*).

37 *Ibid.*, pp. 587–88: "People often say, "It is only a notion,' contrasting the notion not only with the Idea but with sensuous, spatial and temporal, palpable reality as something more excellent than the Notion; and then the *abstract* is held to be of less account than the concrete because it lacks so much of this kind of material. In this view, to abstract means to select from the concrete object for *our subjective purposes this or that mark* without thereby detracting from the worth or status of the many other properties and features left out of account; on the contrary, these as *real* retain their validity completely unimpaired, only they are left yonder, on the other side; thus it is only the *inability* of the understanding to assimilate such wealth that compels it to content itself with the impoverished abstraction. Now to regard the given material of intuition and the manifold representation as the *real* in contrast to what is *thought*, to the Notion, is a view, the abandonment of which is not only a condition of philosophizing but is already presupposed by religion."

38 *Ibid.*, p. 588: " . . . how can there be any need for religion, how can religion have any meaning, if the fleeting and superficial phenomena of the world of sensuous particulars are still regarded as the truth? But philosophy gives a *reasoned* insight into the true state of the case with regard to the reality of sensuous being; it assumes the stages of feeling and intuition as precedent to the understanding in so far as they are conditions of its genesis, but only in the sense that it is conditioned by their *reality*. Abstract thinking, therefore, is not to be regarded as a mere setting aside of the sensuous material, the reality of which is not as mere *phenomenal appearance* to the *essential*, which is manifested only in the *Notion*."

39 "Of course, if what is taken up into the Notion from the concrete phenomenon is to serve only as a *mark* or *sign*, it certainly may be any mere random sensuous particular determination of the object, selected from the others on the basis of any random external interest and of a similar kind and nature as the rest" (*ibid.*).

40 "This original synthesis of apperception is one of the most profound principles for

speculative development; it contains the beginning of a true apprehension of the nature of the Notion and is completely opposed to that empty identity or abstract universality which is not within itself a synthesis. . . . It is in keeping with this standpoint, too, that the Notion without the manifold of intuition is again declared to be *empty* and *devoid of content* despite the fact that it is a synthesis *a priori*; as such, it surely does contain determinateness and difference within itself" (*ibid*., p. 589).

KENLEY ROYCE DOVE

HEGEL'S 'DEDUCTION OF THE CONCEPT OF SCIENCE'

In one of his most memorable epigrams, Aristotle declares it to be the mark of an educated man to seek in each type of inquiry only as much precision or exactness as the nature of the subject matter allows. This remark, which appears at the beginning of his inquiry into 'the practical sciences' (i.e., Ethics and Politics), is designed to forestall the kind of mistakes made by his great predecessor, whom Aristotle criticizes for trying to determine the unity of the *polis* too precisely. He does not mean to suggest that an educated man should avoid the study of works such as Plato's *Republic*; the suggestion is rather that his time would be better spent if he did not attempt to discover scientifically demonstrative arguments in them.

The subject matter of the present inquiry is Hegel's *Phenomenology of Spirit*. Many have been persuaded that the book should be read in the spirit of Aristotle's practical philosophy, that one should not seek scientific precision where it cannot be found, and that Hegel, like Plato, can be read with great profit without succumbing to 'the Platonic Fallacy'. This manner of approaching the *Phenomenology* is extremely attractive, and for at least two reasons. (1) After a century and a half of scholarly writing about the *Phenomenology*, no consensus has been reached either concerning the precise subject matter or the argument of the book. To the extent that agreement has been reached, it has tended to be negative, namely, that Hegel himself altered his intentions about the subject matter or the argument during the course of his actual composition of the book. (2) The subject matter Hegel takes up in the second half of the book, whether or not it is continuous with and develops out of the foregoing subject matter, bears a very strong family resemblance to the subject matter of Aristotle's practical philosophy. Indeed chapter VI, 'Geist', begins with a treatment of a form of ethical life whose basic structure directly parallels Aristotle's juxtaposition of the *oikos* and *polis* in the first book of the *Politics*. All the more reason, then, for reading Hegel's *Phenomenology* in the spirit of Aristotle's practical philosophy. Readers agree that many of the trees are magnificent; but the wood is dark. So why get lost?

Probably the darkest saying Hegel ever made about the *Phenomenology* is to be found in the opening pages of the *Science of Logic*. It is especially

R. S. Cohen and M. W. Wartofsky (eds.), Hegel and the Sciences, 271–281.

vexing since it is one of his very few utterances about the function of the work as a whole. He there refers to the book as "the deduction of the concept of science" and assigns the concept, thus deduced, the formidable role of being the necessary presupposition for his otherwise presuppositionless *Logic.*[1] The *Logic*, on the other hand, takes up in its first chapter a subject matter which seems to bear no resemblance to that of the second half of the *Phenomenology*. It begins with 'pure being'. And if we ask what this might be, the immediate context does not seem to be very illuminating, for we are only told what 'pure being' is not: it is *not* definite and it is *not* mediate. It is "*das unbestimmte Unmittelbare* (*WL* [1812] **1**, p. 22)". In fact, it is said to be nothing more nor less than Nothing.

This famous and irritating beginning of the *Logic* has been the object of a considerable literature.[2] But, as in the case of the *Phenomenology*, nothing like a critical consensus has been arrived at. The *Logic* continues to be read, but its beginning remains an enigma.

I

Given this situation, it would be foolhardy simply to assert that the mysteries of the *Logic* might be unlocked by referring all difficulties of its beginning to the *Phenomenology*, as its presumptive 'deduction'. For, as we have seen, the *Phenomenology* itself is no less problematical than the *Logic*. But perhaps some light might be cast on the question of the *Phenomenology*'s subject matter if the question of its argument were focussed by the problem of the beginning of the *Logic*. The aim of this paper is to explore the conjecture that the *Phenomenology*, when read as a "deduction of the concept of science," will reveal more clearly the nature of its subject matter.

As we have seen, the subject matter at the beginning of the second half of the *Phenomenology*, 'Geist', is remarkably akin to that of Aristotle's practical philosophy. It is also strikingly different from the traditional subject matter of German Idealism, i.e., the '*Ich*'. Now when we look at the pages of the *Logic* immediately preceding the treatment of 'pure being', we find that Hegel explicitly contrasts '*Ich*' and '*Geist*' with respect to the problem of an absolute beginning in science. An examination of this discussion, "Womit muss der Anfang der Wissenschaft gemacht werden?" (with special reference to the recently republished 1812 edition), will suggest an interpretative conjecture concerning the *Phenomenology*.

The section begins with the statement of what appears to be a dilemma. On the one hand, the result of the *Phenomenology*, 'pure knowing', is said to be presupposed by the *Logic*, the *pure* science (*WL* [1812], p. 6). On the other hand, an absolute science must itself *be* an *absolute* beginning, and as such it may presuppose nothing (*WL* [1812], p. 7). If the *Logic* is taken to begin as a science of the absolute, it will therefore be incompatible with the *Phenomenology*. Since Hegel has explicitly affirmed that the *Logic* does require and must presuppose the "deduction of the concept of science," and that the *Phenomenology* is this deduction, the implication seems to be that the *Logic* does not and cannot have an absolute beginning.

An explication of this, at first sight, impossible dilemma is to be found on the next page of the 1812 edition. The paragraph in question has been discussed only rarely in the Hegel literature; and since it is critical to our problem, it will be well to cite it entire:

In der ersten so eben gegebenen Darstellung des Seyns als des Anfangs ist der Begriff des Wissens vorausgesetzt. Somit ist dieser Anfang nicht absolut, sondern kommt aus der vorhergehenden Bewegung des Bewusstseyns her. Die Wissenschaft dieser Bewegung [i.e., the *Phenomenology of Spirit*], aus der das Wissen resultirt, müsste nun den absoluten Anfang haben. Sie macht ihn mit dem *unmittelbaren Bewusstseyn*, dem Wissen, dass etwas ist. – Das Seyn macht so hier gleichfalls den Anfang, aber als Bestimmung einer concreten Gestalt, des Bewusstseyns; ERST DAS REINE WISSEN, DER GEIST, DER SICH VON SEINER ERSCHEINUNG ALS BEWUSSTSEYN BEFREYT HAT, HAT AUCH DAS FREYE, REINE SEYN ZU SEINEM ANFANG. – Aber jener Anfang, das unmittelbare Bewusstseyn, enthält das Ich als bezogen auf ein schlechthin Anderes, und umgekehrt, den Gegenstand bezogen auf Ich; somit eine Vermittlung. – Zwar enthält das Bewusstseyn die beyden Vermittelnden – die auch wiederum die Vermittelten sind, – selbst, weist somit nicht über sich hinaus, und ist in sich beschlossen. Aber indem die Vermittlung gegenseitig ist, so ist jedes Vermittelnde auch vermittelt, somit keine wahrhafte Unmittelbarkeit vorhanden. – Aber umgekehrt, WÄRE eine solche vorhanden, so ist sie, DA SIE NICHT BEGRÜNDET IST, etwas WILLKÜRLICHES und ZUFÄLLIGES (*WL* [1812], p. 8 – emphasis added).

As in the case of Plato's recollection myth, which is said simply to transport the paradox of learning (*Meno*, 80E) to another level, Hegel's remarks concerning the beginning of logical science and its derivation from the *Phenomenology* have been seen as a dodge. In the light of this paragraph, however, we see that Hegel clearly distinguishes between two ways of beginning and two types of subject matter in which these beginnings may be made. The one kind of beginning is absolute, but not immediate. The other is immediate, but not absolute. In the former, immediate consciousness begins absolutely, with the knowledge that something is. It is absolute because consciousness refers to nothing beyond itself. It simply affirms the absolute identity of what is 'for it', pure being, and what it takes itself to be, the pure *Ich*. As the affirmation of a relation, however, consciousness is not immediate but rather the reflective mediation of the *Ich*, identified with what is for it (the object), and the object, identified with the *Ich*. Both terms of this identity are therefore mediating and mediated. And to affirm the immediacy of either by itself would be to remove it from the absoluteness of the identity, to declare it fortuitous, arbitrary, and groundless. Consciousness is therefore the subject of the absolute way of beginning. It begins with an absolute grounding, the identity of its terms: the *Ich* and pure being. And this is the beginning of the *Phenomenology*, the shape of consciousness called 'Sense Certainty'.

The other way of beginning, which is immediate but not absolute, first becomes possible as the *result* of the movement (*Bewegung*) which begins absolutely with consciousness and through which '*Geist*' is "liberated from its appearance as consciousness." As such, it is said to be 'pure knowing' and to have 'pure being' as *its* beginning. 'Pure being' was also the beginning, indeed, the absolute beginning of the '*Ich*' in the *Phenomenology*. But, as we have seen, it was there a term in a relation of absolute identity, the unity of consciousness. At the end of the *Phenomenology* 'pure being' becomes a beginning, the beginning of 'pure knowing', now conceived as Spirit purified of all relation to an object, pure spiritual unity (*geistige Einheit*). Here there is no absolute beginning because there is no absolute ground; spiritual unity is pure immediacy, immediacy as the result of a process of purification. And through this process, consciousness, which began the movement described in the *Phenomenology* with an absolute grounding, seems to have become groundless.

These observations suggest that the key problem for a coherent interpretation of the *Phenomenology* will be an understanding of how spiritual unity emerges through the movement of mediation which *begins* with the absolute identity of the '*Ich*' and its object. Put in its most concise form, the question is how Spirit emerges – and is not simply introjected – in the '*Wissenschaft der Erfahrung des Bewusstseins*'. Is this Science continuous with the '*Wissenschaft der Phänomenologie des Geistes*'?

To this end it will be important to notice that the absolute identity of consciousness *is* absolute only from a standpoint independent of consciousness. Hegel makes this point most clearly in his Jena lectures of 1803–04. "Aber dies *Bewusstseiende*, und das, dessen *es sich bewusst ist*, ist nur für einen Dritten diese Einheit des Bewusstseins, nicht für sie selbst . . . ".[3] In the *Phenomenology* it is clear that this 'third' is the 'Wir', who observe the entire development of consciousness from its absolute beginning to its '*Aufhebung*' in pure knowing.[4] The absoluteness of the beginning of the

Phenomenology must be seen to consist not merely in the identity of the two terms of consciousness, *Ich* and object, for this inner reflective unity is an absolute unity only by means of an 'external reflection', or '*für uns*' (see *WL* [L], 2, p. 18). The '*Wir*' is therefore a constitutive element of consciousness from the point of its absolute beginning and because it *is* an absolute beginning.

It must be noted, however, that the absolute unity of consciousness with which the *Phenomenology* begins is not posited by the 'Wir'. The 'Wir' is rather pre-thematically posited by consciousness as the necessary condition of its claim to absoluteness. The development of the argument in the book may therefore be read as a progressive thematization by consciousness of that ultimate, but at first abstractly posited ground of its absoluteness: the '*Wir*'. The *Ich* of consciousness therefore has two objects, the one with which it identifies itself as a knowing subject, and the other the essential ground on which its claims of ultimate or absolute identity with its first object are based. In its relationship to the first object, the *Ich* takes itself to be absolutely certain of itself. It is absolutely at one with its object. But as the knowing subject in an absolutely scientific mode of knowing, it must also express the truth conditions, the grounds, for its certainty. And thus it begins the process of thematizing its second object. This second object is at first taken to be simply the *Ansich* or the essence of the certainty constituting the relation to the first object. But when consciousness actually expresses what it takes the absolute ground or essence (*Ansich* or *Wesen*) of its certainty to be, the ground expressed reveals itself to be merely an *Ansich* or essence *for consciousness* and not something in itself or absolute. We therefore see that consciousness, as consciousness, cannot *immediately* thematize its ground; it cannot, as Hegel so strikingly puts it, "shoot the absolute out of a pistol." But it does take the now thematized essence or *Ansich* to be new ground for its subsequent truth claims. And this process of mediation in which consciousness progressively thematizes its absolute continues throughout the *Phenomenology*.

The movement (*Bewegung*) of consciousness presented in the *Phenomenology* may therefore be understood as a sequence of stages (the '*Gestalten des Bewusstseins*'), each of which is defined by the fundamental truth claim that consciousness has come to the point of thematizing. But although the *Gestalten* are discontinuous, each being defined by a definite *Ansich*, all share the form of consciousness as such, all are forms of knowing in which a subject takes itself to be related to an object. Viewed in this way, we can also see that this movement will have a terminus, namely, at that stage when knowing no longer has the form of consciousness, when, as Hegel expresses it in the *Logic*,[5] the "opposition of consciousness" (i.e., *Ich* and Object) will have been overcome. At the completion of its voyage of discovery, the last *Gestalt* of consciousness thus comes to discover what it is in truth, namely '*Wir*'. But, as such, the knowing subject is no longer in the form of consciousness, no longer a subject making truth-claims. For at this stage there is no longer any ground for asserting the absoluteness of any relation between subject and object. The '*Wir*', purified of its relation to consciousness, is pure knowing (*reines Wissen*), no longer a knowing grounded in the absolute, but knowing knowing itself. And, as such, it is pure immediacy, the pure and unmediated knowing from which the 'pure science', i.e., *Logic*, begins.

The problematic beginning of the *Logic*, 'pure being' characterized as "*die* un*bestimmte* Un*mittelbarkeit*," does, as we have seen, have the form of a reflective expression. And Hegel, in the section of the *Logic* referred to above, clearly acknowledges this.[6] But whether or not such expressions, at the beginning of the Logic of '*Sein*', illicitly presuppose categories yet to be developed in the Logic of '*Wesen*', is a question that might be viewed in another light if an introductory "deduction of the concept of science" includes an example of the Logic of *Wesen* in the development of consciousness' truth-claims. Then "*die unbestimmte Unmittelbarkeit*" might be read as a result rather than an anticipation.

II

If we grant the foregoing conjecture with respect to the *argument* of the *Phenomenology*, conceived as a "deduction of the concept of science," what light is thrown upon the question of the *subject matter* of the book? How may this *subject matter* be understood to undergo a development conforming to the proof-structure demanded by this argument? More specifically, what does the conjectured development of consciousness, as a progressive thematization and ultimate manifestation of its absolute ground, suggest about the significance of the ground posited by consciousness, the '*Wir*', for us as *readers* of the *Phenomenology*, on the one hand, and for the concept of '*Geist*' on the other?

In the first place, it is clear from the Introduction to the *Phenomenology* that Hegel shows his readers why it would be impossible for them to make any justifiable truth-claims at the beginning of an exposition of science (*Phän* [H, 1937], pp. 63–66). Hegel does not at this or any other point in the *Phenomenology* invite us to begin or observe the beginning of pure science, i.e., *Logic*, but rather to witness the development of phenomenal science, that is, a science begun by consciousness and in which consciousness expresses, gives phenomenal form to, its own absolute truth claims. For it is the very nature of consciousness to be a consciousness-of, of an *object*, and, as a scientifically knowing subject, to set down standards in terms of which it is prepared to utter warrant statements about its objective knowledge (*Phän* [H, 1937], p. 71).

Thus we, as readers of the *Phenomenology*, are explicitly invited to suspend all judgment, to engage in what might be called an *epoché* with respect to our subject matter, knowing consciousness, which by its nature does make truth-claims. In this respect, we, as readers of the *Phenomenology* are from the outset akin to that '*Wir*' in which consciousness was above seen pre-thematically to ground its claim to knowledge of the absolute. But at this stage, our disengagement, our *epoché*, is no more than a disengagement from the standpoint of consciousness. If we, as readers of the *Phenomenology*, are to come to the actual insight that we are not merely akin to but at one with the '*Wir*', the 'pure knowing' with which the *Phenomenology* ends, then this will only come as a result of our detailed comprehension of how consciousness, our subject matter, itself progresses from its absolute beginning to its final sublimation in the 'Wir'. For us readers, the suspension of judgment that we undertake at the beginning of the *Phenomenology* is a mere thought-experiment.

Secondly, let us turn to the question of '*Geist*'. At the conclusion of the *Phenomenology* (*Phän* [H, 1937], p. 564) as well as in Hegel's Jena lectures,[7] absolute knowing is described as "der sich als Geist wissende Geist." May this be regarded as an alternate expression for what we have called the '*Wir*'? If so, then 'Geist' in its final form must be a pure structure of interaction in which all members participate, not in virtue of any determinate characteristics they have as individuals and not in virtue of any claims that they have to the truth, but purely and simply as members. But how is such a pure spiritual unity conceivable? As a unity directly presented to consciousness it is, quite clearly, incomprehensible. As we have seen, consciousness as such cannot thematize the '*Wir*'; it can only come to a oneness with it. But then isn't this state of 'oneness' with the '*Wir*' simply a Hegelian reformulation of the neo-Platonic doctrine that there is only one *noûs*, or a rational mysticism through which individual consciousnesses ecstatically come to unity in a transcendental or meta-consciousness?[8]

Hegel's indication of how the ultimate spiritual unity in the *Phenomenology* is reached suggests that his concept of '*Geist*' is not to be understood in terms of a supra-individual meta-consciousness. Knowing in such a unity would still have the form of consciousness. The final unification spelled out in the last chapter of the *Phenomenology* suggests rather that '*Geist*' is a unity in plurality and a plurality in unity. The unification in which '*Geist*, comes to know itself as '*Geist*' is here presented as a synthesis between the result of chapter VI, Spirit in the form of consciousness, and chapter VII, Spirit in the form of self-consciousness (*Phän* [H, 1937], p. 553ff). Chapter VI, in turn, is treated here, and also in the introductions to chapters VI and VII, as the culmination of chapters I–V in the sense that they show themselves, in retrospect, to be 'abstractions' from Spirit in the form of consciousness (*Phän* [H, 1937], p. 314). The entire book, viewed from its result, thus assumes the following three-fold structure:

A. Spirit in the form of consciousness, including the abstracted shapes of consciousness, Self-consciousness and Reason
B. Spirit in the form of Self-consciousness
C. Spirit knowing itself as Spirit, i.e., as the synthesis of A. and B.

Our question is now whether or not this structure may be seen to correspond to the general argument-structure required for a "deduction of the concept of science."

In part A, consciousness finally comes to show itself as the indeterminate middle term (*Mitte*) of Spirit (*Phän* [H, 1937], p. 477). It reaches this stage in the form of moral consciences that reciprocally *recognize* each other by negating the significance of the determinate being (*Dasein*) in which they appear to one another. This negation Hegel calls 'forgiveness'. In the 'reconciling yes' of moral forgiveness, which negates the opposition of consciousness between the individual moral '*Ichs*', 'Spirit' is present in its absoluteness. It is not, however, Spirit knowing itself as Spirit, because the medium in which these '*Ichs*' are for one another is not merely transparent but void. It is the final stage in the process of negatively purifying the factors mediating recognition. This process explicitly began in chapter IV, when we, as readers of the *Phenomenology*, first came to see that the *Ich*, as self-consciousness, is actually constituted through its being recognized *as* an *Ich*, when we were first presented with the 'concept of Spirit' (*Phän* [H, 1937], p. 140), and began to view the concept of 'spiritual unity' (*Phän* [H, 1937], p. 141ff) as it developed through the movement of recognition, beginning with the radically non-reciprocal recognition of Lordship and Servitude and culminating in the negation of that act constituting all forms of servitude, the ascription of naturally determinate being to the essence of another self-consciousness.[9]

For Spirit to know itself as Spirit, this negative process undergone by Spirit must be complemented by a development through which the phenomenal *medium* of reciprocal recognition is retained, but purified of its natural determinacy and made into a pure receptacle for the generally arbitrary, but in each instance specific, *recognizability* of Spirit. Whereas the spiritual unity in chapter VI came to be constituted by the inward and subjective acceptance of the inner moral conviction of the other, spiritual unity in chapter VII is a unity in otherness (*Phän* [H, 1937], p. 544). Here it is not naturally determinate being as such that is relinquished by consciousness, but rather nature as a form of being, which is known as Evil (*Phän* [H, 1937], p. 544), as the absolute other of its own absolute ground. Since Spirit in the form of self-consciousness has come to experience the defect of naturally determinate being in the form of knowing, this evil character of its phenomenal world is capable of being overcome. But Spirit in the form of self-consciousness still knows itself by means of representations, it is for itself still "spiritual essence" (*geistiges Wesen*),[10] spiritual being from the standpoint of consciousness (*Phän* [H, 1937], p. 526), spiritual unity as grounded in a beyond.

Erst nachdem es [consciousness] die Hoffnung aufgegeben, auf eine äusserliche, d.h. fremde Weise das Fremdsein aufzuheben, wendet es sich . . . an sich selbst, an seine eigne Welt und Gegenwart, ENTDECKT SIE ALS SEIN EIGENTUM und hat somit den ersten Schritt getan, aus der *Intellektual-welt* herabzusteigen, oder vielmehr deren abstraktes Element mit dem wirklichen Sein zu begeisten (*Phän* [H, 1937], p. 559 – emphasis added).

The end of the *Phenomenology* may thus be interpreted as that point when knowing in the form of consciousness reaches a *Gestalt* which is no longer a "determinate *Gestalt* of consciousness," when knowing, as the result of its own immanent development, no longer reflexively posits an independent *Ansich* as the ground of its knowledge. It thus actualizes through the dialectical stages of its experience what was, for us at the outset, a mere thought-experiment, our '*weglassen*' (*Phän* [H, 1937], p. 72) of all standards or criteria. As readers of the *Phenomenology*, we thereby come to see that we *are* the '*Wir*' posited in consciousness's absolute beginning. In the resultant mode of knowing, pure knowing, nothing is determinate, because the reflexive determinations of consciousness have been suspended, and all is immediate, because the medium, the world in which spirit knows itself as spirit, is no longer postulated in an alien realm, but is discovered to be its own property (*GPR*, §§ 41 and 62). The indeterminate immediacy which the *Phenomenology* issues must not be understood as itself a 'we', what might be called a transcendental intersubjectivity; but the end of the *Phenomenology*, conceived as "the deduction of the concept of science", is to be grasped as *resulting from* a unification into the 'Wir', the *immediate* unity of spirit as opposed to the *absolute* unity of the *Ich.*

If the supra-individual shape of Spirit, that is, the *Volksgeist* with which chapter VI begins, were not susceptible of being resolved into pure knowing, Spirit knowing itself as Spirit, then the only appropriate scientific method for discussing the subject matter of a book about Spirit, e.g., the *Phenomenology*, would be the avowedly imprecise method of Aristotle's practical sciences. The foregoing exploration of our conjecture that the *Phenomenology* is a deduction has by no means 'proved' that the work actually carries out such a resolution; it has simply outlined a proof-structure suggested by the *Logic*, which would entail the resolution of spiritual essence into spiritual unity. If the implied reading of the *Phenomenology* also sheds some light on the beginning of *pure* science, "*das unbestimmte Unmittelbare*," then we might also be assisted in our effort to understand why Hegel regarded *science* itself as a freedom from all arbitrary, fortuitous or given determinations and how he comprehended the emergence of determination as a pure movement of "*die Sache selbst*," independent of all reflexive consciousness.

State University of New York
College at Purchase

NOTE

1 The question whether Hegel later, i.e., after 1812, retracted this interpretation of the *Phenomenology* has been the subject of a considerable literature. In addition to the reminder that Hegel explicitly reaffirmed this view in the last year of his life (see the second edition of the first book of the *Science of Logic*), the reader may be recommended Hans Friedrich Fulda's masterful survey of the critical literature in his book, *Das Problem einer Einleitung in Hegels Wissenschaft der Logik* (Frankfurt, Klostermann, 1965).

2 A critical review of nineteenth century attacks and defenses of the beginning of the *Logic* may be found in Dieter Henrich's short essay, 'Anfang und Methode der Logik', *Hegel-Studien, Beiheft* **1.**

3 *Jenenser Realphilosophie*, ed. J. Hoffmeister. 2 vols. (Leipzig, Meiner, 1932), Vol. 1, p. 201. Cf. *Phän* [H. 1937], p. 74.

4 For a more detailed discussion of the question of the '*Wir*' in the *Phenomenology*, see my essay, 'Hegel's Phenomenological Method', *Review of Metaphysics* **23** (1970), 615–641.

5 See *WL* [1812], p. xii; *WL* [L] **1**, p. 30; *Phän* [H, 1937], p. 549.

6 "Die einfache Unmittelbarkeit ist selbst ein Reflexions-ausdruck, und bezieht sich auf den Unterschied von dem Vermittelten" (*WL* [1812], p. 7).

[7] *Jenenser Realphilosophie*, Vol. 1, p. 262.

[8] For a systematic exploration of these themes, without special reference to Hegel, see Philip Merlan, *Monopsychism, Mysticism, Metaconsciousness*, 2nd ed. (The Hague, Nijhoff, 1969), esp. pp. 114ff.

[9] It should also be noted that, before we were presented with the 'concept of Spirit', the unity of consciousness, in chapters I–III, was mediated by us. See *Phän* [H, 1937], pp. 81, 85, 95, 103.

[10] For further explication of this important category in the *Phänomenologie*, see especially pp. 300, 415, 446, 532.

DANIEL COOK

THEORY AND PRAXIS AND THE BEGINNING OF SCIENCE

In the Introduction to his *Naturphilosophie* in the *Encyclopädie*, Hegel begins by examining various 'Ways of Considering Nature' (*PN*, p. 3). He starts with man's *practical* approach to Nature (§ 245), and only after discussing this approach does he turn to man's thinking or theoretical consciousness or consideration of Nature (§ 246). The order is understandable given that man's first relation to nature is an immediate and external one, whereas the latter is viewed as containing the means for the satisfaction of one's own practical urges. Through our practical approach to Nature we transform the particular objects of Nature into means for our own immediate physical survival and well-being. Only then does Hegel turn, in the next paragraph of the *Enyklopädie*, to man's theoretical or thinking consideration of Nature and the Philosophy of Nature *per se.*

Hegel apparently thought it necessary to justify beginning his *Naturphilosophie* in this fashion, for he concludes the opening *Zusatz* with an explanation of this ordering. He makes his standard methodological appeal for the need to begin the conceptual analysis of a science with a general comprehension of the various particular, concrete manifestations of man's encounter with the natural world. "In order therefore to possess the Idea [of Nature], we must traverse a series of specifications through which it is first there for us" (*PN*, p. 6). If we, nevertheless, claim that we wish to approach Nature as *thinkers*, Hegel says

> . . . there are, in the first place, other ways of approaching Nature which I will mention, not for the sake of completeness, but because we shall find in them the elements or moments which are requisite for a knowledge of the Idea and which individually reach our consciousness earlier in other *ways of considering Nature* (*PN*, p. 4).

Only in this fashion, Hegel concludes, will we really comprehend what truly characterizes the philosophical or scientific approach to the realm of Nature: the need, at the outset, to synthesize properly the practical and theoretical approaches to Nature.[1]

In this brief paper, I would like to highlight the relation (dialectical or otherwise) between the practical and theoretical elements in Hegel's

R. S. Cohen and M. W. Wartofsky (eds.), Hegel and the Sciences, 283–286.

treatment of the beginning of Science, of *Wissenschaft.* I take the following three examples for my purposes:

(1) the beginning of the Science of Nature (*Naturphilosophie*), as I began to discuss it above;

(2) the beginning of the 'Science of Logic' (*Wissenschaft der Logik*); and, finally,

(3) the beginning of the 'Science of the Experience of Consciousness' (*Wissenschaft der Erfahrung des Bewusstseins*), the *Phenomenology of Mind.*

In discussing the next way of considering Nature – the theoretical or thinking approach – Hegel sets out to expose the contradictions present in such an attitude as it is usually exemplified in physics and natural philosophy. Such an approach claims to apprehend the world as it truly is and as a science, it does so through the conscious usage of universal and class terms. This is proper since such a scientific approach "is directed to a knowledge of the *universal* aspects of Nature" (*PN*, p. 6). But in trying to comprehend the variegated world of Nature, we change its character. The use of theoretical terms in science has an immediate practical effect which must be duly noted by a philosophy of science. As the *Zusatz* to this section puts it:

> . . . instead of leaving Nature as she is, and taking her as she is in truth, instead of simply perceiving her, we make her into something quite different. In thinking things, we transform them into something universal; but things are singular and the Lion as such does not exist (*PN*, § 246, *Remark, Z,* p. 7).

The theoretical language and logic of empirical science negate the particulars of sense experience as much as the practical approach mentioned above. Only by properly understanding this negative power of the theoretic stance can we achieve a proper perspective for our inquiry into the Philosophy of Nature.

This same view of the practical importance of the use of language and logic is the topic of Hegel's discussion at the beginning of another science – the 'science of logic'. In the opening paragraphs of the Preface to the second edition, Hegel says,

> Into all that becomes something inward for men, an image or conception as such, into all that he makes his own, language has penetrated, and everything that he has transformed into language and expresses in it contains a category – concealed, mixed with other forms or clearly determined as such, so much is logic his natural element, indeed his own peculiar *nature*. If nature as such, as the physical world, is contrasted with the spiritual sphere, then logic must certainly be said to be the supernatural element which permeates every relationship of man to nature, his sensation, intuition, desire, need, instinct, and

simply by so doing transforms it into something human, even though only formally human, into ideas and purposes (*SL* [M], pp. 31–32).

Hegel refers in this Preface to the "natural logic" of ordinary consciousness "as an instrument and a means" ("als zum Gebrauch und als Mittell") for coming to terms with the natural world. Ordinary or natural consciousness view the terms it uses as extrinsic to the world with which it is dealing. Only later does the intrinsic, universal quality of its pragmatic, unself-conscious language and logic become apparent. By further analyzing the categories of Natural thinking, we can then start down the path to scientific or philosophical thinking. What should be noted here, once again, is the practical dimension always present in Hegel's discussion of the beginning of science – even when dealing with apparently purely theoretic activities such as thinking and talking.

If our analysis of Hegel's approach to the beginning of the above 'sciences' is correct, then it should help us to understand the beginning of the 'Science of the Experience of Consciousness' – i.e., the first section entitled 'Consciousness'. This section is often neglected by commentators on this 'Science' on the grounds that it is purely epistemological and theoretical. If it is discussed at all, it is usually only in preparation for a more exhaustive treatment of later sections such as 'Master-Servant' or 'Unhappy Consciousness' in the section entitled 'Self-Consciousness.' The latter are taken to be seminal since they are the first stages in the *Phenomenology* that develop the practical implications of Hegel's 'Science of the Experience of Consciousness'. Yet when we examine the section on 'Consciousness' – 'Sense-Certainty', for example – we find that Hegel's analysis can be seen as a demonstration of the practical implications of the theoretic stance of 'Sense-Certainty'. The universalizing power of thinking and talking of natural consciousness goes beyond its immediate and external relationship to the natural world. Hegel explicitly says that such theoretic activities as the thinking and talking of natural consciousness in 'Sense-Certainty' are practical activities that prefigure the later stages in 'Self-Consciousness' of desire and the need to negate the particularity of the external world (*Phen* [B], pp. 158–159). Hegel gives some fanciful examples of this basic universal experience – e.g., dumb animals gobbling up the sense-particulars confronting them in the natural world. The use of this analogy to convey the negating power present in thinking and talking may be strained, but it does show us that Hegel was intent on having his readers see the practical force of the theoretical activity of consciousness. Interestingly enough, this same example of the animals is used in the *Zusatz* to § 246 of the *Naturphilosophie*, where Hegel exposes the contradictions present in a view of theoretical consciousness which affirms the ultimacy and inscrutability of the natural objects confronting it (*PN*, p. 9).

Thus, I have tried to show that in three different 'beginnings' of science, the theoretic stance of consciousness is viewed by Hegel as in fact representing a very practical relationship between consciousness and the natural world, and that this theoretic activity is indeed the first genuine *practical* step in the *Bildung* of consciousness. However we may choose to define '*Wissenschaft*', one necessary element in Hegel's general conception of this term is the close relationship between theory and praxis in the conceptual analysis of any scientific or philosophical endeavor. To conclude with Hegel's words to his friend Neithammer:

> I am daily convinced more and more that theoretical labor accomplishes more in the world than the practical; if the realm of thinking is revolutionized, reality itself cannot hold out.[2]

Brooklyn
CUNY

NOTES

[1] "In so doing [considering other ways of approaching Nature], we shall come to the point where the characteristic feature of our inquiry becomes prominent. Our approach to Nature is partly practical and partly theoretical. An examination of the theoretical approach will reveal a contradiction which, thirdly, will lead us to our standpoint; to resolve the contradiction we must incorporate what is peculiar to the practical approach, and by this means practical and theoretical will be united and integrated into a totality" (*PN*, p. 4).

[2] "Die theoretische Arbeit, überzeuge ich mich täglich mehr, bringt mehr zustande in der Welt als die praktische; ist erst das Reich der Vorstellung revolutioniert, so hält die Wirklichkeit nicht aus." Cited by F. Nicolin and O. Pöggeler, 'Zur Einführung' in G. W. F. Hegel, *Enzyklopädie der philosophischen Wissenschaften im Grundrisse* (1830) (Hamburg, Meiner, 1959), p. XXII.

L. D. EASTON

THE FIRST AMERICAN INTERPRETATION OF HEGEL IN J. B. STALLO'S PHILOSOPHY OF SCIENCE

Among the Ohio Hegelians – an informal group of writers and social reformers who championed Hegel's thought in Cincinnati from 1848 to 1860, prior to the St. Louis Philosophical Society – John Bernard Stallo was outstanding in mastery of philosophy and public influence.[1] His special interest was philosophy of science, the underlying principles and assumptions of physics, chemistry, biology and sociology and the foundations of knowledge in those areas. Pursuing this interest in his teaching and writing, he concentrated on two main points he found in Hegel: first, the thesis that all experience and the actualities of nature and society are essentially process, becoming, development, evolution; secondly, the thesis that the real nature of things is not to be found in a separate realm, a 'beyond' that transcends experience, but rather in the phenomena themselves taken concretely in their verifiable connections and context.

In applying and interpreting these two theses Stallo also criticized Hegel, thus anticipating his mature philosophy of science that professed to repudiate Hegel altogether. Stallo's relation to Hegel from the beginning was that of critical follower. In the preface to his first major writing in 1848, *The General Principles of the Philosophy of Nature*, Stallo wrote: "I cheerfully acknowledge that for many of my views I am indebted to the study of Hegel's philosophy, although, generally, these views are as independent of Hegel as Hegel is (if it be permitted *magnis componere parva*) of Schelling."

I

Stallo absorbed Hegel's views from 1841 to 1848 while he was studying and teaching mathematics, physics and chemistry at St. Xavier's College in Cincinnati and St. John's College at Fordham, New York. In his native Germany his schooling had included mathematics and languages, and at a teacher's college in Vechta he had been exposed to Kant's view of space and time and the relation of concepts to sense-impressions. But apparently Hegel's writings were among the 'forbidden books' in his boyhood home. A few years after his arrival in Cincinnati in 1839 at the age of sixteen, he began studying Hegel, Schelling, and Humboldt "with great enthusiasm." These studies came

R. S. Cohen and M. W. Wartofsky (eds.), Hegel and the Sciences, 287–299.

to fruition in *The General Principles of the Philosophy of Nature, with An Outline of Some of Its Recent Developments among the Germans, Embracing the Philosophical Systems of Schelling and Hegel and Oken's System of Nature.* By the time the book was published in Boston in 1848, however, Stallo had decided to study law, and the following year he was admitted to the bar in Cincinnati where he became a prominent lawyer and judge.

As a result of the pattern of his studies Stallo's philosophy of science in *The General Principles* was more than a literary or library achievement. It was stimulated and informed by study of current research in physics, chemistry, and biology and by first-hand observation or laboratory work. Hence Stallo was far more concerned than Hegel to control his conclusions by the empirical data of natural science and increasingly criticized Hegel in this respect. To be sure, Hegel had carefully and widely studied the natural sciences. In 1831 he wrote a book on planetary orbits that Stallo quoted at length, he was an active member of three scientific societies,[2] and he kept up-to-date in his reading in the sciences as shown in the *Zusätze* to various sections of his *Naturphilosophie.* But in comparison with Stallo, Hegel's philosophy of nature was primarily a literary achievement, a product of his writing desk. Hence he frequently denigrated experimentation and "merely sensuous notions" in favor of a purely reflective development of an idea or principle, the exact opposite of the trend in Stallo's thought.

II

Stallo's appropriation and application of Hegel's thesis that whatever is truly real or actual is a becoming – process, activity, development – is especially apparent in the first third of *The General Principles* where Stallo develops his own philosophy of science. It is further confirmed in his summary of Hegel's philosophical system in the second part of *The General Principles*, a summary that took up a third of the entire book and twice as many pages as he devoted to summarizing Kant, Schelling, and Oken.

The underlying premise in Stallo's view was that mind or thought is fundamentally identical with the forces that actuate the whole natural world. As he put it:

> The fundamental principle upon which, according to my conviction, all true philosophy of nature rests, is, that the different manifestations of the vitality which bursts forth in nature's phenomena are comprehensively united, centered in the mind; that the implacable rigor of cosmic laws, which sways *extensive matter*, is identical with the eternal freedom of mind in its *infinite intensity.*[3]

Such a principle, Stallo recognized, was opposed to the dominant materialism and "misconceived Baconism" of the time as well as to the dualism in Kant's philosophy between matter and spirit, the objective and subjective. It was, however, the core of 'German idealism' since Kant, and asserted, in contrast to English philosophy, the 'reality of the Ideal.' But mind, thought, or 'the Ideal' is not to be viewed as a fixed, rigid substratum of the phenomenal world. It is, rather, essentially activity, motion, process, development. The ultimate unity of changing phenomena can lie only in what is self-mediated and exists by itself – namely, mind or spirit. In Stallo's words:

> Mind is the absolutely Restless in itself the absolutely Creative, the absolutely Free. Mind is not the blank abstraction, not the *caput mortuum* of the External. The Deity, the absolute Mind, is the absolute intrinsic process – the substance which causes, produces itself, – gazes into its own eye.[4]

The source of Stallo's fundamental principle is apparent in his summary of Hegel's philosophy. The 'German idealism' he had in mind was particularly that of Hegel who had seen that the Absolute is not, as with Schelling, an abstract identity of nature and spirit but rather a development, generation, or process whose motive principle is its inherent self-opposition. It is a process which has itself both for its material and its object or result but in dualizing itself yet remains an identity. Such activity in and for itself, Stallo held, is thought wherein the unit enters into self-opposition to achieve self-recognition, to reestablish its unity, a unity that requires and contains the antithesis. Hence Hegel 'proved' that the reality or 'constant' of the series of phenomenal variations in nature is "the result of the dialectic process of thought."[5]

Hegel's view of mind or spirit as process, activity, development was particularly apparent, Stallo observed, in *The Phenomenology of Mind.* With Hegel spirit is not to be understood as an entity or determinate substance but rather "that being which is truly subject, or, what is the same thing is truly realized and actual solely in the process of positing itself, or in mediating with its own self in transitions from one state or position to the opposite." Similarly, 'experience' is the process whereby what is immediate in sensation and reflection leads to something else and then also returns to itself to yield a whole which is truth for that movement. "The truth is the whole," said Hegel in a familiar passage that Stallo repeated. "The whole, however, is merely the essential nature reaching its completeness through the process of its own development" (*Phen* [B], pp. 80–81).[6]

With such underlying principles Stallo adopted Hegel's overall position on the self-development and manifestation of Spirit in three dialectically related phases akin to the major points of *The Encyclopedia of Philosophical Sciences.* The three phases of Spirit, Stallo held, are:

1. The Spiritual as the absolute origin of all existences, abstractly taken. 2. The exterioration of the Spiritual as manifestation in existence, abstractly taken. 3. The Spiritual as thus sustaining itself, regenerated in its exterioration, or the Spiritual taken in its concrete identity and truth.[7]

In support of his view that process, movement or activity is "the absolutely *primum*, the first principle in existence," Stallo relied on Hegel's purely reflective argument in the *Logic* that being and nonbeing can only be adequately comprehended through their unity in 'becoming' (*Werden*), the model of all subsequent syntheses and thus all other adequate concepts. In addition, however, he introduced some arguments of his own as he attacked materialistic conclusions from physics and chemistry. To those who argued that in the movements of nature there must be absolute rest as the bearer and source of motion he replied:

It is impossible to conceive rest without the concomitant idea of equilibrium, and therefore of motion. It is impossible to construe motion out of rest; rest is an incident to motion and consequently to be explained from the nature of the latter.[8]

"There is," Stallo concluded, "nowhere absolute rest, but motion everywhere." Further, those who hold that there must be some material substance as a fixed, unchanging substratum behind changing phenomena cannot possibly square their view with Dalton's law of compression of gases or Graham's law of diffusion. With such arguments as these Stallo was more concerned than Hegel to control his conclusions by the empirical data of natural science.

The final and longest section of Stallo's statement of his own philosophy of science was entitled 'Evolutions.' It provided details on the "phases and processes" in physical motion, the solar system, light, sound and heat, chemical combination, electricity, vegetable and animal organisms, recreative and procreative mind, and the organization of society. This section was the heart of Stallo's philosophy of science, applying and seeking to confirm his central thesis concerning process, movement, activity. It also revealed the distinctive features of his view and thus how he differed from and criticized Hegel.

Stallo used the word 'evolution' synonymously with 'process,' 'development,' 'origination,' and he thought of it as progressive, 'spiral' movement towards increasing complexity of organization and greater multiplicity of relations. He was well aware that Hegel had used the term '*Evolution*' to refer to the serial progression in nature from the imperfect and formless liquid element through fishes and land animals to man. But Hegel had dismissed this as "a completely empty thought." It explains nothing and in some respects is inferior to its correlative opposite, the Oriental idea of '*Emanation*.' For Hegel the only 'development' and 'origination' is in the Notion which "timelessly and in a universal manner posits all particularity in existence." On this basis he concluded, in spite of his reading in Lamarck and Kant, that species do not change, that organic fossils never really lived, and that the "lightning of life" strikes a complete creature into previously lifeless matter like Minerva from the brow of Jupiter. "The Mosaic story of creation," said Hegel, "is still the best in its quite naive statement that on this day planets came into being, on another day the animals, and on another day man" (*PN*, pp. 20–22, 284).[9]

Stallo, on the other hand, regarded "generative modification of certain species" as empirically substantiated and saw organic fossils as evidences of "the consecutive advance of vegetable and zoological genera." By implication he dismissed the Mosaic story as a form of occasionalism that is as external as the purest materialism to the mediated unity of efficient and final causation. Stallo agreed with Hegel that temporal succession is inapplicable to the true ground of all nature, absolute Mind or Spirit as "absolute intrinsic process." Its relation to its phenomenal manifestations is tenseless as in the relation of whole or part, and Stallo frequently referred to the Absolute, following Hegel, as 'the Whole.' But the essential nature of the Absolute is to manifest and realize itself in the phenomenal and material world characterized by spatiality and temporality. "Consequently," Stallo concluded, "whatever is in the Spiritual as unital intensity is likewise found in material nature, therefore in time and space as succession and co-arrangement."[10] Apparently, then, Stallo's use of 'evolution' as a synonym for 'process,' 'development,' and 'becoming' was not accidental. It signalized his differences from Hegel, integrating process more closely with empirical data of natural science and providing a more coherent position than Hegel had achieved.

Stallo frequently used 'history' as a synonym for 'process,' 'becoming,' and 'evolution.' It has been suggested that Hegel temporalized the realm of spirit in history but generally denied the historical development of nature because of his conservative temperament in a pre-Darwinian age.[11] Stallo also lived in a pre-Darwinian age but had the audacity and originality to try to resolve the dualism between a non-historical nature and a historical realm of spirit in a more coherent, thoroughgoing evolutionism. His exposition of evolutions widely followed Hegel's *Naturphilosophie* in its sequence or subject matter, and even some of Hegel's subtitles, from mechanics and motion through the solar system, light, chemical combinations, and organisms to end in a discussion of sexuality. But then Stallo went on to elaborate evolutions in what Hegel had separated from nature as subjective and objective mind, namely, perception, imagination, memory, reasoning, mental derangements, the organization of society in family, civil society, and state and finally the major phases of mankind's history.

In sum, Stallo held that Hegel's position on process, becoming, and development was abortive. It was not thoroughgoing but was still infected with static substantialism, an appeal to a fixed entity or 'thing' as the basis of explanation. Hegel invited this criticism, Stallo felt, in maintaining that every true philosophy must stand on the platform of Spinozism. Though Hegel saw clearly that the existence of things is their becoming (*Werden*) – hence Stallo could reverently regard him as 'one of the philosophical saviours of nature" – he was not consistent and thoroughgoing in the application of this insight. His construction and his method still belonged to the old school that makes 'the thing' the basis of its deductions and changeless rest the fundamental fact. As a result, said Stallo,

> His consequences are not infrequently fallacious, and his transitions are arbitrary. He is like a man who has invented gunpowder; now he wishes to shoot but unfortunately with the old weapon, the lance, and at the same time to keep the lance in his hand. Hence also the unintelligibility of his language, the jargon of real and ideal "positing," of being *at* itself and *for* itself (*an sich* and *für sich*) etc.[12]

Stallo's more thoroughgoing development of the principle of process as pervasive evolutions was a main factor in his influence on Ralph Waldo Emerson and American literature. Prior to reading Stallo's *General Principles* Emerson was absorbed in the 'grandeur' of the efforts of Schelling and Oken to unite natural and moral philosophy. Beginning in November, 1849, he copied into his journals extracts from *The General Principles* to the effect that "Every individual existence is but a living history . . . . The development of all forms will be spiral. . . . Animals are but foetal forms of man." In subsequent published writings he asserted that "nothing stands still in Nature" and "everything is organic, freedom also, not to add but to grow and unfold." Though Emerson got little from Hegel's writings except a headache, Stallo's extension of Hegel's principle of process into evolutions intelligibly provided him with a foundation sufficient to his interests. In 1873 Emerson observed that while Darwin's *Origin of Species* appeared in 1859, a decade earlier Stallo had written that "animals are but foetal forms of man" and "the lines of our ancestry run into all the phenomena of the material world."[13]

III

Further interpreting Hegel, Stallo defended a second main thesis for philosophy of science, namely, that the real nature of things is not to be found in a separate realm, a 'beyond' that transcends experience, but rather in the phenomena themselves taken concretely in their connections and context.[14] Negatively this thesis involved a rejection of transcendent metaphysics, the view that 'reality' lies in a separate realm that observed phenomena somehow copy or reflect. Affirmatively it asserted that reality or truth is to be found in phenomena, and in this respect it is akin to current philosophies of experience – empiricism, phenomenalism, positivism – which vigorously criticize transcendent metaphysics and seek its elimination from responsible thought. This thesis has already been suggested in the previous discussion of Absolute Mind or 'the Whole' as process. Such a conception outlaws any fixed, unchanging substance or entity that might serve as a permanent reality behind or beneath changing phenomena.

In his *General Principles* Stallo was especially attracted to the way that Hegel overcame the dualisms between idea and reality, subjective and objective, spirit and matter left by Kant's philosophy. By resolutely eschewing 'abstractions' or "some hypostatic notion of universality," Hegel could provide an effective unification of science and metaphysics. As Stallo put it:

> Hegel does not attempt to evolve concrete forms from an abstraction; his "Absolute" is essentially concrete . . . . Since truth is apprehended not as something reposing in the bosom of its own being, but as the "Whole in its development," as the Absolute, not abstractly taken, but also in its phenomenal existence, in its individual exterioration, the system of metaphysics, which formerly consisted of nothing but formalities, must encroach upon the domain of all science. Instead of an establishment of certain forms, merely for construing the various material, form and material now stand in necessary relation; the material – nature, etc. – enter as essentially into metaphysical reasoning as the old formulas. It cannot, therefore, be startling to see that the natural sciences, history, etc. are an integral part of metaphysics.[15]

Following Hegel in his view of perception and observation, Stallo held that external objects are not somehow engraved on the mind but rather their particularities become general qualities and relations. Objects are not something transcendent to their qualities and properties. Rather, "the predicates *are* the subject; they are not glued, plastered upon it, and torn off when the subject is idealized. Everything exists *in* its properties, not *beside* them." These properties, moreover, are basically an ensemble of relations. "Objects,"

Stallo held, "are thoroughly relative; their being is a complex of relations."[16] He indicated that this conclusion was also Hegel's. Summarizing the section of *The Phenomenology of Mind* on 'Observation,' Stallo found Hegel saying that the observed object is its qualities, and since these qualities as differing from one another involve relationships, the unity of the object depends on the observer and "the individual object is nothing more than the relation to other objects." With such a view of perception and observation Stallo paid particular attention to Hegel's conclusions on 'Force and Understanding' in *The Phenomenology* and saw them as rejecting transcendent metaphysics. In Stallo's summary in his *General Principles*:

> We look into the interior of things only through the phenomenon; the interior itself is transcendental, a "beyond" for our consciousness. The transcendental interior, however, reveals nothing whatever to consciousness; no more, to use Hegel's own simile than *pure* darkness or *pure* light reveals anything to the gaze. But the supersensual "beyond" results from mediation; it proceeds from the phenomenon, and the phenomenon is its reality. The Supersensual is but the Sensual taken in its truth, taken as phenomenon and not as a permanent reality, which it has amply proved itself not to be. We behold the play of forces, a continual shifting of determinate appearances, whose truth consists merely in the law which manifests itself there. The *law* is the permanent image of the fleeting phenomenon. The supersensual world is the quiet realm of laws, indeed beyond the world of observation, since this exhibits the law only in continuous change; but it is nevertheless present in the world of observation and its immediate type.[17]

But even as he praised Hegel for fusing the Absolute and phenomena and thus advancing what might be called a 'concrete phenomenalism' in opposition to transcendent metaphysics, Stallo was critical of Hegel in this respect. There were, Stallo felt, numerous abstractions in Hegel's view, points where the Absolute did not exist in and through the phenomena but remained independent of the empirical data of natural science. In short, there were numerous points where Hegel held to a "hypostatic notion of universality" as in traditional metaphysics.

We have already noted how Stallo found an element of static substantialism in Hegel's view and criticized his treatment of process in nature as abortive, as failing to embody the transitions and movement of phenomena. In developing his own view of evolutions in physics, Stallo widely followed Hegel but sometimes digressed to criticize him for failing to fuse idea and phenomena. For example, he noted that Hegel's view of single and double diffraction of light did not account for certain observed deviations or distortions, and the mutual dependencies of refraction and reflection as well as the limit of refraction could not be derived from his theory at all.[18] Though

Stallo endorsed Hegel's general conception of light as reciprocal manifestation or 'showing,' an "exhibition of the Spiritual One in the materially Dual," he found a defect in the development of this idea, a point where Hegel's view was abstract and infected with "a hypostatic notion of universality." Stallo argued that generality and individuality in existence were not be to identified with sun and planets as Hegel had done. Rather, "the solar systems are the first cosmic individuality, not the isolated planets" and the universal as such cannot be the concrete material sun. In short, Stallo found an abstraction in Hegel's view, a point where he failed to follow through on his own premise that objects consist of their relations and their unity lies in a system, in laws *of* and *in* the moving phenomena. Stallo found the same mistake in Hegel's political theory where the king is the incorporated universal corresponding to the sun. The universal, Stallo argued, is rather "nothing but the *pronounced* unity of the individuals," and this implies that only a democratic republic, not monarchy, can be fully "adequate to the idea."[19]

Seventeen years later Stallo's criticism of Hegel's abstractness was more pointed and sweeping. Writing on the foundations of scientific knowledge, he noted that Hegel's view of history did make some provision for "geographic conditions" but Spirit as "developed logic" moved high above and beyond external conditions and held firm against the contingencies it employed.[20] Finally in his book of 1881 on *Concepts and Theories of Modern Physics* Stallo's criticism of Hegel reached its climax. He was not, he said, further developing the doctrines of his *General Principles* though there was much in it of which he was not ashamed. It was, unfortunately, written under the spell of "Hegel's ontological reveries . . . one of the unavoidable disorders of intellectual infancy." In particular, he regarded Hegel's Absolute or Being as a monstrous verbalism. It was a reified *summum genus*, "the specter of the copula between an extinct subject and a departed predicate, . . . a sign of predication which 'lags superflous on the stage' after both the predicate and that whereof it is predicated have disappeared."[21] In this respect Stallo regarded Hegel's thought as the very model of a major 'metaphysical error' in the interpretation of scientific concepts. Through the study of comparative linguistics he had become sensitively aware of the unity of thought and language and the way metaphors are inseparable from language, but in science they must be controlled by precisely observed correspondences, relations of structure, and functional equivalence.[22]

Yet in spite of Stallo's claim to have broken decisively with Hegel's philosophy, the main points of his theory of cognition in *Concepts and Theories* may readily be identified with Hegelian principles he espoused in

The General Principles three decades earlier. The theory of cognition was fundamental in *Concepts and Theories*, and Stallo regarded it as the central point and purpose of the whole book. It undergirded his continuing criticism of materialism, the "atomo-mechanical view of nature." The atomic theory, he allowed, was useful in chemistry as an expository device, but for its most vigorous defenders it was "substantially identical with the cardinal doctrines of ontological metaphysics" and could not in main features be squared with data of observation and experiment. Stallo's theory of cognition was also the basis for detecting 'metaphysical errors' that he illustrated in detail in the last eight chapters of *Concepts and Theories* where he criticized the 'absolute space' of Newton and Euler, the notion of absolutely independent physical objects, and cosmological speculation about "the origin of the universe as an absolute whole in the light of physical and dynamical laws."

The first point in Stallo's theory of cognition asserted that thought deals not with things themselves but with "states of consciousness" and every object of cognition is a synthesis of subjective and objective elements so "all knowledge is relative to the cognizing faculty." He had said substantially the same thing in *The General Principles* where, following Hegel's *Phenomenology*, he saw perception and observation as involving an interchange between conscious mind and apparently external objects and the reality of consciousness in and through them. Secondly, Stallo's theory of cognition asserted that objects are known through their relations and have no attributes except through their relations. In 1848 he based this conclusion on Hegel's view that the 'sense certainty' of objects perceived as 'here' and 'now' involves grasping them as relationships, but in 1881 it was supported from Helmholtz's research on vision. Helmholtz, as Stallo may have known, had attributed the nineteenth century schism between philosophy and science to Hegel's *Naturphilosophie* that seemed "absolutely crazy" to the scientists, an attempt "to construct *a priori* the results of other sciences."[23] The third main point in Stallo's theory of cognition asserted that thought deals only with a selected class of relations, so our thought of things is always 'symbolical.' *Concepts and Theories* identified this point with views of Leibniz and Sir William Hamilton, but in *The Phenomenology of Mind*, which Stallo had summarized three decades earlier, Hegel maintained that abstractions ("things of the intellect") are indispensable to perception. In wider implications of this theory of cognition, which Stallo called 'phenomenalism,' there was a vivid echo of Hegel's coherence theory of truth. "There is no physical reality," Stallo held, "which is not phenomenal. The only test of physical reality is sensible experience." What, then, about the conflicting deliverances of experience? What is 'apparent' in contrast to what is 'real,' Stallo concluded, is merely a partial deliverance of sense mistaken for the whole because the senses are not fully and precisely interrogated, their whole story is not heard. With 'Absolute' in place of 'real' he had earlier ascribed this position to Hegel for whom truth was the "Whole in its development" involving "phenomenal existence" in the findings of natural science.[24]

Concepts and Theories of Modern Physics won increasing recognition as an important contribution to philosophy of science. It was part of an eminent scientific series, went through several printings and translations and was reissued in 1960 by Harvard University Press as a "landmark of intellectual history" with an introduction by P. W. Bridgman. In Europe its importance was appreciated largely through the attention of Ernst Mach, father of the recently prominent 'scientific empiricism.' Mach became interested in the book whose "scientific aims so closely approximated" his own and wrote a special foreword to the German edition. He shared Stallo's rejection of metaphysics in the mechanical-atomic theory, agreed with his view of physical concepts as relations, and concurred with him in rejecting pronouncements on "the world as a whole." Mach also found in *Concepts and Theories* "traces" of Stallo's earlier "Hegelian lines," though he did not specify them.[25] Those lines are apparent in the main points of his theory of cognition as already noted. They suggest an interesting irony in the history of ideas – namely, that Stallo's phenomenalism, anticipating forms of empiricism that have become increasingly prominent in America since William James, was rooted in the ideas of the one philosopher most empiricists have regarded as their arch-enemy, Hegel.

IV

Early and late in his lifetime Stallo saw his philosophy of science as implying and supporting freedom of inquiry, the fullest freedom of mind. He was aware that Hegel had supported freedom of inquiry in discussing the relation of church to state but had grounded it in constitutional monarchy. But Stallo held that only a democratic republic, not monarchy, is "adequate to the idea." Only democracy is consonant with Hegel's premise that the substance of society and history is spirit as reason, and reason, Stallo argued, involves the unity of system and relationship, not the extraneous unity of a single man. Universal reason in society is nothing but law reproducing itself in and from the individual. "The organization of society," he concluded, "is, therefore, *essentially* democratic." Furthermore, social development is a 'dialectical process' in which partial points of view and particular errors are required for the emergence of truth. Such a process requires freedom of thought and inquiry as its essential medium. Similarly, but on different grounds, Stallo saw his mature philosophy of science – his phenomenalism that professed to repudiate Hegel altogether but actually extended major 'Hegelian lines' – as a means of freeing intelligence for growth and openness to experience. Of his *Concepts and Theories of Modern Physics* he said:

Its tendency is throughout to eliminate from science its latent metaphysical elements, to foster and not repress the spirit of experimental investigation, and to accredit instead of discrediting the great endeavor of scientific research to gain a sure foothold on solid empirical ground, where the real data of experience may be reduced without ontological prepossessions.[26]

Thus Stallo's philosophy of science in its beginnings and in its maturity was fully and unswervingly committed to freedom of thought and inquiry, to the free mind.

Ohio Wesleyan University
Delaware, Ohio

NOTES

1 For biographical and historical details see L. D. Easton, *Hegel's First American Followers, The Ohio Hegelians – J. B. Stallo, Peter Kaufmann, Moncure Conway, and August Willich* (Athens, Ohio University Press, 1967), Chs. I-III presenting the nub of the main theses in this essay, but here those theses are developed further, with details of Stallo's criticism of Hegel.

2 Cf. Franz Wiedmann, *Hegel*, trans. J. Neugroshel (New York, Pegasus, 1968), pp. 32–33, 36, 44.

3 J. B. Stallo, *The General Principles of the Philosophy of Nature* (Boston, Crosby and Nichols, 1848), pp. vii-viii. Some sixty pages of this scarce book are reproduced, with original pagination in square brackets, in an appendix to Easton, *op. cit.*

4 *Ibid.*, p. 43.

5 *Ibid.*, p. 333, 345.

6 Cf. Stallo, *General Principles*, pp. 50–51.

7 *General Principles*, p. 44.

8 *Ibid.*, p. 24; cf. *ibid.*, pp. 51, 345, 45–46.

9 *Encyclopädie der Philosophischen Wissenchaften* ed. G. Lasson (Leipzig, Felix Meiner, 1930), §§ 249, 339. (Stallo used the Michelet text and *Zusätze* of Hegel's 'Naturphilosophie.')

10 *General Principles*, p. 44; cf. *ibid.*, pp. 6–7, 11–12.

11 Cf. J. N. Findlay, 'Foreword,' Hegel, *Philosophy of Nature*, trans. A. V. Miller (Oxford, Clarendon Press, 1970), p. xv; *Hegel, A Re-examination* (New York, Macmillan, 1958), p. 272.

12 *General Principles*, p. 50
13 For further details and specific documentation in Emerson's writings see Easton, *op cit*., pp. 45–49 and A. W. Plumstead, and W. H. Gilman, eds., *The Journals and Miscellaneous Notebooks of Ralph Waldo Emerson* (Cambridge, Belknap Press of Harvard University Press, 1975), Vol., XI, p. 200.
14 Cf. J. N. Findlay, 'Some Merits of Hegelianism,' *Proceedings of the Aristotelian Society* **56** (1956), pp. 7–9.
15 *General Principles*, p. 337.
16 *Ibid*., pp. 131–34.
17 *Ibid*., p. 360.
18 *Ibid*., p. 74.
19 *Ibid*., p. 72; cf. *ibid*., pp. 158–62, 517–18.
20 Stallo, *Reden, Abhandlungen und Briefe* (New York, Steiger, 1893), p. 112.
21 Stallo, *Concepts and Theories of Modern Physics*, ed. P. W. Bridgman (Cambridge, Mass., Harvard University Press), p. 178; cf. *ibid*., pp. 6–7.
22 Cf. *ibid*., pp. 10–11; Stallo, "Speculative Science," *Popular Science Monthly* **21** (1882), pp. 151–53. In *General Principles* Stallo did not summarize Hegel's views on language in *Philosophy of Mind* (trans. William Wallace from *The Encyclopaedia of the Philosophical Sciences* (Oxford, Clarendon Press, 1894), § 459), views that resembled his own mature conclusions at some points. Cf. Easton, *op. cit*., pp. 215–19.
23 Cf. W. C. Dampier-Whetham, *A History of Science* (New York, Macmillan, 1929), p. 313 quoting from Helmholtz, *Popular Lectures on Scientific Subjects,* trans. E. Atkinson (London, Longmans, Green, 1873), p. 5.
24 Cf. Stallo, *Concepts and Theories*, Ch. IX, p. 216; *General Principles*, pp. 131–35, 354–57; *Phen* [B], pp. 168–78, 80–81.
25 Ernst Mach, 'Vorwort' (1901), J. B. Stallo, *Die Begriffe und Theorien der Modernen Physik*, trans. H. Kleinpeter (Leipzig, Barth, 1911), pp. x-xiii.
26 Stallo, *Concepts and Theories*, p. 4; cf. *General Principles*, pp. 158–67, 517–18 and *PR*, §§ 270, 275–80.

PART THREE

DIALECTICS AND LOGIC

YVON GAUTHIER

HEGEL'S LOGIC FROM A LOGICAL POINT OF VIEW[1]

My title is not intended as a pun. Rather it should be interpreted as a question: how to deal with Hegel's logic from a logical point of view? I mean here by 'a logical point of view' the point of view which adheres to the body of logical truths that we can formulate in classical propositional logic and first-order predicate logic. That is not necessarily a conservative stand, only a careful one. The question indicates that one should be prepared to adopt a critical point of view concerning Hegel's logic. The fact that most logicians are not interested in Hegel's logic is a sign that such a critical attitude is not uncommon. Three years ago, I asked Professor Quine how one could go about Hegel's dialectical logic. He simply answered that one would have to change the laws of logic in order to make sense of Hegel's logic.[2] I doubt that there is anyone who would be ready to support Hegel to such an extent as to abandon the *corpus* of our logical laws. I do not think either that one has to drive to extremities to extract some logical sense from what Hegel called *Die Wissenschaft der Logik*.

1. What kind of logic is Hegel's logic? It is certainly not formal logic as we understand it, with a limited number of primitive terms, axioms and rules of inference. If there is such a thing as a formal system of Hegelian logic, it is certainly buried under a rich coat of nonlogical notions. A different and much easier answer to our question would amount to saying that Hegelian logic is a transcendental logic; transcendental logic would then be the study of the *a priori* structures of logical thought in contradistinction to formal logic as the study of the laws or operations that obey the logical structures of the human mind. Kant had already delimited the provinces of formal and transcendental logic:

> Eine solche Wissenschaft, welche den Ursprung, den Umfang und die objektive Gültigkeit solcher (Verstandes – und Vernunft –) Erkenntnisse bestimmte, würde transzendentale heissen müssen.[3]

For Hegel transcendental-speculative logic reaches even further: what he calls objective logic is nothing less than metaphysics in the traditional sense (*WL* [L], **1**, p. 46); Hegelian logic is therefore transcendental-metaphysical.

R. S. Cohen and M. W. Wartofsky (eds.), Hegel and the Sciences, 303–310.

Transcendental or speculative logic deals with the most general features of thought. Such an undertaking, from Kant to Husserl, is amply justified by its philosophical bearings.[4] Philosophical logic, for being less ambitious, is endowed with more indistinct traits. Metalogical considerations of any sort do not have, however, direct connection with Hegel's logic. Heidegger was probably closer to Hegel's intent when he designated Hegel's logic as an 'Onto-logik' in his work *Die onto-theo-logische Verfassung der Metaphysik.*[5] I prefer to call it simply an onto-logic, meaning a logic of being, instead of using the old ambiguous term 'material logic'. Since we do not have a formal logic, formalization, if it is attempted at all, must be a specified procedure. When we want to formalize or axiomatize a given theory, mathematical or else, we look for the barren structures of the theory and give the essential relationships between the found structures.

With Hegel's logic, we are not given at first an axiomatic skeleton, an uninterpreted or a semi-interpreted language, but a fully interpreted one. We are faced with the interpretation and we have to make our way to the abstract framework. Obviously, if we could uncover the abstract structure totally, there would not be any major problem. Unfortunately, we have to work with an interpretation of the interpretation. Here one can see that our inverted course does in no way raise the question of instantiation. We start with the instantiation and we have to reascend to the pure structure or framework.

There is a second characteristic of Hegelian logic: besides being transcendental or ontological, it is dialectical. Dialectics, as everybody knows, does not have a unique meaning. Many more or less conflicting models of dialectics can coexist. There are more ways than one to formalize the notion of dialectics.[6] The added difficulty of interpreting dialectics in the Hegelian sense involves a thorough examination of Hegel's intended meaning.

2. It is important to notice that dialectical logic in the Hegelian sense is, in principle, amenable to a formal treatment. Of course, the self-movement of the content (*die Selbstbewegung des Inhalts*) does not yield as such to formalization. What we can formalize is again the structure of that movement and it happens that the dialectical movement is endowed with a structure or rather structures.

Hegel himself tells us in the Introduction to his *Logic* that "die Methode ist das Bewusstsein über die Form der inneren Selbstbewegung ihres Inhalts" (*WL* [L] **1**, p. 35). If Hegel warns us that we cannot separate content and method, he does not deny that the method has a form, what I call a structure.

It is clear that the method in question here is the dialectics. That there are different forms or models of dialectics is again not to be doubted. I am alluding here not only to Plato's dialectics, but to post-Hegelian dialectics, the Marxist and the Neo-Marxist models. Even within the restricted range of Hegelian dialectics, there are many models possible, for instance, the dialectical models of Weil and Kojeve and others.[7] Leslie Armour has given recently yet another model of Hegelian dialectics in his *Logic and Reality* [Assen, 1972]. From the logical point of view many variations of dialectics can be conceived, for instance the models of Günther, Asenjo, Kosok and my own models. It is not surprising then that logicians like Beth[8] and Lorenzen[9] express a strong scepticism about the interest and the validity of dialectical logic, notwithstanding the efforts of contemporary Soviet logicians. To put it bluntly, the dialectical method is not a logical method and Hegel's endeavour from a logical point of view does not deserve the name 'logic'.

3. How can we escape such a devastating criticism? It will not suffice to say that most logicians do not know anything about Hegel's logic. If it is objected that Hegel's logic is at the same time and indissolubly an ontology or a metaphysics,[10] the critics can add to their arsenal the charge of irrelevance. Hegel's dialectical logic thus seems unsalvable. It would not do either to argue that Hegel's logic has no similarity at all with classical or orthodox logic and that is what explains most misunderstandings. There are, to be sure, nonclassical systems of logic, e.g. many-valued systems or intuitionistic logic, but their spirit of logical revisionism does not affect the foundations themselves, but specific logical laws, like the law of excluded middle. I do not plead here for logical exclusivism, I am only dispelling illusory hopes or expectations.

What are we left with, if we still want to pursue a formal approach to Hegel's logic? If we exclude non-classical or unorthodox approaches and I do not see what in Hegel's logic would compel us to abandon classical logic – the subsequent analysis of Hegelian logic tries to justify that statement – there is a variety of standard methods that can be used. It seems to me, for example, that one could do for Hegel's logic what Chomsky has done for transformational grammars, to propose a recursive treatment. Of course, natural languages and ontologies or ontologics are subject to algorithmic limitations that make it difficult to transcend the level of trivial generalities, but with the help of recursive analysis, one could define, for example, the generative structure of the dialectical scheme, the invariance properties of dialectical determinations or moments, the continuity of dialectical passage (*Uebergehen*) from one moment to the others, the formal conditions of the circularity of the system, and so on. Another closely related approach is the combinatory approach. I have made some suggestions in that direction in a previous paper.[11] Here one concentrates on the logical operations of Hegel's system and attempts to show their reducibility to some fundamental constants, their inter-connection, the cyclic permutation of logical categories. One could then obtain a relational logical algebra in the sense of Curry.[12] Both approaches, the recursive and the combinatory ones, share a common idea: the algebraic perspective. The general algebraic approach seems to me to be a fruitful undertaking. Boolean structures, group structures, graph structures, lattice structures, all could contribute to a better understanding of Hegel's logic. The question is how to apply those abstract structures to Hegel's particular form of logic. Uses and abuses proliferate. One has only to remember the intricacies of quantum logic. The problem of Hegel's logic is not without similarity with the situation in quantum logic. However, quantum logic has to do with a more limited (non-simultaneously measurable observables), if not less ambiguous, state of affairs. The solution devised for quantum logic, the modular lattice, has only a faint affinity with the *logicisation* of Hegel's logic. Hegel's logic, as was pointed out above, is constituted by an interpretation which covers up, supposedly, an axiomatic skeleton. I indicated how a very close analysis of Hegel's work was necessary to avoid substituting an alien model to Hegel's model of dialectical logic, since as Asenjo rightly emphasizes, "But here there are too many existing applications of dialectic that are far from convergent."[13]

In my opinion, attempts to formalize Hegel's logic all suffer from an insufficient interpretation of Hegel's writings, mainly *Die Wissenschaft der Logik.* Formal treatments of dialectics, despite their intrinsic interest, have to be judged on their relevance for Hegel's logic and at the same time for their logical fertility. I have explained earlier that formalisation in the case of Hegel's logic has a special meaning: formalisation does not exhaust the content of Hegelian dialectics, it aims at disentangling the structural features of the dialectics or, as Hegel puts it, the form of the self-movement of the content from the content itself. In particular full axiomatisation in the usual sense is excluded. On the other hand, I insist upon that point again, the mathematical-algebraic-structures used in the formalization must have a logical meaning. Very often a mathematical structure serves to represent a logical relation, but it remains an empty representation or an 'analogical' abstraction; it gives then only a formal image without any real gain as far as explanation or foundational depth is concerned.

4. Algebraic logic, in the sense of Tarski or Halmos, is the discipline that endeavours to represent logical relations by the means of algebraic structures without renouncing logical meaning and, at the same time, foundational relevance. Algebraic logic as a representational theory has not yet succeeded in integrating the whole of classical orthodox logic; it appears rather paradoxical then to propose a logico-algebraic treatment of Hegel's dialectical logic. But, in my view, it is the most logically fruitful, if not the only valid one.

Of course, what I have said about such an approach in my paper on 'Dialectic Logic and Algebraic Logic' gives a very dim idea of that fruitfulness. But my intention in that paper was to indicate that a closer analysis of Hegel's text was necessary, if a better formal treatment is to be expected. An analysis of the following chapters of *Die Wissenschaft der Logik* would provide us with a larger basis, but we have got already with the first chapter a formal inductive structure that is indicative of the whole. For our logico-algebraic-treatment – here we use the simpler form of polyadic algebras – we need a Boolean algebra (with possibly an enriched structure of additional laws or operations, as is suggested in my other paper), a set for category, subcategories, etc., a function T for transformations, e.g. cyclic permutations and the existential quantifier $\exists$. With the Boolean algebra we have propositional logic, with the existential quantifier, we have (first-order) predicate logic. If we wish to introduce new algebraic operations, it does not mean that we have to change propositional logic, only to expand its domain. But if we want to add new operations and expand the Boolean framework, then we have to resort to a multiplicity of algebraic varieties, lattice structures and group-theoretic structures. The problem with those structures is that they can be as empty as metaphysical notions. Mathematics is not in itself a warranty of validity, we know that from set theory and especially from the theory of large cardinals. Mathematization does not necessarily yield significant results. I think that to obtain logically significant results, we have to keep a constant control of our mathematical apparatus. By multiplying mathematical structures '*sine necessitate*', we simply lose in foundational meaning what we gain in mathematical scope. That is the lesson, in my view, of algebraic logic, a lesson that dialectical logicians should not neglect.

Once we have understood that *structural* representation does not add to the fundamental stock of logical notions, we still can draw from *functional* representability. That's a resource not only in algebraic logic, but in many fields of logic, among which many-valued logics. I mean by functional representation, a mathematical model which emphasizes process and continuity rather than moments, levels and structures. In the case of Hegelian logic, the function T for transformations or cyclic permutations could be introduced besides the function for sursumption as defined in my other paper; other transformational operations, for example, hierarchisation of positions (of the different categories), the circular completeness of the system, would be given a natural formulation in that manner. I can only indicate such an orientation; it is only after significant progress is made in that direction that one will be able to judge the merit of the approach. My aim is to provide an epistemological justification for the idea of a formal approach to Hegel's logic.

It should be clear, after what I have said, that I privilege a rather cautious approach. I think there is enough to do with the logic we have already got, especially from the mathematical point of view, that we should not hurry to invent parallel logics. The example of many-valued logics, mentioned above, is instructive. Many-valued logics, in spite of their conceptual interest, have not contributed essentially to a better understanding of logical thinking, at least up to now.[14] It appears to me that higher-order predicate logic, for example, is capable of mathematical developments beyond comparison with any logical variant. A constructivist viewpoint which favors foundational fruitfulness over conceptual *free enterprises* inclines to a monistic attitude, which is hardly objectionable on philosophical grounds[15] Areas of fertile work in contemporary logic, if we except such borderline fields as modal logic and inductive logic, point to an overwhelming constructivism (recursive and metarecursive function theory, subsystems of analysis, hyperarithmetical hierarchy, functionals, etc ...) while the realism of set-theoretical high cardinalities appears more and more problematic. The onto-logic of idealist dialectics in its Hegelian sense would be close, in my opinion, to the realistic trend and as I have pointed out repeatedly, less susceptible of foundational interest, unless one is prepared to reinterpret idealist dialectics as a general theory of language.[16]

5. The interconnectedness of logic and language in a Hegelian perspective may be a more promising study than the tentative formalizations that have been proposed. Dialectical logic, after all, may be more appropriate for a hermeneutical enterprise interested primarily in the methodological problems of the social and human sciences and I would suggest considering analyticity and dialecticity or hermeneuticity as the opposite poles of scientific inquiry.[17] I do not go as far as to say that axiomatics and dialectic are irreconcilable, but it strikes me that dialectical thinking is at its best when it deals with complex phenomena that are refractory to an analytical treatment. This new

form of the traditional opposition between '*Naturwissenschaften*' and '*Geisteswissenschaften*' is probably not reducible methodologically. The conflict of methods does not lead to dualism, it forces a more harmonious vision of the hierarchical system of knowledge. Hegel certainly strove toward such a vision. His own systematic endeavour, however, did not make much room for the complementarity of methods or approaches. That is due partly to a failure of his time. The contemporary philosopher lives in a different world. Logic does not require any more a ready-made ontology, physics does not need a metaphysics and philosophy has to learn more than to dictate.

Nevertheless, theoretical justification and fundamental speculation remain the perennial task of philosophy and in that Hegel is still a master. Integral understanding of the whole of experience and reality must be pursued, even if it is at inflationary costs. Theories of logic share that ambitious and risky search, to give a structure to totality and a global meaning to our particular points of view. For those who cannot be satisfied, fortunately or unfortunately, with a down-to-facilities pragmatism *à la* Quine,[18] there is still the intellectual adventure of Hegel's logic, from a *dialectical* point of view.

University of Montreal

POSTSCRIPT (NOVEMBER 3, 1978)

Instead of attempting to revise my paper, I'll sketch very briefly my present position on dialectical logic. The logico-algebraic approach described in the text has been pursued by Dubarle with the help of what he calls hyper-Boolean algebra (cf. D. Dubarle and A. Doz, *Logique et dialectique* (Paris, Larousse, 1972)); for my criticisms, see *Dialogue* **13** (1974), pp. 203–205). There is also the formalization proposed by Rogowski (see J.L. Gardies, *La logique du temps* (Paris, P.U.F., 1975)). Since I have not been working on the subject for some years, I shall limit myself to a few remarks. Dialectics could still be a useful tool in social-historical analysis, even though it cannot be considered as an alternative logic, in the sense of intuitionistic logic; the essence of dialectics seems to rest on the interpretation of negation – I have dealt with a notion of 'local' negation, but in a different context (see my 'Intuitionistic Logic and Local Mathematical Theories', *Zeitschrift für mathematische Logik und Grundlagen der Mathematik* **23** (1977), 411–414 and *Fondements des mathématiques* (Montréal, Presses de l'Université de Montréal, 1976)).

Local negation, in my view, would allow for the dialectical passage

'*Aufhebung*' which I translate by 'sursumption', from particular moments or domains to the comprehensive or integrative structures, as in *The Science of Logic*. The process from the local notions to the global ones would be accomplished by negating only locally (in algebraic terminology, complements would be relative or pseudo-relative complements as in Heyting algebra).

NOTES

[1] This paper should be considered as a preface or an introduction to my paper 'Dialectic Logic and Algebraic Logic' circulated before the Congress 'Hegel and the Sciences'.

[2] In his book *Philosophy of Logic* (Englewood Cliffs, N.J., Prentice Hall, 1970) Professor Quine does not even mention Hegelian logic in his chapter 6 on 'Deviant Logics'.

[3] *Kritik der reinen Vernunft* (Hamburg, Felix Meiner, 1956), p. 99.

[4] See P. K. Schneider, *Die Wissenschaftsbegründende Funktion der Transzendentalphilosophie* (Freiburg/Munchen, Karl Alber, 1965).

[5] *Identität und Differenz* (Pfullingen, Neske, 1957), p. 56ff.

[6] See F. G. Asenjo, 'Dialectic Logic' in *Logique et Analyse*, no. 4, 1965, pp. 321–326.

[7] See *La dialectique*, Actes du Congrès de Nice (Paris, Presses Universitaires de France, 1969).

[8] *Dialectica* 6, 1948.

[9] See Lorenzen's remarks at the end of Gotthard Günther's paper 'Das Problem einer Formalisierung der Transzendental-dialektischen Logik' in *Hegel-Studien*, Beiheft 1, Heidelberger Hegel-Tage, 1962, pp. 65–123.

[10] See Georges Noël, *La logique de Hegel*, 2nd ed. (Paris, Vrin, 1967), p. 19.

[11] 'Logique Hégélienne et formalisation', *Dialogue* 6 (1967), pp. 151–165.

[12] See H. B. Curry, *Foundations of Mathematical Logic* (New York, McGraw-Hill, 1963), chapter 4.

[13] Asenjo, *op. cit.*, p. 325.

[14] See A. A. Zinoviev, *Philosophical problems of many-valued logics*, revised edition, ed. and trans. by G. Kung and D. D. Comey (Dordrecht, Reidel, 1963).

[15] Cf. my paper in the Canadian philosophical review, *Dialogue*: 'Logique mathématique et fondements des mathématiques' (*Dialogue* **10** (1971), 243–275).

[16] See my book *L'arc et le cercle. L'essence du langage chez Hegel et Hölderlin* (Paris, Desclée de Brouwer, 1969).

[17] See remarks in the same direction by Zinoviev, *op. cit.*, pp. 112–113 and 117–118.

[18] Quine, *op. cit.*, p. 102.

M. KOSOK

THE DYNAMICS OF HEGELIAN DIALECTICS, AND NON-LINEARITY IN THE SCIENCES

INTRODUCTION

The specific question to which this paper is addressed is the relevancy of Hegel to contemporary science. However, in order to show *how* Hegel is relevant, it is extremely important to *read* Hegel from the *perspective of awareness* that modern science is in the process of developing, in order to see whether there are any formulations explicitly or implicitly expressed by Hegel that touch on the central issues of contemporary science. As will become evident in the paper, the problem of *Non-Linearity* in the sciences is regarded here as an all-encompassing question, and one which furthermore is inseparably related to the problem of formulating a *dialectic* logic of relations. However, the dialectic nature of Non-Linearity proceeds from an awareness of relations that is not usually found in the approach most interpretators of Hegel take. Hegel's system presents an awesome structure of categories and it is very easy to get lost *in* them, if one has not grasped the essential dynamic that flows *through* them, a dynamic, which, from a non-linear perspective, turns out to be a most valuable insight in Hegel's system. We shall therefore reapproach Hegel's system – his *Science of Logic* in particular (although it is not limited to this) – from the perspective of the dynamic of movement which *generates* his categories, stating in fact that it is this very movement that constitutes the *content* of the categories generated, and not the particular categories themselves, regarded as identifiable and analysable terms. For, as we shall see, it is precisely this identification process, singling out terms or elements from their dynamic of relation and movement, which destroys not only the dialectic of *relation*, but the very *terms* themselves.

Now it is precisely this last statement – namely the assertion that in *no* way can Hegel's system or logic be understood through an identification procedure – which is to be taken as *radical* and revolutionary in the fullest sense possible. Usually, any approach to Hegel's logic at one point or another involves an identification process in which some central notion is converged upon – such as Being, Becoming, the Absolute, dialectic movement – and then regarded as an orientation perspective for analysis. This holds true for both the traditional non-formal approach in which Hegel's logic is looked

R. S. Cohen and M. W. Wartofsky (eds.), Hegel and the Sciences, 311–347.

upon as an ontological self-development, and modern formal approaches in which the logic is regarded as an algebra of terms-in-relation. For what results, in either case, is the presentation *of* a set of qualities determined in some simple or complex way, *to* an observer, reader or consciousness, in which the dialectic of awareness co-relating subject to object is hypothesized into a fixed state: Hegel's logic is then a simple product, result or object of consciousness, and not at the same time the subject of consciousness in its *activity* of awareness. It is true that one result of awareness consists in identification, but – as we shall see – this process of quality-localization is but a more complex form of non-linear dynamics in which all elements are always elements *of* movement and never elements *in* movement, there not existing in *any* way elements – as such, not even as illusions, or fictions. Consequently, we are asking the reader to *experience* Hegel's logic and non-linearity as *sheer-movement.*

In order to see how *this perspective makes Hegel relevant to modern science, and furthermore* that *this perspective was also Hegel's very own, requires a return to the originating gestalt with which Hegel begins his* Science of Logic.

I

As complex as Hegel's philosophy appears, there are a few fundamental statements appearing in the beginning of his Science of Logic *(from his preface to and including the first chapter) which clearly and unambiguously set forth his entire gestalt. Thus, Hegel states that "the headings and divisions which occur in this system* (*his* Science of Logic) *are designed* in themselves to have no other significance than that of a Table of Contents" (*SL* [J/S] **1**, p. 66). These divisions and headings

> do not belong to the content and body of the Science, but are compiled by external reflection, which has already run through the whole of the scheme, and hence knows and indicates in advance the sequence of its phases, before these introduce themselves in the subject itself (*SL* [J/S] **1**, p. 65).

Hegel then comments, however, that ". . . the *necessity of connection* and the *immanent origination of* (these) distinctions must show themselves in the discussion of the subject matter, for they are part of the self-development of the concept" (*SL* [J/S] **1**, p. 66). And what *is* this "necessary connection"? Hegel goes on to say that "That by means of which the Concept forges ahead is the . . . . . Negative which it carries within itself; it is that that constitutes the genuine dialectical procedure" (*SL* [J/S] **1**, p. 66).

Now this all important 'Negative', which the concept carries within itself, which forges the true dialectical movement, is introduced by the very bold statement that "The one and only thing for securing scientific progress is knowledge of the logical precept that Negation is just as much Affirmation as Negation . . . " (*SL* [J/S] **1**, pp. 64–65), and this declaration reappears in the form that "neither in heaven nor on earth is there anything not containing both Being and Nothing" . . . because Being and Nothing "are absolutely distinct and yet . . . inseparable" and furthermore "This unity of Being and Nothing, as being the primary truth, is, once and for all, the basis and element of all that follows: therefore, besides Becoming itself, all further logical determinations such as Determinate Being, Quality, and in short all philosophical concepts are examples of this unity" (*SL* [J/S] **1**, pp. 95, 97). In summary, Hegel refers to this both singular *and* universal paradox of being and nothing or affirmation and negation (or immediacy and mediation) characterizing at once his whole science and the dialectic, as the '*aufheben*' or transcending process in which "a thing is transcended only in so far as it has come into unity with its opposite . . . " (*SL* [J/S] **1**, p. 120), a transcendence process which Hegel in his preface refers to as the 'immanent soul" of the content of knowledge and the "*schema of movement* of all concrete knowing" (*SL* [J/S] **1**, p. 37).

Thus, all the various complex identities, divisions and headings Hegel displays in his system are *in themselves not* the content: it is the self-developing Negative as a 'scheme of movement' which is the only content, and *as* a schema of movement it must serve as the basis or 'soul' of all concrete knowledge. What one must grasp, therefore, is this *schema* or *dialectic movement* in its singular and universal nature if one is to judge the significance and relevance Hegel has for the sciences of knowledge as we know them today, and *not* get lost in an analytic dissection of the various categories or identities which Hegel clearly dismisses as being besides the point. Indeed this very distinction that Hegel makes of the categories in-themselves as not being the content but rather their *necessary connection and origination*, should warn the reader that any attempt to hypothesize this movement-of-identities into a set, sequence, or algebra of well-defined terms is doomed from the start, and thus should the reader disdainfully display a proof that all sorts of self-contradictions appear when analyzing the movement of distinct terms as separate entities in relation, or that the movement of terms can be reduced to a *given* pattern of relations, then he would merely be displaying his ignorance of what the point of the whole dialectics is in the first place.

Grasping *directly* the essential point being made, however, takes us on a voyage totally removed in spirit from what our usual conception of logic or structure is, if our training and conditioning rests upon a heavy dosage of what is usually referred to in the sciences as 'linear-thinking' . . . and unfortunately our most cherished responses and habits of consciousness are so weighed down with automatic 'linear-response' that we are not aware most of the time of the type of assumptions and simplifications introduced by this mechanism of linearization . . . . it *appears* so 'natural' that it *dis*appears as a problem. And yet, it is precisely *this* problem of linearity that not only characterizes the paradoxes and boundaries of all the modern sciences and humanities, but at the same time becomes the *means*, when properly resolved, by which Hegelian dialectic reappears transformed as a revolutionary perspective of integration and transcendence *for* the sciences and humanities as they are developing today.

Both the problem of linearity and Hegel's paradoxical unity of being and nothing or affirmation and negation can be grasped with the understanding of what is involved in the activity of 'identification,' or simply put, what one means by the 'identity' of that which our awareness is concerned with. To identify means to bring into focus a certain element, quality, fact, condition . . . a certain event, state of events or experience in such a way as to be able to refer to it in some categorical way. Without identification, our awareness is in a state of immediacy which is *not defined in relation to* categorization. This, however, means that a genuine state of immediacy cannot be characterized as either including or excluding any mediation or categorical state lest this immediacy is 'itself' then perceived as an identified state and hence not immediate. Thus, true immediacy is not an irrational/indeterminate anti-categorical state of dissolution in opposition to a rational/determinate categorical state of resolution, since both involve a specification or positing of specific qualities. Indeed, a state of direct-presence and immediacy presents a logical paradox right at the beginning for not only cannot it be regarded as a well-defined element capable of being identified with a *given* 'truth value,' but this very 'lack of value' cannot be regarded, upon reflection, as its identity either. Identity of any kind *sets limits* and boundaries, while a state of immediacy is a trans-bounded 'totality' or 'singularity', meaning that it cannot be regarded as either determined positively by the presence of a boundary or negatively by the absence of a boundary, for both presence and absence, inclusion or exclusion of limits or boundaries are mediations or modifications delimiting and defining that which it stands in relation to.

Even the notion of trans-bounded, if identified or localized by appearing in opposition to non-trans-bounded, cannot describe immediacy, yet immediacy cannot simply be dismissed as 'something' not-characterizable altogether for not only is this comment about immediacy itself an identifiable state, but the very awareness of the *paradox* which immediacy as a trans-bounded state *must* present, makes immediacy not an *arbitrary* variable simply left for the poets to sing hymns to.

Immediacy or direct awareness *is* . . . which is to say that *as such* it *appears* as a kind of trans-bounded or trans-identifiable 'totality' or 'singularity' that must be in a state of being-experienced in order to be-known, *any* mediated reference to it failing to capture or delimit it into an identity. Immediacy means the complete co-incidence of experience and the knowing of that experience – of object and subject – and thus concrete and non-questionable – any question appearing as a mediation or negation of immediacy by localizing one aspect in opposition to another, creating identifications and categories. And yet consciousness or awareness also identifies, focuses, makes categories and distinctions – namely it mediates immediacy. What then can one say about the relation between immediacy and mediation – or direct presence and its identification without merely throwing up our hands and calling this a meaningless problem not recognizing that this very dismissal in-itself is an identifiable *response* and hence not a way out of the problem?

This question takes us to the heart of our problem – namely that of grasping both the Hegelian *dialectic* and the modern notion of *non-linearity* as manifestations of awareness experienc*ing* or form*ing* identities *out of immediacy*: a non-dialectic logic and linearity on the other hand works only *within* a system or structure of *already-given* identities and is *not* concerned with their activity of appearance, creation or formation . . . they are concerned with what has-been . . . the past . . . and not with 'what-is' . . . with images, projections, categories and memories *qua produced*, and not even *with* images, projection and memories as an active and on-going process and thus an integral aspect of what is. In science, to linearize means to *separate out* into manipulative variables distinctions which constitute a given gestalt – namely distinctions which are perceived as constituting a totality to begin with and not distinctions which are first fixed and defined as isolates and then as *separate* distinctions, fit together as 'parts' into a 'whole'. Linearization is not only useful as a means of reducing the complexity of a given totality of appearance, but also necessary *if* one wants to define, measure and predict – i.e., if one wants to be present as an active and creative element in the world. But then one has to be *self-aware* of *what* identification and linearization means in order not to be blinded by the beauty of our own products – i.e., by *hubris*. All identification is a process that has to appear out of a pre-identified (i.e. pre-defined and pre-measured) state and is not God-given to Man. Thus identification into separate categories is a conditioned and mediated presence, and not in any way absolute, unconditioned, essential truth or the answer and response to anything other than the specific conditions that generated the identities out of a non-identified state of immediacy displaying identification and conditioning as the very price one has to pay in order for awareness to be active. However, we are so used to manipulating, both intellectually and emotionally *categories* or 'abstractions' of experience that one can become insensitive to our existence as an immediate awareness in which our habitual categories (such as logic, matter, self, world, freedom, necessity etc.) are neither identified as given or counter-identified as not-given, but in a 'fluid' state of identities 'being-given' and 'being-formed', and for that matter, re-formed, unformed, transformed and thus expressive of a state or condition that is *radically open* and not fixed . . . and hence not fixed by the very mediations, identities and words being used to describe what is happening. To be radically open does not merely mean to be free as opposed to being determined, for in this form *both* are closures of meaning bouncing each other. To be radically open means not even being bounded by *having* to not have determination, form, delimitation and structure . . . ; to be radically open means experiencing no position or counter-position already determined and fixed in a certain relation of inclusion or exclusion . . . it means experiencing one's being-in-the world as a singular totality being-evolved and transformed in boundless multiplicity and thus not in any way a simple detached entity contemplating other detached entities playing a cosmic game of hide and seek. Words, symbols and identifications have a truly hypnotic spell, freezing into a mold that which is being-experienced into 'an experience' that is separated from the activity of experiencing. But a little reflection or awareness of this condition further reveals that any attempt to then separate out the condition of immediacy as a 'pure-state' *from* the identification process as a categorical or 'impure' state is but *another* example of the categorization of immediacy, and in fact one of the most damaging for, through this particular maneuver, the genuine presence of immediacy becomes eclipsed by the very attempt to 'save' it as something separate from identification – not recognizing that immediacy or that 'which is' cannot itself be *delimited* by manipulation or categorization in any direct or indirect way, but rather must express itself as a totality or singularity in and through whatever process or state *is present*, and this includes the various states of identification, categorization and abstraction taking place.

Actually immediacy or any paradoxical identity-transcending 'state', not being bound by the self-affirming law of identity that A is A, means that such a state need *not* be self-affirming but also could be self-negating, which in turn implies that a trans-identity state can appear *both* trans-identical and non-trans-identical. Thus immediacy, precisely because it is *not* an identity, is free to appear either in pure non-identity form, or in impure identified form, displaying categories of mediation without contradiction, for it is only *identities* that must remain consistently self-affirmative in order to be meaningful. In our blindness, we tend to regard the law of identity that A is A as also being applicative to that which is non-identical, not recognizing that non-identical immediacy, to *be* immediate, cannot be externally delimited by mediation and identity reducing immediacy to a mediation-by-contrast called 'mere immediacy', but must be a totality capable of expressing *both* immediacy and its negation into mediation and identity, making all mediation a self-mediation *of* immediacy.

Put in simple terms, one *cannot* 'stop the world and get off' for even one instant – and in what suspension capture the world by an identity – for *any* activity of awareness, be it defining or measuring, thinking or acting, making abstractions, meditating or producing goods, is a *state-of-presence* which *qua* state of presence is what one *means* when one uses the terms concrete, existential, active, dynamic or immediate to describe that 'which is'. In modern scientific terminology – as distinct from the classical position – one recognises that the very *act* of observing involves, forms and transforms both the observer and the observed – and to *first* postulate a so-called 'metaphysical' sub-structure of hypothetical and hypothesized identities is to fix our minds into a *detached-reflective* mode manipulating already formed elements that no longer display a state of *immediate presence*. The task is, therefore, to become aware of the *phenomenology of reflection* and identification *as it* manifests itself without pre-empting it into an already *given* identity structure. Thus the very appearance of a reflection condition within immediacy implies that what *originally* was present – namely a pre-reflective element-of-immediacy, which we shall refer to as e, is now a determined or *posited* element $+e$ ('plus e') co-existing, however, in relation to its mutually determined context of reflection $-e$ ('minus e'), *both* the posited element $+e$ and its context as its negation $-e$ being necessary to *define* a *singular* boundary of identity or form $\pm e$ *of which* $+e$ and $-e$ are its inseparable distinctions. If one were simply to 'remain' with the original undefined state of immediacy e, one could not in any way refer to it, and hence the initial e 'as such' is not in any way an identity. Recognition of e is literally a *re*-cognition of e and hence a *transformation* of *e* into a state of determination and mediation: awareness *is* transformation. However, mediation is never one-sided, for any posited or determinate immediacy, given now as $+e$ co-exists only in relation to a counter-determined state $-e$ in which *both are mutually* co-defined and thus mutually *in co-balance qua mediation*, no mediated state negating the totality of immediacy *within* which mediations appear: identification is always a *relational* activity and not a matter of simply picking *out* a singlet without also at the same time revealing a relational boundary condition, expressive of an immediate totality. Thus, starting from a pre-mediated immediacy *e*, the phenomenology of reflection discloses a state of paradoxical mutual-mediation $\pm e$ to appear in which every distinct identity only co-exists in relation to a counter-identity, every action in relation to its reaction, every assertion in relation to its negation, every subject in relation to its object, every 'matter' in relation to an 'anti-matter', such that the boundary is *not* constructed *out* of the distinct elements regarded as *already* separately conceived or experienced and *then* brought into a linear compound or 'synthesis', (which *would* be a contradiction *of* terms) but rather the state of boundedness or transition (i.e. 'becoming') between any element and anti-element (i.e. between any notion of being and not-being) *is* what is *immediately* present as a *non-linear singularity or totality* whenever a simple or 'pure' immediacy – called here *e* – reveals an identification process. Thus $\pm e$ or the *self-mediation of e into* a plus-minus mutuality is the self-transformation of *e* or immediacy and we can refer to it as e' or the *totality* of the 'elements' $+e$ and $-e$ *and* the co-relation $\pm e$ as a *singularity* self-generated from an initial singularity. The first order determination of immediacy *e* creates a *cycle* or triad of immediacy returning as e'. Naturally, e' as the new singularity can in turn recycle and exhibit e'', which in turn can reveal e''', – without end – *all* of which are further self-mediations all within the dynamics of an *e*-ing process of immediacy, each newer level not being an *addition to* the previous one, but rather a *deepening of* that which is *being-revealed*, in *each* case the revelation of mediation coming in mutually bounded pairs (e.g. $+e'$ and $-e'$) and in each case, the increase of the complexity of mediation involving a *transformation* of the *type* of structure being-revealed.

We have thus presented the essence of both Hegel's dialectic – namely his scheme of movement – and the essence of non-linearity itself, seeing that *both* are a dynamics of the phenomenology of awareness in its state of reveal*ing* distinctions. This clearly reveals that dialectics is *not* a linear progression *from* a given thesis $+e$, *to* an anti-thesis $-e$, and *then* to a synthesis $\pm e$, but rather a non-linear unfoldment-of-what-is, wherein all being is be*ing*, and all becoming is self-becoming and transformation. *Not* grasping the e, e', e'' e-ing state of immediacy *within* which the + and – functions develop, reduces dialectics to an absurdity, and makes non-linearity in the sciences a mystery. Hegel, unfortunately, did not explicitly spell out the whole dynamics of such a non-linear self-mediation process, and as a result it became easy to get lost *in* a myriad of categories instead of becoming *free* of them. Thus the boundary-transitional nature of Becoming, resting as it does on *both* the *distinctness* of $+e$ and $-e$ and the *inseparability* of $+e$ and $-e$, was not sufficiently brought out as an immediate transition from Being appearing simply as e, and here is where a marriage of Hegelian dialectics with non-linear topology in mathematics would be extremely fruitful, for each is involved – from a different perspective – with the same problem. Hegel's triad of Being, Nothing and Becoming, when analysed in detail *amounts* to the same logic as presented here (i.e. this is *not* a *sequence* of Being, then Nothing and then Becoming) but the *singular* dynamics of its origination is somewhat obscure.

Now the importance of such a dynamics of immediacy, being the unfoldment of what-is and revealing the coming-to-be of identities in a non-linear topology or context-of-relations, is that it does *not* present one with an identity structure predicated on a *given* structure, but rather is totally void of specificity other than the experience of immediacy. To the degree to which one *does* experience himself-in-the-world as a dynamic immediacy, together with the forming and transforming of his identities and categories of immediacy, then to what degree will this dynamics reveal itself in and through the *identities* appearing and not appear as an external structure or mold into which to put or place an assortment of categories, *if* what has been presented *does* 'correspond' to the life and structure of immediacy. This in turn implies that the 'test' of *meaningfulness* such a dynamics of immediacy or dialectic has for knowledge in general and for the sciences in particular *initially* rests on the degree to which one is capable of regarding his total being-in-the-world – namely his feeling, ideas, actions and thoughts in terms of such a dynamic of self-transformation. Otherwise, this scheme of movement merely becomes but a more complex identity structure as linearized into a pattern of relations as any other system. This means that all the various identities and boundary conditions continually confronting one's existence must be grasped in their paradoxical state of transition and levels of transition, including of course any reflected notions *of* the dynamics of immediacy itself when regarded *as* a structure among other structures in the world of dynamic appearance. The paradoxical state of dialectic self-becoming is thus not only *about* immediacy, but also, paradoxically, *within* immediacy, and one must not therefore conclude that the present form of analysis, using as it must linear words and symbols historically conditioned, is the *last-analysis.* However, the beauty – if you will – of a *paradoxical* dynamic structure, is that its *own* transition *as* a structure of identifiable elements is *not* a contradiction to its power of revelation, but is indeed exactly that which *must* be expected if it does indeed reveal the continual self-negating and transforming nature of *any* identity! Paradoxical non-linear dialectics is never an *end* (i.e. an identity finally given), but always a *beginning*, and hence is a logic of creativity and revolution whose activity is the revelation in depth of awareness – and thus self-awareness, for as we have seen, all mediation and reflection is always self-mediation and self-reflection – the *world* of dynamic appearance being the *content* of this awareness.

II. DEVELOPING DIALECTICS AND NON-LINEARITY AS A DYNAMICS OF IMMEDIACY

The problem now in front of us is 'what to *do*' with the dynamic of immediacy once we have – *if* we have – experienced it. However, if we *do* something *with* it, one must be prepared for a linearization of the entire dynamics to express itself, for immediacy is always a state of self-actualisation and totality, while simple *directional* activity implies a narrowing or focusing of perspective – whether that narrowing is called 'looking at the elements or parts' (+ and –), looking at the whole (±), or *any* identifiable aspect of transcending, transitory totality. Of course, such a linearization is itself but an expression of a non-linear totality, but if this totality is lost sight of, expressing itself only indirectly or 'unconsciously', then linearization *also* expresses self-alienation – a type of self-negation in which any original paradoxical totality appears *visible only* in terms of its negation and not directly as such. And the negation of paradox through linearization is none other than being caught up in the trap of 'diction' vs. 'contradiction' which Gödel's theorems beautifully express as the fate of all linearizable (arithmetizable) logics.

Thus, if we 'take' the dynamics of immediacy from e to e' and linearize it, then the three inseparable distinctions appear as separated distinctions obtaining $+e$ as A or the standard assertion, $-e$ as not A or the standard negation, in which case $\pm e$ or the boundary element would merely appear as an indeterminate 'middle' which if retained in a system would lead to a contradiction (being both $+e$ and $-e$ or A as diction and not A as contradiction), or if rejected from a system would lead to incompleteness (i.e. the transition state being excluded). Once linearized, $+e$ and $-e$, being and nothing, or any element and its context, any particle and its space or field, any ego and its world cannot be 'brought back together again' without producing either an overlapping (inconsistency) or underlapping (incompleteness) of identities that are each held to have their *own* separately determined boundaries and thus void of genuine relation. We see, here, how Gödel's incompleteness-inconsistency opposition follows *directly* from the linearization of the fundamental paradoxical triad of dialectic. Thus a three value, *n*-value or infinite value logic does *not* yield a dialectic logic – dialectic movement dealing with self-mediated *transition* states of immediacy and not identity values such as true and false, or for that matter *any* kind of value including the value of 'indeterminacy' as long as these values can appear arithmetized or linearized in any direct or indirect or even 'partial' way as a sequence 1,2,3,4. . . . Of course, for the purposes of simplification one does introduce linearizations and intermediate values, but these become delusions and not clarifications if one forgets the state of immediacy and existential presence that our whole totality is an expression of, within which identity appears as a process that is being-formed and not merely appearing as a product already present (like the numbers 1,2,3,4 . . . , regarding numeration linearly as a given-structure and not dynamically as self-related transitional values whose 'discreteness' is not absolute but only provisional).

The only 'road' or 'Tao' that the dynamics of immediacy follows is *sui generis* and only those who are an integral aspect of such a state of experiencing can discover directly or indirectly aspects of its self-expression. We shall follow the singular e, e', e'', e''' . . . *e-ing* process or state (*not* to be confused with either a simple sequence or set) first in our already expressed variables of + and – interrelatedness, and then tie this dynamics of immediacy with the Hegelian logic, at the same time showing the intrinsic relation this logic has with the formulation of non-linearity in modern scientific theoretics. Once one grasps the initial dynamics of immediacy from e to e' as the unfoldment of three inseparable distinctions, now written in the form $e' = (+e \; -e \; \pm e)$, expressing what e must appear like when referred to or identified within a state of immediacy, then one can also see that e'' would be a further self-transformation of that immediacy now appearing in the form $e'' = (+e' \; -e' \; \pm e')$. Although these *look like* a discrete leveling of similar separate phases, they are only indeterminate singularities whose dynamic interrelatedness is yet to be revealed. Thus, e'' would now have to appear as a two-dimensional triadic transition *of* a triadic transition state and hence a *nine*-term expression of relations occurring simultaneously on two distinct but inseparable levels: e'' thus appears *qualitatively* distinct from e'. This can be expressed thus[1]:

$$e'' = \begin{array}{c} +e' \\ \hline -e' \\ \hline \pm e' \end{array} = \begin{array}{c|c} +\ (+e\ -e & \pm e) \\ \hline -\ (+e\ -e & \pm e) \\ \hline \pm\ (+e\ -e & \pm e) \end{array} = \begin{array}{cccccc|c} + & +e & + & -e & + & & \pm e \\ - & +e & - & -e & - & & \pm e \\ \hline \pm & +e & \pm & -e & \pm & & \pm e \end{array}$$

In general, one obtains (mathematically) a progression e, e', e'', e''' generating an ever-increasing matrix of transition states, such that as the number of *levels* or dimensions 'n' increases according to the numbers 0,1,2,3 . . . n, the number of *terms-in-transition* increases correspondingly according to the progression 1,3,9,27 . . . 3^n, (i.e. triads *of* triads *of* triads . . . etc.), realizing of course that such a *numerical* localization is in linear form – as any *explicit* symbolic relation must be. The only way in which *not* to lose sight of the non-linear dynamics of immediacy, is to grasp whatever is present in terms of a functioning totality – i.e. as a topology of dynamic relations. (Thus, in e'', the 9-term totality reveals 4 oppositional terms $++e$, $+-e$, $-+e$, $--e$; 4 partial transitions $+\pm e$, $-\pm e$, $\pm +e$, $\pm -e$, and 1 complete transition – i.e. a transition of a transition $\pm\pm e$, all 9 acting as a singular two-dimensional dynamics). In the form presented, we see that as the number of dimensions increases to the denumerable infinity 'n' in the 'limit', the number of transition terms increases to the indenumerable infinity '3^n' in the limit, these two infinities being but the initial modality of infinitization that the e-ing state can reveal. In effect, the *infinite* transition condition of the dynamics of immediacy is but a way of expressing the fact that all elements involved 'in' a transition-state are not 'elements-in-transition', but rather 'elements-of-transition', meaning that the elements as such are not primary, but rather the transition-relation is – which in turn means that any expression which formulates a transition relation *between* two elements on one level (level n) must also reflect this transition *within* each of the elements on a deeper level (level 'n plus 1' for example) lest the elements appear separable into *complete* parts independent of their relation, making transition only an 'external' matter and not the very dynamics or 'soul' of immediacy. However, this is but a way of demanding an unending infinity of transitional relations, for either transition is being-experienced as a singular totality *without* comment, in which case one does not single or linearize out *anything*, or one has made a comment which singles out some level or level of relations 'n', which as a linearization is incomplete without a deeper one 'n plus 1', which in turn is incomplete if singled out without still a deeper one, etc. Thus, should *any* immediacy appear reflected or determined as an identity, it must be transitional and thus revelatory of *sub*-identities-of-relation (e.g. e into e'), there not *being* any ultimate identity, sub-identity or meta-identity *reflected upon* that is not *itself also* transitional. *Transcendence is total transition* appearing at any stage – i.e. total 'clarity' of being or 'enlightenment', be it called the Absolute of Hegel, the Void of Zen, the non-localizable I and Thou of Buber, the radical negativity of Nietzsche, or in modern scientific terminology, the non-linear totality of inseparable distinctions expressive of the primary immediacy of experience.

Reviewing the entire dynamics of immediacy once more as a *singular* experiencing state of *self*-mediation, having now 'spelled out' the particular modality of linearization used here as its expression (namely the infinity of multi-dimensional matrices of transition-elements), one can get an intuitive feeling of the entire process if one regards e as a signification of the singularity of immediacy in *any* form; e' that singularity appearing as a first order boundary relation; e'' appearing as a second order mutuality or boundary relation of elements . . . each element *in turn* explicitly exhibiting a boundary relation of mutuality, and so on . . . the number of boundary relations present being a function of the complexity of depth a state of immediacy displays. Indeed, turning to Hegel, we can see that this is exactly how he introduces transitions that move *beyond* the first order transition called Becoming. In describing the two moments of becoming called *arising* and *passing away*, Hegel states that "Becoming thus contains Being and Nothing as *two* such unities, *each* of which itself is a unity of Being and Nothing . . . One of these is Being taken immediately and as related to Nothing (i.e. passing away); the other Nothing taken immediately and as related to Being (i.e. arising). . . . In this manner Becoming is a two-fold determination" (*SL* [J/S] **1**, p. 118).

In terms of our matrix presentation of dynamic immediacy, Becoming as the first transition state is $\pm e$, or e' as a totality of relation between Being determined or posited as $+e$ and Nothing counter-posited as $-e$ (*both* of which originate and are expressions of the original expression of pure or sheer immediacy e, simply called Being). Now a second-level reflection starting with e', would give $+e'$ as a posited or determined state of becoming, which would in turn reveal $-e'$ as a counter-determined state of becoming, meaning that no longer do we have a *simple* boundary relation between $+e$ and $-e$ expressing the immediate form of transition called e'. Thus $+e'$ would be a *posited* or *direct* form of relating $+e$ to $-e$ in e' or becoming in contradistinction to $-e'$ as the *counter-posited* or *inverted* form of relating $+e$ to $-e$ in e': whereas

$+e'$ represents the original state of direct relation between being and nothing, each one *posited as itself*, $-e'$ would reveal an inverted counter-relation in opposition to this posited state and this would mean a state in which being *appearing as nothing* would be in relation to nothing, *appearing as being*. Thus, $-e'$ as 'counter-becoming' would be a relation of terms *each* one of which appears as its *opposite* in order to counter $+e'$ in which each term appears only as *itself*. Hence, if $+e'$ represents, for example, an energy exchange relation between charges (of electricity), $-e'$ would represent an *inverse* energy exchange relation between the *same* charges, tending to counter-act and stabilize a transition state into two transition states moving in opposite directions. In terms of a general topological dynamics, whereas e' in effect appears as a relation due to a simple balance of two opposite *elements* as a result of a self-differentiation of e into $+e$ and $-e$, e'' now appears as a balance of two opposite *relations*, appearing as a result of a self-differentiation of e' into $+e'$ and $-e'$, producing a singular relation-of-a-relation. This, of course, means that in e'', not only must there exist $+e$ and $-e$ elements, producing a posited relation, but *each* in turn must also mirror its opposite in order to manifest a *counter*-relation *between* them, which, at the *same time*, mirrors the *original* relation of transition between them, also *within* each. Thus, intuitively, a second level two-dimensional relation of dialectic opposition reveals not only X and Y, let us say, as oppositions, but also its inverse as $X(Y)$ and $Y(X)$, such that there are *four* $X-Y$ relations: the direct relation between X and Y, the counter-relation between (Y) and (X), the relation within X (between X and (Y)) and the relation within Y (between Y and (X)). Finally, there is the entire structure as a whole, which is nothing but the $X-Y$ relation now taken not as a simple one-dimensional unity, but as a self-differentiated unity – i.e. a unity of unities, or a unity between X and Y each of which is also an XY unity. (This nine-term analysis is an intuitive *simplification* of the formal two-dimensional, nine-term matrix appearing on page 322.) Clearly, higher order matrices would then not only develop further unity within unity within unities, but each quantitatively higher order dimension would require a higher order topology of relations in order to grasp all possible relations as a singular dynamic whole: a three dimensional reflection would have to spell out 27 terms, simultaneously related in a *three*-dimensional space thus displaying *qualitatively* newer modalities of existence not previously visible. Thus the dynamics of immediacy not only represents the initial 'triad' of dialectics non-linearly, but subsequent triadicies continually expand the dimensionality or *perspective* of the dynamics present and hence the *nature of the terms* one is in relation to, revealing a *power* expansion which in the limit generates an *in*denumerable number of terms in an infinite space. In no way, therefore, is dialectic a term-by-term expansion: it is a self-evolving unfoldment of possibilities of relations, any *specific* structure reflective of the richness of the paradoxical immediacy or totality present.

Returning to Hegel's analysis of the Logic, we can now observe the progression of *dimensionality within which* the terms or categories express meaning, without which his logic is reduced to a linear term-by-term unfoldment, or an a-logical presentation of generally related intuitive notions. For Hegel, $+e$ $-e$ $+-e$ form the first triad of Being, Nothing and Becoming. However, this *entire* triad, as an immediate totality e' in turn develops as a triad of triads $+e'$ $-e'$ $+-e'$ which he calls Being (Being-as-such), Determinate Being, and Being-for-Self, which in turn, as a singular totality e'' is called Quality. Thus Being-as-such or $+e'$ is actually the initial 3-term state of becoming posited as a fully immediate state, Determinate Being or $-e'$ is then its negation into counter-becoming or anti-becoming, dealing with the determinate structures of stabilized becoming (such as finitude and infinity) as a counter 3-term state of mediation emerging *from* the 3-term state of becoming, while Being-for-Self or $+-e'$ is the unity of becoming with anti-becoming (or the unity of non-determinate 'universal' transition and movement with determinate structures of particularity) yielding the final 3-term *self*-determinate state which is the unity of indeterminacy and determinacy.

All in all we have nine terms for a second order dynamics of identification e'' called Quality. The zero order state e is pure immediacy as being; the first order determination e' is being-determined-*as* becoming its own other through negation. However, and this is extremely important, since the relation between positive and negative or being and non-being within becoming is not strictly symmetric, e' or becoming is the determination of the *immediacy of being* referred to as e: the negative or any reflection or mediation is *not* an *originating* primary state capable of appearing in a *simply* immediate form. All determination is self-determination *within* being. (In mathematics the negative is defined as a *reflection* from what is posited as positive: it is a relation-state to begin with, such that -1 *times* -1 is a double-reflection back to the positive direction. Positive and negative are asymmetric-symmetries of relation: $+1$ times $+1$ is $+1$, but -1 times -1 is *not* -1 . . . it is also $+1$, indicating that all symmetric alterations that occur between + and − cycles are asymmetrically *based* on the positive.)

Thus e' is a development of a relation between being and its negation *out of the immediacy of being* or e. It is hence only within e' that the negative makes its *first* appearance. This means that e'' as the development of e' will now start with both being and negation as positive immediacies in $+e'$, and only then develop *each* of these as *mediated* negative terms in $-e'$, wherein the original positive and negative immediacies within $+e'$ each now reveal their complimentary opposites (i.e. being revealing negation and negation revealing being through mediation). However, since $+e'$ or *posited becoming* is also the asymmetric movement of being *into* negation, $-e'$ is then the counter asymmetric movement of *mediated* being *into mediated* negation – or negation back into being: $-e'$ is thus the anti-state of counter-becoming balancing the original state of becoming $+e'$, such that $\pm e'$ or e'' is now *both* the symmetric balance of becoming and antibecoming, and the *asymmetric* movement of being into becoming-its-negative (e into e') which has now in turn expanded with a *return* movement of negation back to being, for only in e'' do we *first* find a *double-negative* term ($--e$) representing the *return* movement of being from negation. Thus e, e' and e'' develop an *overall* cycle from immediate being (e), through being-revealing-the-becoming-of-negation (e') to the becoming-of-negation, *re-revealing* the presence of being as a *return* state (e''). In e'' being is revealed as the *end*-state of all becoming, because now being reveals *itself* through negation as self-being or being-for-self. Quality as the full dynamics of e, e', e'' is thus the *permanence* of being or immediacy *through* any determination, limit, negation or becoming. The *nine*-term *matrix* e'' is thus the full self-determination of being or e and it is this *matrix* of double-transition which is the essential *building unit* of Hegel's logic, as much as any *triad* of single transition is the interconnecting *movement or energy* which constitutes the component aspects of such units. Specifically, the *details* of the dynamic unit of Quality e'' can best be revealed by returning to our initial cycle or triad e' where we can see the *cycle itself* recycled, first *determined* as an immediate *positive* cycle, then as the mediating *negative* cycle. (i.e. the *cycle* e' will be recycled as $+e'$ $-e'$ and $\pm e'$ all within e'', just as e originally appeared as $+e$, $-e$ and $\pm e$ within e'.) In the immediate or positive cycle, *all* the forms present are now in immediate form, which means that the original *self-mediated* state of becoming $+-e$ containing + and – in mutual identification now *collapses* as Hegel often puts it, into only an immediate form of itself: the Being and Nothing or + and – within Becoming or e' appear as indifferent immediates or distinctions which have *not yet* explicitly developed each opposite within itself (i.e. e' is now *part* of itself as e'' which *is* to develop). This leaves us now with the mediating or negative cycle. Here all the forms present become mediated: the positive appears mediated *into the negative*, and the negative mediated *into the positive*, each immediate or apparently indifferent term revealing the other within and *relative* to itself. Consequently the identity between the positive and negative which was originally *totally* immediate in the cycle e', and which in the *positive* cycle appeared as a paradox of *two* immediacies in a *co*-immediate state, has now become self-separated in the negative or mediating cycle because each immediacy is now turned to the opposite *within* it as a derived opposite, and is not *directly* or immediately in relation with the opposite 'itself'. Each opposite appears self-bound and not other-bound. In Hegel this means we have moved from the initial or positive triad of Being Nothing and Becoming (itself called Being) to the triad of Determinate Being. Relative to *this* triad, Being or the positive form of the first cycle appears as a *determined* Being called Something or Finitude (mediated *self*-referral), while Nothing or the negative appears as a *counter*-determined Other or Non-finitude (the bad infinity of open-ended *other*-referral): the 'Something' is actually Being acting in a delimited negative way, i.e. it is a *bounded system*, while the 'Other' is actually negativity acting in a positive way – i.e. it is the *context* or the determinate space within which the system as a something appears determined: Determinate Being is *System in a Context* (Diagram, p. 345).

Now the unity of the Something and Other (the two fundamental opposites on the negative level) is naturally not the state of immediate becoming or the co-*immediacy* of Being and Nothing, but rather the mediated state of Anti-Becoming, or the co-*mediated* state in which a determined Being and a determined Negation relate to each other as apparently *external* elements.

This, however, immediately collapses, for the system and its context *in unity* generate a synthesis called True Infinity – each determinate Being (as a something) in revealing within itself Negation, in turn is but a revelation *of* the negative *itself qua* Other, which relative to *its* domain in turn is the revelation of Being: i.e. being externalized *within* negativity is now a return *to* immediate Being. This, hence, leads us to consider the *unity* of the positive or immediate cycle of Becoming and the negative or mediating cycle of Determinate Being, for not only does a system and its context now *ex*-ist as distinguished *in*dependent and *self*-related entities, but each in turn has revealed the dependency and other-relatedness previously expressed *directly* in the pure state of Becoming.

In the self-mediated cycle, then, *each* opposite, both the positive and negative, now express *a total cycle* of return, for each opposite is (1) both its *immediate* presence and hence in a direct relation *to* its opposite as a co-immediate (2) a revelation of its complementary opposite as a derived or mediated presence *within* itself separating each opposite *from* its immediate relation to its *actual* immediate opposite and now (3) a return to itself *through* its own opposite – i.e. a return as a self-mediated immediacy *having explicitly* developed within itself its own opposition. This means, however, that the opposites, as immediates are *once more* in a direct relation of unity *with* each other as in Becoming, but at the same time, each is also a unique monad *of* the other within it. Hence each opposite *itself* is in a state of Self-Determination and a *unity* of opposites, while at the same time the opposites *together* are *also* in a state of *direct* unity: This is the Being For Self stage in which the unity of being and nothing is also *a unity of unities:* Being, having emerged into a determined Something or system, returns to itself as the One of Attraction; Nothingness – having emerged into a counter-determined context, also returns to its negativity as the Many of Repulsion. The Unity of the One and Many, or attraction and repulsion is the *overall* unity for the e'' cycle. It is a unity in which the original element of immediacy called e or Being, having become a *transition* state e' and a revelation of its negative as Nothingness, has *in turn* generated a higher order negation, in which *both* Being and Nothingness, as *two* co-immediacies, have *each* become negated within themselves, only to reveal a counter-transition to the e', in which Being not only disappears into Negativity, but any Negativity *as* an immediate element itself in turn reveals Being. Hence with e'', being or e is in a double transition state, *explicitly* returning being to itself through any negation which has emerged from it. Being as containing all determination and negation *within* it is Quality.

Now in mathematics, the movement from e to e' to e'' is structurally similar to the movement from a function to its first derivative, and then to its second derivative, which is *not* as such a linear one-dimensional sum of two derivatives, but a derivative *of* a derivative, or a state of transition-in-transition. In physics, position, velocity as the first derivative (like becoming) and acceleration as the second derivative (the motion itself in movement) likewise express three states of space-time relation important to describe fundamental *types* of movement possible. Position is simple 'being', velocity is simple 'change-of-position', and hence position vs. velocity merely describes the simple opposition between non-action and action. However, in order to describe inter-action, or a state of non-action (like a body at rest) which is actually in a sub-state of action (like sub-units in movement), then the acceleration of *changing* velocities (and hence forces, dynamically speaking) are required, for now not only is there simple movement in one-direction, but there is also the movement of this movement changing it into an opposite direction – and continually re-changing it if this accelerated sub-movement is stabilized into a body (e.g. stabilized into harmonic oscillating motion). Consequently, simple position can only express *mass* or inertia – a state of rest. Then simple velocity or movement expresses *momentum* – a state of action or activity. Finally, a state of interaction describes a physical system of units in dynamic relation, in which *both* non-activity of rest (for the system as a *whole*) and the activity of movement (for *every part* of the system) can coincide to express a condition of *dynamic-equilibrium*: viz., a state of mass expressing its movement as *internal energy*. Thus with only position, there is *no motion* evident at all, with velocity there is only *explicit motion*, but with accelerations and changing velocities, a system can display *implicit motion* in the form of stored energy (e.g., rest-energy, potential energy).

Now in mathematics, not only are there an infinite number of derivatives possible and hence states of transition, similar to the e, e', e'' infinite matrices of transition, but complex number theory likewise shows that the type of number-structures possible also – theoretically – build up in complex matrix form – giving us the 'hyper-complex sequence' (real, complex, quaternion, caley etc.) continually expressing more and more subtle *dimensionalities* (i.e. qualitative aspects) of number. In physics there are, of course, an infinite number of derivatives of motion possible, and the *qualitative complexity* of interaction depends upon the number, ordering and interrelation *of* these states of movement. Thus the *dynamics* of the logic of Hegel, when expressed in terms of the e, e', e'' e-ing process of transition not only accounts for the dimensional non-linear complexity of his logic, but has immediate relation in mathematics and physics to the structures with which they present their modalities of transition.

One can now move beyond e'' or the state of Quality as such and generate e''' as a three-dimensional transition state involving $+e''$ or Quality posited against $-e''$, Quantity, generating $+e''$ called Measure – which is a triad of 9-term matrices, giving us 27 basic terms for the state of Measure. The entire triad e''' is called Being (as a totality). If Quality is the self-determination of Being, then Quantity is the self-determination of its opposite – namely Non-Being or the 'other' of Being. Thus Quantity is the externalization of Quality, and Measure is the unity of both Quality as 'internal Being' and Quantity as 'external Being', giving us e''' as the totality of Being. Dynamically, this means investigating the state of transitions which a self-integrated state of Becoming (i.e. e'') can display, or the complexities of three transition states simultaneously interrelating. Finally, in Hegel's logic, only one more reflection is needed, and the entire structure is displayed: $+e'''$ as Being, $-e'''$ as

Essence and $\pm e'''$ as Notion gives us the fourth-dimensional state of transition called e'''' or the Logic itself revealed in four reflections and 81 basic transition terms (i.e. 3^4 is 81). Dynamically, this means that *starting with* e'' or *Quality* as our basic *initial* term (the state of self-integrated Becoming), one can generate a *second two*-dimensional matrix with the 'sequence' e'', e''', e'''' similar to the previous sequence, e, e', e'' which *ended* with e''. Thus e'''' is to e'', as e'' is to e. This means that e'''' or the entire logic of Hegel is structurally the Qualification *of* Quality, or a display of a meta-Quality state (e'''') *each* term of which is a positive or negative state of Quality (e'') or self-integrated Becoming similar to the way in which Quality itself (e'') was a meta-Being state (i.e. the return of Being to itself or Being-for-itself) in which each term was a positive or negative state of Being (e). In this way, one can also relate the two intermediate e' and e''' states to each other. Thus, whereas $+e'$, $-e'$ and $+-e'$ stood for Being (as such), Determinate Being and Being-for-Self, $+e'''$, $-e'''$ and $+-e'''$ stand for Being (as a totality), Essence and Notion – making the *last* triad but a higher order version of the former. Indeed, the *even* number dimensions, e, e'', e'''', are all forms of doublets or doublets of doublets (of transition) and hence all in a form in which any given transition or negation is coupled to another transition and negation, producing a negation-of-the-negation or return to Being-type-of-structure, while the *odd* number dimensions e', e''' are all forms of singlets of negation or transition, and hence in a Non-Being or Negation type of structure, not forgetting however that each succeeding higher even or odd dimension is not a mechanical repetition of the previous one, but rather is a development of the previous one in conjunction with an intermediate opposite.

In developing Hegelian dialectic as a dynamic topology of transforming dimensionality, one can visualize this process geometrically (i.e. linearize it into a specific model) by considering it as an evolution of spaces – from the point (zero dimension), to the line, the surface, the volume and higher order spaces, in which one obtains (1) a complete one-to-one correspondence between all the topological *elements* present in a particular space and all the dialectic terms present in a particular dimension of transition – and (2) a complete correspondence between the topological *relations* present in that space, and the dialectic relations present for a particular dimension. Thus for 3^n, the first value is $n = 0$, giving us 3^0 or the value 1. This is both the element e, and the singular point. For 3^1, we obtain the value 3, and this corresponds to the first dimension e, in which $+e$ and $-e$ correspond respectively to the two end points of a one-dimensional line, while $\pm e$ corresponds to the line itself – i.e. to the line of co-relation determining the two end points or polarities. Thus, the original zero-dimensional *space* – namely the *singular point* – now appears within the one-dimensional space as a double-*term* or double *sub-space* (i.e. as two points) giving expression to a new singular space – namely the line itself. For 3^2 we obtain the value 9. As previously pointed out, a two-dimensional dialectic matrix has *four* opposition elements ((++*e*, +–*e*, –+*e* and ––*e*), four partial transitions (+ +–*e*, – +–*e*; and +– +*e*, +– –*e*), and one complete transition (+– +–*e*). The four oppositions correspond to the *four* points of a square-surface, the four partial transitions correspond to the four lines connecting these points, and the one total transition corresponds to the two-dimensional square-surface itself whose boundary is now composed of four one-dimensional lines simulteneously in co-relation. (Looking back at the matrix on page 322, together with its analysis on the pages following, will enable one to intuit the correspondence outlined. Besides the obvious properties mentioned here, there are others to be considered, e.g. the role of the diagonals, which all give additional expression to the topology of dialectic relations.) Thus in moving from one- to two-dimensional dialectic, the elements in relation are no longer points, but lines (i.e. relations themselves) and furthermore, the singular space they generate is *not* according to the same kind of structure found in the more simple one dimensional dialectic of a triad. A two-dimensional dialectic has to be grasped as a singular 9-term relation of relations generating simultaneous transitions in a two-dimensional space.

For a three-dimensional dialectic, 3^3 gives 27 terms. This corresponds to the cubic volume which has 8 points, 12 lines, 6 planes and one singular volume – namely the entire space itself as a three-dimension transition space. We see that the complexity of transition relations jumps *qualitatively*, each time the *previous* space becomes a sub-space or element within a larger singular space. For a four-dimensional dialectic, 3^4 gives 81 terms. This corresponds to the 'tesserac' – i.e. a four-dimensional cube having 16 points, 32 lines, 24 surfaces, 8 volumes and one four-dimensional volume that is bounded by the 8 volumes (as the three-dimensional volume is bounded by 6 planes). Hegel's logic, as a 3^4 or 81-term dialectic generated through four major transition states, which function *simultaneously* in co-relation, is thus mapped topologically into a 'tesserac' – i.e. a four-dimensional cube, which gives a singular expression to these 81 elements of transition in a singular gestalt! Small wonder that it is difficult to imagine as a totality, let alone higher order spaces into 5, 6, . . . etc., etc. dimensions not even considered.

Two immediate observations appear. First, the number of singular *points* or opposition elements (of zero dimension) present in each space generate a 'sequence' 1, 2, 4, 8, 16 . . . or 2^n within the overall 3^n 'sequence', showing us that Hegel's *dia*lectic is in effect the infinitizing of the basic 'two-value' (i.e. 2^n) nature of paradoxical oppositions, giving us an *indenumerable* number of *inseparable* elements (2^n is mathematically the power of the *continuum* of points) – i.e. illustrating that dialectic logic is actually a *singular* 'value' continuum of transition points appearing *through* a paradoxing process of *self*-opposition in which all distinctions are inseparable (but not *in*distinct). Ordinary 2-value logic is hence a first order approximation to a dialectic logic, as is a 4-value modal logic a second order approximation. In general any *n*-value logic is an approximation within the continuum of inseparable transition points. Secondly, a topological logic demands that the terms or elements present in any perspective are *not* all of the same *power* – i.e. certain terms are 'point-like', others, line or surface-'like' – indicating that one must be aware of the dimensionality of the terms present as expressions of different stages of transition-relation – i.e. of different stages of self-oppositional relation. In conclusion, we can see that dialectic is expressible through *non-linear, topological continuum* states of transition and not linear algebraic atoms of fixed identity, and the model here presented is itself but a simple-geometry of such a dialectic, which, as a dynamic of immediacy, cannot be bound to any one linearized presentation of non-linearity.

The above topological presentation of dialectic structure as a non-linear transition complex both generating and functioning through a multi-dimensional 'space' (or rather 'space-time' since it is a *dynamic*-of-becoming presenting not only symmetrical-'spatial' forms of +–*e* mutuality, but asymmetrical-'temporal' modalities of transcending-mutuality e, e' . . .) has been hitherto completely neglected in any analysis of dialectic, but is here regarded as its most significant essence – and furthermore one which opens up its immediate relevancy to the sciences and humanities. Consequently, we can now indicate two crucial aspects necessary in any non-linear or dialectic process: (1) its 'symmetric' modality of expressing all elements as +–*e* mutual transition elements functioning within *qualitatively* distinct states of *dimensionality*, thus eliminating the notion that dialectic builds up elements in a singular linear term-by-term progression: elements appear in qualitative 'jumps' of 1, 3, 9, 27, 81 *grouping* of elements according to a *power* expansion 3^n which reflects the changes of perspective needed whenever there is a change in the quantity of +–*e* transition states existing in a non-linear complex (e.g. e''' or 27 terms involving *three* simultaneous +– transition or boundary states) – and (2) its 'asymmetric' modality of expressing an 'infinite' or 'open-ended' transformation process e, e', e'', e''' . . . that is consequently *free* of restricting the results of this transformation to any *final* identity. Thus dialectic non-linear logic is a logic which is *beyond* any relation to an intrinsic *ontology*, being instead a *logic of phenomenology* (i.e. the 'ology' of *phenomen*-ology), and thus a logic dealing with the specific characteristics of non-linearity *as* they appear through the dynamics of phenomena, all ontological states being then linearized and identified references to the various mediations and determinations of self-mediating immediacy.

With regard to relating these two necessary aspects of a non-linear dialectic to Hegel, the first aspect demands that the various categories generated in any portion must be grasped in relation to their *specific dimensionality* as the determining context of movement determining their nature *as* a moment-of-transition (and not a moment-in-transition). Thus, as already explained, the entire Logic of Hegel has 81 basic terms (plus many sub-terms which are always capable of appearing through the infinity of reflection possible on any particular element or groups of elements in their dynamic of immediacy): these 81 terms are, however, generated through only 4 fundamental changes of perspective, due to the fact that there are only 4 simultaneous transition states in co-relation, e'''', expressing 3^4 or 81 terms arising out of the non-linear triadic process of constant self-triadicity (i.e. triads *of* triads *of* triads . . . . etc. etc., taken four times). Hence one has to *first* be aware of the 'type of space' present – in reading Hegel – within which the 'points' express themselves as 'monadic' reflections of that dynamic space – the space referring to the dimensionality, the points to the categories, and the 'monadic' relation referring to the dialectic between point and space (which is the same type of relation, which Leibniz as the inventor of the Calculus – and the 'monad' concept – attributed to the *mathematical* concept of point and space in the dynamics of differentiation and integration). Thus Dialectic-non-linear logic or 'language' is to standard linear logic or language, as the Calculus (or what is called 'analysis') is to algebra: the first is dynamic-generative and the second is static. Consequently, in analyzing categories in Hegel's section of the Notion, for example, it would be *destroying* the non-linear dialectic of the categories – and hence the very categories themselves (categories being terms *of* but not *in* dialectic movement) *if* they are merely *linearly* read as a term-by-term elaboration, and not seen as *sub*-terms, and indeed sub-terms to the fourth power (3^4) of the original immediacy (i.e. e or Being) expressing its dynamic *now* as a fourth order self-mediated immediacy (i.e. e' or Becoming). To arbitrarily start at some middle point in the Logic without full and explicit awareness of the generating movement of immediacy *it is expressing*, is to either degenerate the terms into linear, algebraic abstractions, or to regard *that* particular starting point *once again* as the original starting point, expressing Being, Becoming, Quality etc. – all of which are but lower order transition states already transcended and '*aufgehoben*'. This, however, demands a knowledge of the 'dialectic-matrix' of multi-dimensional transition states (as, e.g. outlined in simple form on page 322) seen as a dynamic *topology* (Diagrams, pp. 345, 346).

But, non-linearity and topological relations likewise express themselves through their opposite – namely linear algebraic localizations of identity – non-linearity being the scientific expression of dialectic *paradox*. Hence, while non-linear dynamic generation is the *content* or 'substance' of the elements or 'terms' *of* movement, linear identifications are the *form* (i.e. literally 'outline' of determination) expressed by these elements, causing them to appear as categories or 'words' of experience: this is generating '*logos*' or mediated categorical-relation out of '*eros*' or dynamic immediacy, in order to give a higher order expression to the self-generating '*telos*' of self-mediated immediacy – i.e. the dynamic immediacy or becoming which now includes within itself its own 'concept' or category of becoming. Consequently, linearization and categorization represents the *particularity* of the specific historic (space-time) act, giving *finite form* or localization to the trans-historical' content of immediacy in contradiction to the 'historical' form of mediation: as dialectic opposition or mutuality, *both* are *simultaneously* an expression of the singularity of non-linear immediacy. If one gives 'priority' to non-linear trans-historicity *in opposition to* historical form taking on specific localization, then not only is non-linear trans-historicity as a dynamic immediacy destroyed, but what replaces it is the notion of an a-*historical* 'eternity' of non-temporal existence in contradiction to historical temporality. Dialectic immediacy as trans-historicity then appears as a type of alienated *idealism* taking place in a purely non-human impersonal dimension, of which the world as a *materialism* is but a 'poor' unauthentic copy – *at best.* Unfortunately, Hegel's linearization of his non-linear dialectic expresses itself in many cases as an idealism of 'pure thought' which seems to reduce the *essence* of awareness or consciousness to a spiritual process which is *cleansed* of all impure 'material' localizations historically determined, while on the other hand, the *existence* of awareness – i.e. in Nature – *then* introduces (in opposition to the essence of awareness as Logic) this essence as a *mechanical* mirroring into an alienated existence. Consequently, when Spirit issues from Nature as the *reality* of awareness and the transcendence of alienated objectivity, we shall see that either (1) Nature or objectivity is *only* cancelled but not preserved or (2) *only* preserved but not cancelled. Transcendence is dialectic paradoxical immediacy, so if Nature or objectivity (i.e. sensual, material interaction) is regarded as an altogether other-ness not already present within Logic or subjectivity, then there indeed is a mysterious 'jump' from one domain into another, *with* which the idealists feel at odds, and *to* which the materialists can rightly point as a flaw. Spirit, as the mutuality of a pure Logic and an alienated otherness called Nature is then a diction vs. contradiction polarity and not a paradox, i.e. Nature *within* subjectivity can either (1) appear only as a vanishing moment – the otherness *cancelled* – and hence a mechanical return to the logic of subjectivity without being genuinely integrated or (2) remain within the subjectivity – preserved – *as* an alienation in the heart of awareness – and most likely combinations of both. Nature, objectivity, or space time historicity as a particular *determined* here-now localization must be the *authentic* self-otherness to which subjectivity as a *dynamic* essence or Logic (an *Eros*) or *field-of-presence* must have reference, *as* its form (i.e. its *Logos* of determination) in the modality of *events of existence* such that in reality, Spirit or *Geist* is a field-event mutuality – a phenomenology of subject and object in which *neither as an identification* has any meaning *without* the other. Consequently Marx attempted to reintegrate sensual materialism into the dialectic process of Hegel – and this perspective as a whole is best outlined in his famous *Theses on Feuerbach*, where he criticizes *both* onesided idealism *and* materialism as destroying the genuine mutuality between the *active* subject and the object-of-action. Unfortunately – and dialectically – this reintegration of authentic objectivity with the spirituality of Hegel likewise has tended to become a counter one-sided distortion in the hands of the materialists.

The genuine *content* of dialectic is consequently a dynamic of immediacy ($e, e', e'' \ldots$) which on the one hand is, as-a-*totality*, ever-present as a trans-temporal immediacy (i.e. the whole dynamics *itself* expressed *as* an ever-present *singularity* e), while on the other hand, its reflection can take the *Logos* or *form* of a triadic mutuality ($+e$, $-e$, $+-e$), the $+e$ now being the identifiable or linearizable dynamic of immediacy *posited* as a *given essence* or subjectivity (i.e. sheer immediacy now pictured as a Logic of *field* of relations); the $-e$, now the counter-identifiable mediation counter-posited as negativity, determination, objectivity ('*gegenständlichkeit*') or the *events* of *existence* (i.e. mediation as Nature); and $+-e$ as the self-mediated mutuality of essence and existence, subjectivity and objectivity called *reality* or Spirit (*Geist*). Naturally in this modality, Hegel's Encyclopedia of Logic, Nature and

Spirit are *all* three modalities of a higher order dynamic of immediacy whose three *initial mediations* (i.e. Logic as $+e$, Nature as $-e$ and Spirit as the mutual mediation $+-e$) are *each already* the entire dynamic of immediacy – i.e. each is a reflection of the entire e, e', e'' . . . relative to one *type* of fundamental mediation. Logic regards the entire +, –, ± self-mutuality process relative to the positive modality of *presence* (i.e. a *field* of presence regarded as a *given* universe of possibility), Nature the same process relative to the *negative* modality of *localized events* of existence and extension, while Spirit is the same process relative to the *transitive* modality of mutuality between the field of presence and its events. Such an interpretation of both the dynamics of immediacy and Hegel's system requires a detailed investigation into the dialectic of infinity – for it presupposes that the entire non-linear totality expressed here as the e, e', e'' . . . e-ing state or process *can* itself be experienced and grasped as a singularity relative to which higher order e-ing states of immediacy in turn appear. Furthermore, it opens up the possibility of an infinity of infinities . . . *ad infinitum* of dynamic transition states, each dimension of infinity opening up newer interpretations of the fundamental $+e$, $-e$, $\pm e$ mutuality expressive of any e to e' transition. Naturally, all such e to e' transitions would represent a form of the original Being-Nothing-Becoming triadicity. However, if it is permissable to speak about levels of infinite totalities, then each higher order level would require a qualitatively more developed presentation of the original triadicity. It is suggested here[2] that Logic, Nature and Spirit in Hegel's Encyclopedia represents such a higher order form of Being, Nothing and Becoming – a form in which the *implicit* phenomenology of subject-object co-relation inherent in the first order dynamic of immediacy (expressive of such terms-of-movement as Being, Becoming, Quality etc.) now becomes an *explicit* phenomenology, the subjective aspect seen as Logic, the objective as Nature, and their co-relativity as Spirit. With such a *form* of relations, the terms or elements appearing in all three modalities would be of the *same* quality: not only does nature or objectivity appear as a linearizable localization into definitive determinations appearing in space and time, but logic or the subjective field of relations likewise appears linearizable into definitive determinations historically conditioned in space and time (and similarly for Spirit).

The point of such a presentation is to illustrate the following fact of dynamic immediacy: the dynamic *has* to be total and singular at any stage, and should one reflect *on* this totality *as a condition* of presence (i.e. as a *field* of presence or Logic) in opposition to that which is present (the particular *events*), then *both* the immediacy as a field and its opposite –

namely the events of mediation are *now co*-mediations within a larger totality and immediacy. Hence one cannot separate out the dynamics of immediacy as a fixed point of immediacy – i.e. as a given *logic*, relative to which the *world* appears as its content without at the same time mediating that immediacy into a larger and as-yet-unfixed or un-mediated state, within which it now functions as *but* a co-relative variable. Hence, once Hegel decides to *localize* his Logic of dynamic immediacy by mediating it relative to Nature, then *both* are but co-relative sub-mediations within a larger dynamic – now given as Logic, Nature and Spirit. At no time can one come up with a *particular logic* of relations – spelling out a definitive structure that is *merely* a subjective thought un-conditioned by objective existence: to *specify* (i.e. linearize out) a field of presence or 'subjective space' of essences, is to counter-specify a co-relative state of existential events or 'objective space', both of which are co-determined. Consequently, logical relations must reflect their objective conditioning as referring *to* events-in-a-state-of-interaction, and natural events must in turn – if specified out as 'facts' – refer to the subjective field of presence to which they appear as facts. In reality, *Geist* or Spirit is then the mutual co-relativity of fact and essence, each only linearizable in relation to the other.

What has been thus far outlined for a non-linear presentation of dialectic has only taken into explicit consideration the first aspect of non-linearity previously mentioned – namely the appearance of all elements within a dynamic *dimensionality* or space. We now turn to the second aspect – namely the *infinite* nature of such a dimensionality, which has already appeared in the first aspect as a potentiality – in other words – as a *way* of giving expression to orders or levels of dynamic immediacy that give *explicit* expression to the dialectic phenomenology of subject-object relation. We can now justify the usage of such a method of presentation – showing the actuality or reality of dialectic infinity – not only as a notion *within* a dynamic of immediacy but as a quality *of* that dynamic itself. Just because there is no *final* transition state in the *e*-ing process, and furthermore, because all of these stages of transition themselves form a singular *inseparable* dynamics (e, e', e'' do *not* form a sequence of separate terms), this dynamics neither *con*verges to a *final* limit point, nor *di*verges into *a plurality* of isolated steps or positions. One can say, instead, that it *trans*-verges, representing a self-mediating totality whose singularity expresses itself through mutuality of opposition and vice versa. Rather than ending up in a static *being* (i.e. a posited Absolute), or not-ending up at all, but merely generative of an endless state of *becoming* (i.e. an anti-Absolute, Absolute), the dynamic of immediacy is a movement of *self-becoming* which as a *singular* state of immediacy is as totally present at any one state, as it is at any other, but in terms of the *depth of relations visible* at any stage of awareness, shows a definite leveling of complexity, or dimensionality. Thus, dialectic movement is neither a mere a-causal immediacy of indeterminacy, nor a mere mediated, causal process ever-striving toward a determined limit which does not actually exist. Dialectic movement is always an expression of self-movement and self-transcendence, meaning that it expresses mutual causation or self-causation of all its elements simultaneously effecting and being-effected and hence giving rise to either shallow or rich expressions of such a self-totality of movement.

Consequently, dialectic movement is a self-infinitizing whole, which can either be regarded from the perspective of 'the whole as a process', or 'the process as a whole'. The process as a whole is the reality of the singular aspect of self-infinitizing, while on the other hand, the whole as a process gives reality to the multiplicity this whole expresses, such that each element reflects the whole, and at the same time the whole is a reflection and response to each element: universality of the whole is co-determinate with the uniqueness of each element, whole and element (or space and event) being but co-relative aspects of a functioning totality which cannot be reduced to either a plurality of atoms or a monolithic gestalt.

Returning briefly to Hegel, one can now state that his Logic and his *Encyclopedia* (and his Phenomenology as a dialectic of consciousness) must all be re-viewed in terms of an un-ending infinity of *possible* terms, and that the 3^4 terms appearing as the major categories of his logic are consequently only the *way* in which this self-infinitizing totality expresses itself in *4* major changes of perspective. Indeed, there is nothing in Hegel's works that suggests the necessary appearance of a *terminating* absolute at any particular point in its development. Rather what *does* terminate the discussion at a finite place (i.e. after a finite number of steps), is the very table of contents given in the beginning which comes from external (i.e. linear) reflection, and which represents in Hegel's *specific terminology* the extent of historical and cultural development thus far taken place. If we keep to Hegel's own presentation of his overall gestalt as outlined in this paper at the beginning – *not* delimiting and linearizing the self-movement *into* a particular set of identities historically conditioned, then one is left with a dynamic 'schema of movement' as the phenomenology of awareness which, precisely *because* it is trans-identical and beyond being bound to any linearized ontological system of ontology, is then *open* to act as the *logos* or logic of *any* modality of appearance or phenomena appearing (i.e. appearing as events) within the history of awareness.

III. CONCLUDING REMARKS: TOWARDS A MODERN VERSION OF HEGEL'S ENCYCLOPEDIA OF THE SCIENCES

Having described the dynamic of immediacy not only as a trans-historical presence, but also as a historical process – regarding Hegel's *Encyclopedia* essentially as an expanded phenomenology of subject-objectivity, we can now utilize the *Encyclopedia* as a guiding principle in bringing about a modern dialectic of the sciences, which will at the same time be a non-linear 'rejuvenation' of the *Encyclopedia*.

Thus, the *Encyclopedia* regards the science of man to be an integral activity of consciousness, displaying its distinctions dialectically – and consequently generating domains of science, each of which views the *entire* universe relative to a *particular* mode. The major triadicity is of course the science of Logic as the field of presence (subjective space), the science of Nature as the events-present (objective space), and the science of Spirit as the field-event mutuality (transitive subject-object space). The central issue involved is the 'de-ideologizing' of the sciences, (i.e. the transcendence of their one-sided identifications) and the creation of a dynamic integrative perspective which is *functionally* relevant for the development of the particular sciences. Non-linearity becomes, consequently, the essential process to discover and elucidate within the sciences, reflecting as it does, the transitive and hence integrative nature of all historically determined categories, identities and concepts utilized in the scientific acts of definition and measurement. The project thus amounts to viewing science as a process of formation and transformation in which *all* elements are seen to be elements-of-transition – including their identifications. This in turn requires an investigation of the various 'boundary questions' continually plaguing and threatening the essential identities of the various sciences. In the mathematical-logical disciplines, we find the problem of 'undecidability'; within physics and natural phenomena, there exists the corresponding problems of 'indeterminacy', while in the psycho-social sciences of Spirit as man, there is the question of the 'ambiguity' of human identity. However, instead of seeing these as 'problems', dialectic rather grasps these conditions as the very expression of the overall *paradox of immediacy*, and consequently as a *positive*-sign of health: any closed and fully determined system would be taken to be a sign of decadence and degeneration. In the three overall domains of science, one can point to the *three fundamental problems of identity* (the same three Nietzsche points to in *Beyond Good and Evil*!): the status of 'the symbol', 'the material object', and 'the ego' or self. In modern science, the classical atomicity of regarding existence essentially in terms of a linear composition of identifiable units – such as words, objects and persons (and thus groups of words, objects and persons giving us language, nature and society) is precisely what has been put into question by the 'undecidability' of symbols, the 'indeterminacy' of objects and the 'ambiguity' of the ego. From a dynamic of immediacy, however, the paradox of transition inverts one's 'normalized' (i.e. one's linear identity) awareness by becoming aware of the non-existence of *any* kind of '*Ding-an-sich*' or singular point of reference such as the symbol, object or ego.[3]

Thus, for example, it is Gödel's theorems themselves which show that any arithmetizable (linearizable) logic is bound to be either inconsistent or incomplete and hence 'undecidable'. Furthermore, this undecidability arises *exactly at the boundary* between any one symbol and its context or negation. Thus, in attempting to give expression to the coupling between a term and its negation – expressing therefore a boundary condition of co-relation – this co-relation leads either to an inconsistency if accepted into a logic, or an incompleteness if excluded. Only by regarding the *boundary itself* as a dynamic transition *generating* its terms, instead of first regarding the terms as pre-given and then *put into* co-relation, can one truly transcend the limitation of Gödel's theorems. Hence the so-called Liar's Paradox, which prompted Gödel's formulations, is seen to be but the *immediate* state in which all symbolic terminology must give expression to: transition generates self-opposition between mutually conditioned elements such as true or false, present or absent (being or nothing). Any symbol, word, term or designated element at once presents and denies – it presents itself *through* denying its context, and it denies its context *through* presenting itself, hence *conditioning* the presence of a *term* by the very de-lineation it has *with* its context. Thus if one learns that the truth or presence of a term is conditioned by its negation, then simply negating the term will *not* be a transcendence of that term but only a reference to the 'other-side' of the mutually co-determining boundary, and hence a *negative* referral *to* that identity. (To say that "I am *not* going to the movies" is only a reference to the movies as a possible but lacking content and does not of itself refer to anything else.) Thus all negation is *self*-negation, meaning that it expresses the intrinsic paradox between any mediation and its *co*-mediation. Rebellion as an act of denial is but the mirror-image of conformity: *revolution* or real-transformation must grasp *both* what *is* and what *is not* as a singular self-related totality – revolution must be a dynamic of *immediacy* which transcends mediation *by including both* mediations in a single perspective that transcends one-sided linear ideas and 'ideology'. However, by then regarding this immediacy, revolution, or transition *itself* as an exclusive position in opposition *to* mediation, non-revolution or non-transition, is to destroy its transition. Transition, *once identified*, must itself be in transition in order to express itself. Thus *levels* of transition and movement (and paradox) *must* express themselves through alternating mediations of both transition and non-transition-movement and rest, paradox and non-paradox (i.e. identity and contra-identity or contradiction). There *is* no problem to either the Liar's Paradox, the 'class of all classes' or the infinity of infinities, for in non-linear transition dynamics, *all elements are self-referential* and hence self-transcending by their very nature.

This now takes us from logic to physics or natural phenomena, for there the very formulation of how particles exist in the so-called physical vacuum (as element and context) is but a material counter-part to the symbolic formulation of boundary-transition states. Thus the physical vacuum is not just empty, but is characterized by energy-levels and potential processes; charges as particles are defined in co-relation with their space as fields and so on. As scientists are discovering, physical 'contact' and interaction *itself* is not just a linear interaction (as Newtonian, relativistic *and* quantum mechanical interaction is usually described), but involves complex dimensions of activity in which the ordinary conservation principles of momentum and energy are put into question. Indeed, conservation principles, symmetry relations, elasticity (restoring forces), constancy of action (such as the speed of light for relativity and Planck's constant for quantum mechanics) and closed systems (second law of thermodynamics) are *all linear* expressions of interaction which hold true only to the degree to which one can *assume* a fixed identity or parameter governing a *state* of movement. The state of movement *itself* is always open to *its* context of interaction, which can be *ignored* in order to obtain focused *knowledge* about a given structure, should this context not *appear* significant: knowledge is based upon ignorance – i.e. *ignor*-ance, while wisdom on the other hand (as the 'pure *fool*' instinctively knows) must transcend the *Hubris* of its own possessions – i.e. accumulated knowledge or 'facts', by recognizing the limitation *of* limitation – i.e. the *mutuality* between any delimited element and its context: wisdom is thus the awareness of *double*-negation (double-and-mutual mediation or limitation), and hence the self-negation of paradox, *revealing the totality of immediacy*.

Detailed analysis on the physical level reveals, consequently, that *any* kind of fixed *constant* of motion is actually a parameter which itself can function as a *variable* within expanded domains, and are therefore first order approximations of a general non-linear state in which no particular identity need be conserved (or need be non-conserved) – but only relatively conserved. (Indeed, one can rephrase the logic of non-linearity by stating that the contradiction between *constant* and *variable* is replaced by a range-presentation of *parameters*, a parameter being a constant of motion or relation relative to *one* system, but a variable in turn when that system is *itself* subject to change relative to a more comprehensive system. Thus parameters of motion give expression to transitory states of stability, and stabilized modes of transition and thus *levels* of material interaction (e.g. electronic, chemical, biological). Thus relativity and quantum mechanics, as sophisticated as they are, still give rise to confusion and complexity (and contradictions) because the non-linear complexities they do reveal, are revealed in a linear way, *hiding* the richness of relation instead of exposing it. Indeed, from a non-linear point of view, quantum mechanics and relativity give expression to the mutuality between whole and part (space and elements) – relativity stating that no absolute whole or space exists, the whole existing only relative to each element, while quantum mechanics in turn states that there exists no absolute part or determined element, each element or particle existing only in relation to the whole of space (as a wave-function). This mutuality, however, is lost sight of by the linear presentations *of* this dynamic between whole and part.[4]

Finally, coming to the science of the Spirit, or the domain of psycho-social existence – the fundamental problem comes down to clarifying the boundary between the ego and the world, or between self and nature – i.e. the relation between subject and object itself. From a *linear* perspective, ego and world, mind and matter, feeling and fact, soul and body (*ad nauseam*) must all lead to ambiguities, for with such a co-relation, Gödel's theorem on the coupling of contraries (i.e. identities regarded as separate though related) *must* demand either inconsistencies or incompleteness. Thus, any given self appears without itself being a world (i.e. *body*-less) and the world it faces appears without selfhood (i.e. a world absent of 'other' selves), if we regard subject and object as separable distinctions, resulting in ambiguity and alienation. On the other hand, should the boundary of coupling between self and world *itself* be regarded as the immediacy present, then a body-less self or a self-less world would be impossible: *as* a transition state, each exists only relative to the other and 'within' the other as a co-determining presence. Thus, every I or self is automatically a world-conditioned self – i.e. a body, and the world in turn is a self-conditioned world – i.e. one displaying self-hood – and consequently a *world* recognizable as a society of selves. Consequently, the dialectic between self and world on a higher dimension or level reveals the fact that the self itself is a self-world sub-relation, and the world a world-self-sub-dialectic. As a result, the relation between ego and nature or self and world is *not* only a relation *between* them, but also *within* each – meaning that on the one hand, a self-world relation is a *world-world* relation of naturalism (between self *as* a world or body *and* the world), *and* a *self-self* relation of humanism (between a given self and an 'other' self or selves-*in*-the world): the appearance of the self *as* a world or body is thus co-relative to the *world* appearing as a self. The self-world relation is automatically both a physical one and a spiritual subjective one, any ego being subject to both natural and social 'laws' of existence. Hence the psychology of the self or ego involves an intrinsic natural and social perspective – and the *unity* of the two.[5]

Indeed, the *self-infinitizing* relation between self and world is precisely the domain of philosophic-religious experience, meaning that psychology, social-dynamics and spiritual awareness are all but different manifestations of the same dynamic.

Now a rejuvenated Hegelian *Encyclopedia* of the sciences would have to give expression to both (1) the dynamics of Logic, Nature and Spirit as expressed here, but now developed in detail – i.e. as *levels*, sub-levels and meta-levels of coordinated transition states displaying an $e - e' - e''$ topology (as briefly indicated above) and (2) the co-relation between subjective-logic, objective-nature and mutual spirituality – a co-relation demanding a strict *phenomenological* approach showing *how* the dialectic *within* each domain at the same time is an expression of the dialectic *between* them, and vice versa.

We have therefore – in conclusion – mapped out both a 'general' and a 'special' Hegelian dialectic, viewed in terms of non-linearity and a dynamic of immediacy. The general or 'universal' dialectic is the dynamics of immediacy itself – regarded as a trans-historical experience that is totally void of identification. The special or 'particular' dialectic, however, is the self-manifestation of the 'universal' trans-historical immediacy *through* its historical appearance, revealing identities and focused consciousness, such that both the universal and particular (immediacy and mediation) are *co-simultaneously* the very *way* in which the universal dynamic appears as a self-mediating and self-negating process. Actually, *all* analysis, presentation and description is historical . . . *all* identification is mediation and opposition, including the very reference *to* a dynamic of immediacy. Consequently, the *only way* in which to grasp the 'purity' of the universal dynamics is to recognize the fact that *anything visible* or *anything grasped* is *but* an identity which is conditioned and partial – no matter how lofty, spiritual or idealistic it is designated or intuited to be – functioning within the void of silence – the singular totality of immediacy, *whose total absence as an identity itself is the only way in which it functions as a total-presence:* life is the *absolute* paradox of being.

(For a summary of Hegelian dialectic in terms of a non-linear matrix topology, please see the charts on pp. 345–347.

Fairleigh Dickinson University
Rutherford, New Jersey

NOTES

[1] See my articles, 'The Formalization of Hegel's Dialectical Logic', *International Philosophical Quarterly* **6** (1966), 596–631; and 'The Dialectical Matrix', *Telos*, no. 5 (1970), 115–159.

[5] See 'The Formalization of Hegel's Dialectical Logic', *op. cit.*

[3] See my 'The Dynamics of Paradox', *Telos*, no. 5 (1970), 31–43.

[4] See my articles on paradox and on the dialectical matrix in *Telos*, no. 5, cited above.

[5] See 'The Dialectical Matrix', *op. cit.*

CHART – PART I

Hegel's Logic as a 4-Dimensional, Nonlinear 'Matrix-Topology'

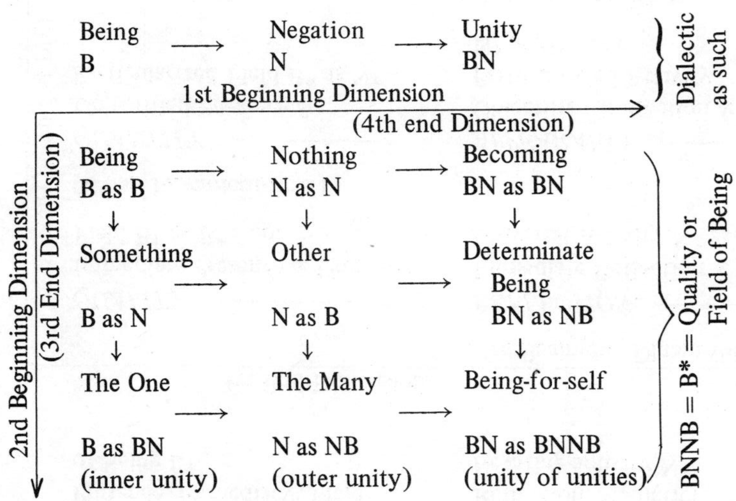

From Being (B) to Quality (B*) in 2 dimensions: $3^2 = 9$ terms-of transition; a Dialectic-Matrix *simultaneously emerging*, non-linear in two-dimensions
From Quality (B*) to Absolute Idea (B**) in 2 *additional* dimensions: $3^2 \times 3^2 = 3^4$ or $9 \times 9 = 81$ transition-terms of a Dialectic Matrix-of-a-Matrix simultaneously emerging, nonlinearly in 4 dimensions creating the entire *Logic*: all Being *is*, Be-*ing*: Dialectic as sheer self-reflexive Being

CHART – PART II

BEING → Presence, Immediacy, Field of Being B*	*ESSENCE* → Reflection, Mediation, Negating activity N*	*NOTION* Self-Reflection, Self-Mediation, Self-Acting-Field of Being B*N*

} *Logic* as such

. . . .

1st End Dimension →

(4th Beginning Dimension)

(3rd Beginning Dimension)

QUALITY → Immediate Presence & Universal Field B* as B* . . or (B* as ↓ *Immanent-field*)	*REFLECTION* → Immediate Reflection & Universal Activity N* as N* ↓	*SUBJECTIVITY* Immediate Self-Reflection & Universal Self-Acting-Field B*N* as B*N* ↓
QUANTITY → Objectified Presence & Particularized Field B* as N* ↓	*APPEARANCE* → Objectified Reflection & Particularized Activity N* as B* . . or (N* ↓ revealing B* as *Appearing-field*)	*OBJECTIVITY* Objectified Self-Reflection & Particularized Self-Acting-Field B*N* as N*B* ↓
MEASURE → Integrated Presence & Total Field B* as B*N*	*ACTUALITY* → Integrated Reflection & Total Activity N* as N*B*	*IDEA* Integrated Self-Reflection & Total Self-Acting-Field B*N* as B*N*N*B* . . or (B* revealing B* through N*: the *Self-Appearing-Field*)

↓ 2nd End Dimension

} B*N*N*B* = Absolute Idea or Sheer Being (B**)

	Logic ⟶	*Nature* ⟶	*S*	
			Spirit	
	Sheer Being as Intension B**	Sheer Being as Extension N**	Sheer Being: Self extending Intension B**N**	*Meta* Logic Encyclopedia
Trans- and *Pre*-logic	*Spirit* as ⟶	*Spirit* as ⟶	*Spirit* as	Hegel's "System of Knowledge"
	Immediacy	*Mediation*	*Self-mediation*	
	Hegel's *Phenomenology*	Hegel's *Science of Logic*	Hegel's *Encyclopedia*	
	S ·	S →	S ⮌	

The *Self-Dialectic of Spirit: Hegel as Nonlinear Dynamics*

H. C. SABELLI

MATHEMATICAL DIALECTICS, SCIENTIFIC LOGIC AND THE PSYCHOANALYSIS OF THINKING

As an experimental scientist, my interest in *dialectics* lies in *its use as a logic of science* and I chose this particular aspect of it in order to include a discussion of the work of Kosok and Gauthier at the Boston Symposium. At that time, I proposed to utilize dialectics as an intuitive model from which to construct mathematical structures applicable to evolutionary theory and to the logic of real thinking. I have not changed the article for publication: its message, I feel, is still valid and the task, unaccomplished. But, to accurately reflect my present viewpoint, I have added a brief discussion of *dialectics as one of the natural patterns of thinking.* The naturalistic observation of dialectical thinking may contribute to our understanding of philosophical dialectics; conversely, it suggests a possible contribution of dialectics to psychological science.

The establishment of the groundwork for the *development of mathematical dialectics* is the great value of the recent work of M. Kosok, Y. Gauthier, F. G. Asenjo, G. Günther, and a few others. Such an attempt transcends the mere formalization of the Hegelian system because mathematical dialectics as a logic of science must also be rooted in mathematics and in contemporary science. Therefore, it implies an actual revision and critique of Hegelian dialectics.

Kant proposed that discipline achieves the status of science only with the use of mathematical methods. Mathematization thus liberates science from empty speculation, dogmaticism and blind empiricism. In spite of its limitations, mathematical logic based on non-dialectic Aristotelian logic is used in computation, in mathematical research and as a logic of science. Such is not the case with dialectics that has not been so formalized. Thus, dialectics must become mathematical dialectics.

On the other hand, symbolic logic needs to incorporate dialectics since, as Hegel indicated in his famous dictum, rationality is the adaptation of thought to reality. Natural, historical, and psychological processes, as described by modern science, do not satisfy the postulates of classical non-dialectic ontology; this is in part described by what Kosok calls the 'non-linearity of science'. More generally, since the mathematical model of natural processes has required the development of many different mathematical structures,

R. S. Cohen and M. W. Wartofsky (eds.), Hegel and the Sciences, 349–359.

mathematical logic cannot be limited to the logical interpretation of lattice theory as a calculus of propositions. Group theory, topology, etc., can also be given a logical interpretation. Such extended mathematical logic is equivalent to the formalization of some aspects of dialectic logic. In fact, dialectics is defined as the science of the most general laws of motion of matter, history, and thought (logic); mathematics is the formal science that constructs abstract models for natural processes including the most general laws of motion. *Thus mathematics is mathematical dialectics* if given an ontological and logical interpretation.

WHAT ARE THE ELEMENTS OF A MATHEMATICAL DIALECTICS?

In *The Science of Logic*, Hegel rejects the concept that logic makes abstraction of the content of thought. Thus, the elements of a mathematical dialectics, concepts or propositions, must be taken simultaneously in extension and in intension. Set theory can be used to develop a logic of the connotation of concepts in which simple attributes can be taken as elements and compound attributes as sets. *The lattice of attributes is not isomorphic to that of classes (in contradiction to Boole's first law).* For instance, there is an infinite number of attributes corresponding to the extensional empty set. Moreover, as the definition of a given class is more sharply defined, the definition of its connotation becomes less well-defined and vice versa. For instance, the well-defined attributes 'liquid' and 'solid', 'acid' and 'base', do not correspond to well-defined classes; a given substance can act as an acid if and only if there are other conditions under which it will act as a base. Extension and connotation are related by an *uncertainty principle* equivalent to the relation between position and momentum in Quantum Mechanics. Thus, a concept taken both in intension and extension satisfies Kosok's requirement that an element of a dialectic calculus cannot be well defined. Let us then define a concept as an ordered pair $< a, \hat{a} >$ *the first element of which is a connotation (set of attributes) a and the second member in the corresponding class* $\hat{a}$ *of objects.* We now say that concept $< a, \hat{a} >$ implies $< b, \hat{b} >$ if and only if $a < b$ and $\hat{a} < \hat{b}$. This *double implication* solves many of the paradoxes of implication that plague classic Boolean logic and as a consequence, Hempel's theory of confirmation.

One needs to consider the connotation of concepts to formalize the Hegelian categories because all such categories are co-extensively universal. The interpretation of concepts in connotation allows for and requires the logical interpretation of more powerful mathematical structures than set theory. In particular, attributes can be so defined as to allow for stronger forms of negation than complementation. For instance, physical dimensions are so defined as to allow for a propositional logic based on group theory.

NEGATION: SYMMETRIC VS HIERARCHICAL

One of the advantages implicit in the use of mathematical models is that they clarify our ideas. There are, in fact, different conceptions of dialectic negation, but this multiplicity is often obscured by the ambiguity of the language. In Kosok's conception of dialectics, negation is symmetric. Reflection upon any immediate unformed element *e* is not a single element but rather a pair of elements *+e* and *−e* emerging by mutual opposition. In other words, the initial, immediate term serves as a base for both mediated terms, positive and negative. The terms are like the three vertices of a triangle standing on one vertex. Assertion and negation are equally potent. The relation between them

Kosok's system:

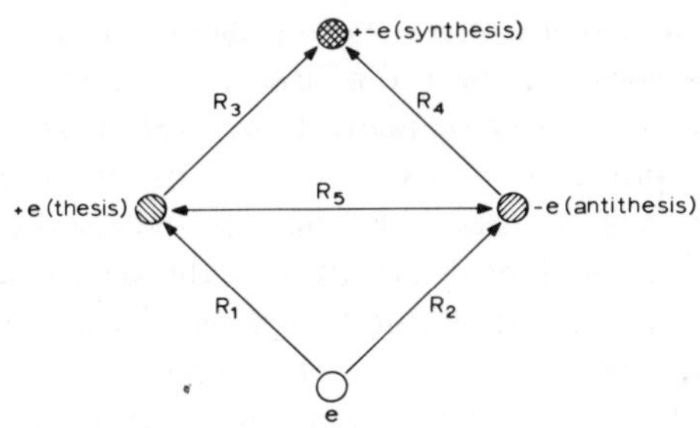

Relations symmetric R_5
non-symmetric R_1, R_2, R_3, R_4

Sublation

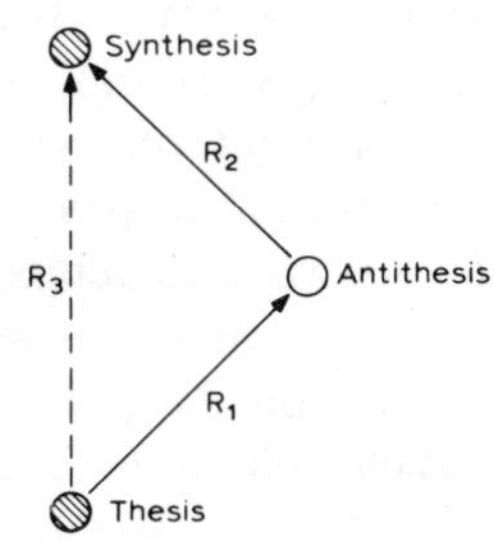

$R_1 = R_2$ = sublation relation $\downarrow$
R_3: order relation $<$

Fig. 1.

is symmetric and they are both sublated by their synthesis, which is intermediate in quality between the two. This hierarchy of elements does not imply evolution in time. The immediate inferior term *e* coexists with the superior mediated terms $+e$ and $-e$ (Figure 1, top).

In classic mathematical logic, negation is complementation, that is to say, absence. Kosok also interprets dialectic negation as an absence; he exemplifies negation by the pair: "I am going to the movies", "I am not going to the movies". This is not sublation, which implies a hierarchical order. Moreover, group inverses may be interpreted to formalize negation as an absence, thereby introducing one of the most powerful mathematical tools in the formalization of dialectics.

Already Gauthier mentioned the use of group theory in the formalization of Hegelian logic. Group inverses obviously abstract the characteristics of symmetric negation. Group theory allows many values in addition to an element and its inverse, thereby eliminating the principle of excluded third, the excluded middle. Moreover, it allows for contradictory statements (elements that are their own inverses). Group theory, however, emphasizes the reversibility of processes, the 'return' of the synthesis to the thesis, by allowing the existence of an identity element. It does not model dialectic negation. The assertion of both a proposition (a) and its denial (a^{-1}) implies the assertion of neither in group theory. In contrast, the assertion of a hypothesis (thesis) and its refutation (antithesis) implies the denial of the former, the assertion of the latter, and the synthesis of a new hypothesis accounting for both the former hypothesis and its counterexample.

Symmetric negation emphasizes the mutually exclusive character of the two terms of a contradiction. Hence, dialectics is described as the need for two consistent (non-contradictory) but incomplete descriptions of the world. Kosok proposed a 'principle of non-identity': "It is impossible to have both the law of contradiction and the law of the excluded middle"; hence, a description of reality can be consistent but incomplete, complete but inconsistent. A more essential form of negation is the hierarchical non-symmetric negation known in dialectics as sublation and in philosophy of science as refutation or falsification. It implies evolution in time and similarity between the terms of the contradiction. The opposition of the two terms of the contradiction is transient, their unity is essential and becomes permanent in a consistent and complete synthesis. The relation between thesis and antithesis and in the same manner the antithesis is sublated by the synthesis. Each term creates new properties by conserving, transcending and superseding the other terms.

Dialectic movement is represented as a helix because evolution has two components: (i) a reversible change in which thesis and antithesis symmetrically negate each other and the synthesis represents a return to the thesis, following its definition as a negation of the negation; and (ii) a non-reversible process for which all three stages represent a movement in the same direction.

ALGEBRAIC MODEL OF SUBLATION: HELIX THEORY:

The notion of sublation can be abstracted by a mathematical relation defined by the postulates of asymmetry (Figure 1, bottom):

P1 For no pair a and b, $a \downarrow b$ and $b \downarrow a$

and indirect transitivity:

P2 If $a \downarrow b$ and $b \downarrow c$ and $c \downarrow d$, then $a \downarrow\!\!/ c$ (direct intransitivity) and $a \downarrow d$ (indirect transitivity)

$a \downarrow b$ will be read 'a is sublated by b' and $a \downarrow\!\!/ b$ will be read 'a is not sublated by b'. The first postulate may be read to mean that if a proposition a is refuted by another proposition b, then b cannot in turn be refuted by a. The second postulate then states that if proposition a is refuted by proposition b and, in turn, proposition b is refuted by proposition c, c actually does not refute a. In fact, it is quite obvious that the refutation of a refutation is a confirmation. If, in turn, c is refuted by another proposition d, d is a refutation of a. From postulate 1, one can immediately demonstrate that sublation is non-reflexive. As the binary relation of order defined in lattice theory formalizes verbs such as 'to be', 'to imply' and 'to include' because it is reflexive, antisymmetric and transitive, the binary relation of sublation formalizes verbs such as 'to sublate', 'to become' and 'to refute', for instance, the refutation of a hypothesis by a contradicting observation. Thus, this axiomatization of dialectics may be interpreted as a calculus of falsification (as defined by Popper) similar to the calculus of confirmation developed by Hempel to formalize inductive logic.

Sets of elements ordered by a sublation relation will be called Helicoids. Figure 2 shows the graphic representation of some finite Helicoids. Each dot represents an element (for instance, a proposition) and each arrow represents a sublation relation. One can easily demonstrate as a theorem that a set of elements ordered by a sublation relation is divided into two sets which we will call colors. These two colors are disjoint subsets, and moreover, the union of these two subsets is equal to the total set. For those of you familiar with

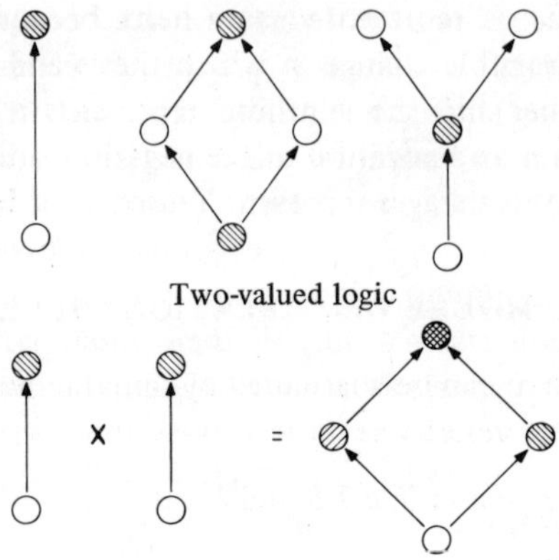

The direct product of two two-element Helicoids corresponds to Kosok's *R* (*e*)

Fig. 2.

mathematical terminology, a Helicoid is the algebraic representation of bipartite oriented graphs. Of course, one can also develop algebraic systems for oriented graphs of more than two colors to represent other types of periodicity such as those found in Mendeleyev's table of chemical elements. One can prove that each subset (color) is internally ordered: if *a* is sublated by *b* and *b* is sublated by *c*, then the relation holding between *a* and *c* satisfies the postulates of the order relation $<$ 'weaker than'. As in Boolean logic, we can take order to represent 'implication' and we say that '*a* implies *c*'. Thus, each element implies the antithesis of its antithesis. For instance, the statement of a hypothesis implies that any of its refutations in turn can be refuted. This is in contrast to classic logic in which the negation of the negation is identical to the initial element.

In two-valued logic, the value 'false' is weaker than the negation of 'true'. Thus, 'false' is sublated by 'true' according to our mathematical definition of sublation. Hence, two-valued logic is simply a two-element Helicoid and the present model of mathematical dialectics may be considered as a generalization of two-valued logic. This serves as a validation of our model and reflects the fact that logic is contained in dialectics.

Note in Figure 2 that the direct product of two-element Helicoids (= two-element chains) is a four-element set which corresponds to Kosok's *R* (*e*). Helicoids have important algebraic properties. One can define binary operations similar to those of classic mathematical logic to represent the connectives 'and' and 'or'. These connectives, however, differ from those of classic logic because truth values need not form a totally ordered lattice (as in the two-valued propositional calculus). Moreover, a Helicoid may have any number of truth values (including infinite) and such truth values alternate in two colors; elements of one color are consistent between themselves and contradictory to those of the other color.

Mathematical logic refers only to the composition of unchanged atomic propositions into a compound statement. Atomic propositions may be changed in their meaning as a result of their combination into compound statements in a similar manner as in nature the interaction of two elementary particles generates new ones. The synthesis of hypotheses and their refutations is essential to the formulation of new hypotheses. An example of the synthesis of two conepts into a new one is the synthesis of rest and motion in inertia.

A Helix is a Helicoid in which we can define a unitary operation antithesis and a binary operation synthesis, to model such non-linear combinations of concepts, and more generally to abstract the dialectic laws of movement. To present the mathematical theory of helices at this point would require considerable space. An article describing the model as well as its use in the interpretation of empirical data has been published (H. C. Sabelli, 'A Pharmacological Approach for Modelling Neuronal Nets' in *Biocybernetics*, Vol. IV, H. Drischel and P. Dettmar, editors (Jena, VEB Gustav Fischer Verlag, 1972)).

Let me now touch some more general points raised by the work of Kosok and Gauthier.

First, *the relation between dialectic and non-dialectic logic:* Most modern logicians have simply ignored dialectics. Conversely many dialecticians have regarded mathematical logic as sterile. The best description of the worst conception of the relation between dialectic and formal logic is given by Zinoviev:

> Dialectics by its very essence cannot be subjected to formalization by way of a formal calculus; it has other tasks than logic, and solves them by other methods. . . . Dialectics did actually point to the fact that two-valued logic has limitations, once it excludes situations of the kind $\exists(x)\ P\ (x)$ and $P'\ (x)$. But this limitation is one which makes science viable.

This statement recognizes the actual meaning of the non-formalization of dialectics: science is outside the domain of dialectics. I conceive dialectics as a set of heuristic rules which guide us to search for probable (but not certain) phenomena.

Any attempt to develop a mathematical dialectics is a search for a synthesis of symbolic logic and dialectics. The nature of their relationship, however, is yet to be established. The logical principles of non-contradiction and excluded middle are often interpreted as absolute laws, preventing, eliminating or forbidding contradictions throughout the objective world and not only in its model in consciousness. The dialectic principle states that real processes include contradictions within themselves which, however, cannot be permanent but lead to changes that solve the contradiction. The contradiction between hypotheses and counterexamples leads to the synthesis of new non-contradictory concepts governed by the logical principle of non-contradiction. Thus, the principle of non-contradiction is a particular case of the principle that contradiction causes a change that abolishes the contradiction; non-dialectic logic is a part of dialectic logic. Both contradiction and non-contradiction are real; the initial contradiction causes change leading to non-contradiction.

Kosok, on the other hand, places *immediacy* beyond the domain of diction and contradiction. The opposition of subjectivity, immediacy and essence, to objectivity, mediacy and categorization, is rejected by Hegel. In his *Science of Logic*, he points out that "Neither in heaven nor on earth, in the spirit or elsewhere, is there anything not containing the immediate and the mediation".

Secondly, an important contribution of Kosok's paper is his emphasis on the notion of structure, a central concept for structuralism and general systems theory that needs to be incorporated into classic dialectics. Likewise, the probabilistic view of the world developed by modern science may generate – when incorporated into the Hegelian framework – a new 'probabilisitc dialectics' which may account more clearly for the existence of change and contradiction.

Like Sartre and other modern dialecticians, Kosok rightly notes the increase in structural complexity with time (formalized by his expanding matrices) and the existence of two alternating phases in dialectic development, a phenomenological or analytical phase (exemplified by the need of classification and specialization in science) and a synthetic phase of convergence. But I think that he incorrectly stresses the symmetry between the two terms of a contradiction. Mathematical logic is governed by the laws of duality. The so-called postulates of symmetry are so widely accepted in physics that many as yet unobserved phenomena have been formulated for reasons of symmetry. However, symmetry breaks down in the most important parameter: time. The Hegelian categories are highly asymmetric.

Much more work will be needed to actually formalize dialectic structures and probably *different mathematical configurations will be required to abstract the different dialectic categories.* Moreover, the consideration of order and structure is not sufficient. Even more important is that of time-dependent, changing orderings and structurations. The process of ordering is both logically and existentially prior to the static order that it generates. Thus, Kosok points out the *need for a dynamic mathematical structure to formalize dialectics.* Order is abstracted by lattice theory; ordering has not as yet been formalized mathematically. However, if one accepts that all reality is dialectical, then all the basic mathematical structures (groups, lattice, topology) must be given a logical interpretation.

Third, *is the dialectic principle a natural law or a law of thinking*? Gauthier proposes that "there is only one real dialectics, the dialectics of language". Many authors, including many Marxists such as Sartre, Lefebvre and Deborin have interpreted dialectics as belonging to the spirit or to human history. As a scientist, I am attracted to dialectics because I see evolution and contradiction as the main features of both natural and human processes. In the introduction to the *Science of Logic*, Hegel rejected this notion that contradiction exists only in the thought but not in the 'thing-in-itself'. Black and white, he says, make grey both on the wall and on the painter's palette.

Finally, let me point out that the value of mathematical models of dialectics and of dialectic logic itself, is their application to the logic of science. It is often contended that symbolic logic provides exact rules of inference, whereas the attempt to guide our thoughts by the laws of dialectic logic in the absence of sufficient concrete information can often lead to false conclusions or even to the ability to prove one proposition and its opposite. The merit of symbolic logic lies in its usefulness (e.g., Boolean algebra in circuit design). In fact, symbolic logic itself is far from consistent (as witnessed by the set theory paradoxes). Exactness alone is no praiseworthy merit. After all, a clock that does not run will be the only one to tell the exact time twice a day, and "He alone won't betray in whom none will confide. And the nymph may be chaste that has never been tried" (Congreve, *Love for Love*).

A characteristic of the thinking brain is its ability to suffer delusions, illusions and hallucinations. The usefulness of a logic of science is measured by its fruitfulness as a heuristics for hypothesis formulation and testing rather than by the exactness of its conclusions in the absence of experimental data. Scientific hypotheses are based upon and tested by experiment, observation, practice.

The dialectic notion of *the unity of content and form implies that the axiomatization of dialectics is neither impossible nor trivial.* On the other hand, scientific theories (including dialectics) are neither static nor definitive;

hence, a mathematical calculus can only formalize some aspects of dialectics.

ADDENDUM: DIALECTICS AND THE PSYCHIATRIC ANALYSIS OF THINKING

Dealing with the concreteness of human thinking rather than with simplified psychological or logical constructs, psychiatry necessarily speaks dialectics. Psychiatrists often do so unknowingly, as Monsieur Jourdain spoke prose, but the validity of their contribution to a multidisciplinary study of thinking cannot be overlooked. Piaget has, in fact, revolutionized the study of thought via his concept of experimental epistemology. His pediatric studies have shown an ordered sequence of development in which different mathematical structures successively serve the normative role of an organizing logic. Group structures, for instance, characterize certain stages of development, while the overall evolution of thinking is obviously more complex than Hegelian dialectics. Within this framework, *dialectic thinking is one among many natural patterns of thinking* (e.g., abstract-mathematical, mechanistic, autistic, free-associations, etc.). It seems likely that biological evolution has selected such multiplicity as being more advantageous than a single pattern of thinking. Within his view, the rejection of any one of these thinking modes may be a misguidedly premature attempt to correct evolution. A naturalisitc study of dialectic thinking may offer significant contributions to philosophical dialectics.

Conversely, there is a wide range of applications of dialectics to psychiatry, particularly because contradiction is an essential component of many different types of thinking processes. For instance, the brilliant analysis made by I. Lakatos of the logic of mathematical discoveries via the contradiction of 'Proofs and Refutations' illustrates the dialectic nature of creative thinking; and of course, one of Freud's main contributions was his discovery of the 'illogical' but not random nature of the 'primary process' thinking observed in dreams, psychoanalytic free-associations, errors and slips of the tongue, etc. Because dialectics deals with contradictory thinking – which non-dialectic logic excludes – it may have a clinical application in psychiatry in the interpretation of the covert and often contradictory meaning of thinking and behavior. For instance, the dialectic notion that opposites imply each other provides a 'dictionary' for probable (but not certain) psychoanalytic interpretations (e.g., irritable ↔ irritating ↔ irritated, dominant ↔ submissive; anger ↔ fear). Freud discovered that in dreams things are often represented by their opposite and most psychiatrists utilize their expectation of associations between contradictory moods and attitudes as heuristics for emotional insights. Likewise, the Hegelian concept of the master as the slave of the slave (the asymmetric reciprocity of sublation) finds wide applications in family therapy, suggesting to search for the covert domination which others exert upon the overtly dominant family member. The law of the reciprocal transformation of quantity and quality serves as a warning against the logical fallacies which may be involved in '*a potiori*' arguments (to focus on the extreme and most clearly identifiable phenomenon of a group to analyze the whole range of related processes), which, following Freud, have become characteristic of psychoanalytic reasoning. The dialectic notion of change as the result of the interaction between internal and external contradictions supersedes the sterile polemics of nature vs nurture, intrapsychic vs interpersonal, etc., etc. 'Interactionism' is a modern name for this Hegelian thesis. A scientific revolution (from reversible mechanicism to the irreversible dialectic evolution of thermodynamics) occurred when physicists turned their eyes from celestial bodies to the concrete pursuits of human labor. In a similar manner, philosophical dialectics may find new avenues of development by focusing on actual thinking. Contemporary science shows the fruitfulness of combining experimental, axiomatic and theoretical approaches: for instance, theoretical physics enriched both experimental physics and mathematical geometry. Likewise, dialectic theory may serve to develop mathematical logic into a mathematical dialectics, and to provide a theory ('psychodialectics') for the psychoanalysis of interpersonal dialogue. Philosophy has already had a major impact in psychiatry as illustrated by the phenomenological approach of Jaspers (now used widely in diagnosis) and the use of dialectic concepts by Erikson, Piaget and others.

IVAN SOLL

COMMENTS ON KOSOK'S INTERPRETATION OF HEGEL'S LOGIC

Kosok's interpretation as well as philosophic endorsement of Hegelian logic and dialectic does not even pretend to rest on a painstaking examination of Hegel's logic in its entirety, but rather upon what Kosok considers to be "a few fundamental statements appearing in the beginning of Hegel's *Science of Logic* . . . , which clearly and unambiguously set forth his entire gestalt" (p. 312). Examining the handful of sentences Kosok quotes I do not find anything like an entire *Gestalt* of Hegel's logic or dialectic set forth, not even set forth unclearly and ambiguously. But rather than quibble about what can or cannot be 'found' in these eight sentences or sentence fragments, I would like to deal directly with Kosok's conception of Hegel's logic.

Kosok asks us to view Hegel's logic as 'sheer-movement,' and he claims that, "it is this very movement that constitutes the *content* of the categories generated, and not the particular categories themselves, regarded as identifiable and analysable terms" (p. 311). This call to adapt a dynamic view of Hegel's logic – which so formulated rings somehow virtuous but extremely vague – is given more content by coupling it with a prohibition against what Kosok terms "the identification process, singling out terms or elements from their dynamic of relation and movement . . ." (*ibid.*). "To identify (Kosok tells us) means to bring into focus a certain element, quality, fact, condition . . . a certain event, state of events or experience in such a way as to be able to refer to it in some categorical way." Furthermore: "Without identification our awareness is in a state of immediacy which is *not defined in relation to* categorization" (p. 314). Kosok takes what he calls a 'state of immediacy' to be central to Hegelian logic and dialectic. In fact he offers the definition that "dialectic is the dynamics of immediacy" (p. 343).

Since our awareness is in a 'state of immediacy' only if there is no identification, and identification is defined as the structuring of our awareness by means of categories, the state of immediacy Kosok indicates seems to be a state of awareness in which there is no structuring of consciousness through the application of concepts, no division or grouping of experience into similar parts or aspects, no identification or even discernment of particulars in experience or of particular kinds of experience. The state of awareness indicated seems to be what philosophers in the Kantian tradition have either,

R. S. Cohen and M. W. Wartofsky (eds.), Hegel and the Sciences, 361–364.

like Kant himself, held to be an impossible state of affairs, or like C. I. Lewis, a rare and evanescent one, or even if more common and stable, an inchoate demiconsciousness, "nothing but the dream of an incomprehensible present" (to use Josiah Royce's formulation) – and according to all versions not a state to be considered human consciousness in its full form. On the other hand, Kosok's talk of a state of immediacy is reminiscent of various laudatory descriptions of states of enlightenment or exaltation induced by faith, pharmacy or philosophical meditation. Kosok seems to fit neatly into this second tradition when he claims that immediate awareness is itself ineffable. But his is not the hackneyed and trivial claim that certain kinds of experience cannot be *completely* captured in language. His is the far more radical claim that 'immediate awareness' cannot be characterized *at all*, identified in any degree, or even referred to. His general argument is that any category – no matter how apt – used to characterize this state of awareness functions, like all concepts and categories, only by having a contrastive set and thus characterizes immediate awareness by contrasting it to what it is not or, in Hegelian terminology, through the mediation of what it is not. Yet immediate awareness is defined as a state free of distinctions and contrasts. Thus immediate awareness cannot even be characterized as immediate "lest this immediacy is 'itself' then perceived as an identified state [i.e. in opposition to a mediated state] and hence not immediate" (p. 314) and "even the notion of trans-bounded, if identified or localized by appearing in opposition to non-trans-bounded, cannot describe immediacy" (*ibid.*). Moreover he says "immediacy cannot simply be dismissed as 'something' not-characterizable altogether" (p. 315) for this again contrasts immediacy to something which is characterizable. Again, I suppose, because simple reference relies on contrast, "If one were simply to 'remain' with the original undefined state of immediacy . . . , one could not in any way refer to it . . ." (p. 317).

This general line of argument rests, I believe, on a confusion between being able to characterize or refer to the kind of state Kosok calls 'immediate awareness' and being able to do the same while remaining in such a state. If one were in such a state of awareness, it would be impossible to characterize or refer to it – that is, it would be impossible to characterize or refer to a state of immediate awareness and remain in that state. One could of course refer to or characterize a state of immediate awareness by freely using categories that have contrastive sets. In so doing one would be performing actions Kosok does not allow to someone in a state of immediate awareness, but why must the characterization of such a state be self-characterization and any reference to it be self-reference? All Kosok's arguments even tend to show is the impossibility of self-reference and self-description by someone in a certain state while he is in that state, yet he writes as if he has shown that nobody can ever characterize or refer to this state. What Kosok has done is analogous to arguing that a man in a coma cannot describe his facial features while in the coma, and taking this as tantamount to showing those facial features to be intrinsically indescribable. But the man can come out of the coma and describe what his facial features were even at the time of the coma and certainly someone else can describe those features. What does the fact that Rembrandt cannot paint a self-portrait while holding a golden helmet in one hand and an Edam cheese in the other have to do with the possibility of his posing full-handed for Franz Hals or his posing the objects and painting a similar self-portrait with the aid of a mirror and memory.

In light of this criticism what remains of "the *paradox* which (according to Kosok) immediacy . . . *must* present" (p. 315). If immediacy is a state of consciousness in which no self-reference or self-characterization can be made and – let us remember – no reference or characterization of any kind can be made, then anyone desiring to engage in these activities of 'identification' must move out of a state of immediacy to perform them. If one, in fact, cannot eat his cake and have it too, does that constitute a paradox?

Kosok thinks that what he takes to be the inevitability of this alleged paradox "makes immediacy not an *arbitrary* variable simply left for the poets to sing hymns to" (*ibid.*). Instead he sings hymns of his own. Hegel, however, did not sing such hymns. As Kosok points out, Hegel associated the notion of immediacy with the beginning of his logic and with Being, its first category, but Kosok neglects to point out that Hegel went on to stress the intellectual and conceptual poverty of the beginning of any philosophical enterprise and the conceptual vacuousness of the category of Being. Furthermore in the *Vorbegriff* to the *Encyclopedia*, Hegel devotes an entire section to a disdainful rejection of 'immediate knowledge,' i.e., of claims to non-conceptual, intuitive knowledge by post-Kantian romantics like Jacobi. For Hegel, the knowledge presented at the beginning of the logic like the knowledge offered by the intuitionists is immediate in that it lacks the painstaking conceptual development of the logic itself. Hegel stresses the virtue and necessity of this development, which he often describes as a process of 'mediation.' The same theme of the necessary mediation of knowledge through the careful analysis and development of concepts is stressed throughout the preface to the *Phenomenology*, where a state of awareness approximating Mr. Kosok's immediacy is famously lampooned as "the night in which all cows are black."

Hegel, in contrast to Kosok, generally uses the notion of immediacy pejoratively. How Hegelian then can a theory be which views dialectics as the dynamics of immediacy?

Kosok's answer to this question would, I suppose, lie in his claim that: "Hegel *starts* with immediacy and *remains* within immediacy, all mediations being self-mediations of immediacy." Is this a deep dialectical truth or double-talk? In what sense does Hegel 'remain within immediacy' in his logic? Though Hegel associates immediacy with Being, the first or immediately encountered category of the logic and claims that all later categories of his logic are further determinations of Being, he also says that these categories 'mediate' the original category of Being. Though Kosok's state of immediacy is defined as free of categorization, identification or mediation, he quixotically tries to fit Hegel's complex analysis of many categories and their relations into his categoryless state of immediate awareness by claiming that the mediations are mediations of immediacy. But is this not like arguing that a certain artist's designs are monochromatic because, though he uses different colors, he applies them to a canvas that is monochromatic at the outset? Kosok might want to rejoin that Hegel remains within immediacy because his mediations are *self*-mediations of immediacy. But even if the categorization and mediation in Hegel's logic are somehow self-generated, how does that in any way alter the fact that the body of Hegel's logic is structured and thus mediated by categories.

Kosok is here stressing an idea that, unlike some of his other ideas, is actually to be found in Hegel, namely, that the logic is self-generating. Leaving aside the question of whether *Hegel's* logic is really self-generating, the logic offered by Kosok is not. The state of immediate awareness as delineated by Kosok does not inexorably generate any further state. Only an attempt to characterize, identify or refer to it generates further states or further logical complexity, and this act of referring or characterizing is, on Kosok's own account, extrinsic to what Kosok defines as a state of immediacy.

However interesting one finds Kosok's dialectical logic, its deviations from Hegel's logic are as striking as its similarities. To follow Kosok's advice and leave behind Hegel's myriad of categories or the Hegelian analysis and dissection of the various categories may well be to leave behind any logic deserving to be called Hegelian.

University of Wisconsin
Madison

BIBLIOGRAPHICAL NOTE

[The contributors have referred to various editions and translations of Hegel's texts. The relevant bibliographical data are listed below, as well as the abbreviations used for citation in the text. In addition, the standard translation of Kant's First Critique is listed at the end. – Ed.]

COLLECTED WORKS

1. *Werke.* Edited by P. Marheineke, J. Schulze, E. Gans, L. Henning, H. Hotho, K. Michelet, F. Förster . . . 19 vols. in 20. Berlin, Dunker and Humblot, 1832–1887.
2. *Sämtliche Werke* . Edited by G. Lasson. 18 vols. [Some titles went through more than edition.] Leipzig, Meiner, 1905–1940.
   In the course of publication, J. Hoffmeister began to edit some of the volumes of this series in a 'critical edition'.
3. *Sämtliche Werke*. Jubiläumsausgabe. Edited by H. Glockner. 26 vols. Stuttgart, Frommann, 1927–1940. 3rd ed. 1949–
   A 'facsimile' edition based on the original *Werke* which began publication in 1832.
4. *Sämtliche Werke.* Neue kritische Ausgabe. Edited by J. Hoffmeister. Hamburg, Meiner, 1952 –.
   Based on the Lasson edition.

INDIVIDUAL WORKS

GP — *Vorlesungen über die Geschichte der Philosophie.* Edited by H. Glockner. 2 vols. (*Jubiläumsausgabe* **17, 18**, 1928).
The first edition of this work (in three volumes) was edited by Michelet and published in the *Werke* (volumes 13–15) in 1833–36; a second, emended edition followed in 1840. For English translation see *HP.*

GPR — *Grundlinien der Philosophie des Rechts.* Berlin, Nicolai, 1821. For English translation see *PR.*

HP — *Lectures on the History of Philosophy*. Trans. E. S. Haldane and F. H. Simson. 3 vols. [London, Kegan Paul, Trench, Truebner, 1892–1896]. London, Routledge and Kegan Paul; New York, Humanities Press, 1955. Translation of *Vorlesungen über die Geschichte der Philosophie* (*GP*), based on the second emended edition of 1840.

R. S. Cohen and M. W. Wartofsky (eds.), Hegel and the Sciences, 365–367.

L [W, 1892]	*The Logic of Hegel.* Trans. W. Wallace. 2nd rev. and aug. ed. Oxford, Clarendon Press, 1892. Translation of *Logik*, Part I of the *Encyklopädie der philosophischen Wissenschaften*, 1830.
L [W, 1975]	*Logic.* Trans. W. Wallace; with a foreword by J. N. Findlay. Oxford, Clarendon Press, 1975. Translation of *Logik*, Part I of the *Encyklopädie der philosophischen Wissenschaften*, 1830.
Logik	*Logik* (Part I of the *Encyklopädie der philosophischen Wissenschaften*). Edited by L. Henning. (*Werke* 6, 1840). For English translation see *L* [, 1892] and *L* [W, 1975].
PG	*Philosophie des Geistes* (Part III of the *Encyklopädie der philosophischen Wissenschaften*). Edited by L. Boumann. (*Werke* 7, 1845).
Phän [G]	*Phänomenologie des Geistes.* Edited by H. Glockner. (*Jubiläumsausgabe* **2**, 1927). For English translation see *Phen* [B] and *Phen* [M].
Phän [H, 1937]	*Phänomenologie des Geistes.* Edited by J. Hoffmeister. 4th ed. (*Sämtliche Werke* (Lasson ed., critical ed.) **2**, 1937). For English translations see *Phen* [B] and *Phen* [M].
Phän [H, 1952]	*Phänomenologie des Geistes.* Edited by J. Hoffmeister. 6th ed. (*Sämtliche Werke* (new critical ed.) **5**, 1952). For English translations see *Phen* [B] and *Phen* [M].
Phen [B]	*The Phenomenology of Mind.* Trans. J. B. Baillie. 2nd rev. and corr. ed. London, Allen and Unwin; New York, Macmillan, 1931. (Reprinted several times.) Translation of *Phänomenologie des Geistes* (*Phän* [G, H]).
Phen [M]	*Phenomenology of Spirit.* Trans. A. V. Miller, with analysis of the text and foreword by J. N. Findlay. Oxford, Clarendon Press, 1977. Translation of *Phänomenologie des Geistes* (*Phän* [G, H]).
PM	*Philosophy of Mind.* Trans. W. Wallace, together with the *Zusätze* in Boumann's text (1845), trans. A. V. Miller; with foreword by J. N. Findlay. Oxford, Clarendon Press, 1971. Translation of *Philosophie des Geistes* (*PG*), Part III of the *Encyklopädie der philosophischen Wissenschaften*, 1830.
PN	*Philosophy of Nature.* Trans. from Nicolin and Pöggeler's edition (1959) and from the *Zusätze* in Michelet's text (1847) by A. V. Miller; with foreword by J. N. Findlay Oxford, Clarendon Press, 1970. Translation of *Naturphilosophie*, Part II of the *Encyklopädie der philosophischen Wissenschaften*, 1830.
PR	*Philosophy of Right*. Trans. T. M. Knox. Oxford, Clarendon Press, 1942. Translation of *Grundlinien der Philosophie des Rechts* (*GPR*).
SL [J/S]	*Science of Logic* [1929]. Trans. W. H. Johnston and L. G. Struthers. 2 vols. London, Allen and Unwin, 1951. Translation of *Wissenschaft der Logik* (*WL* [L]).
SL [M]	*Science of Logic.* Trans. A. V. Miller, foreword by J. N. Findlay. London, Allen and Unwin; New York, Humanities Press, 1969. Translation of *Wissenschaft der Logik* (*WL* [L]).
WL [1812]	*Wissenschaft der Logik.* 2 vols. Nuremberg, Schrag, 1812–1816. For English translations see *SL* [J/S, M].
WL [L]	*Wissenschaft der Logik.* Edited by G. Lasson. 2 vols. (*Sämtliche*

Werke **3**, **4**, 1923; 1934, reprinted 1948). For English translations see *SL* [J/S, M].

CPR Immanuel Kant: *Critique of Pure Reason*. Trans. N. K. Smith. London, Macmillan, 1929. (Reprinted several times.)

INDEX OF NAMES